CLINICAL ASPECTS OF INTERFERONS

DEVELOPMENTS IN MEDICAL VIROLOGY

Yechiel Becker, Series Editor
Julia Hadar, Managing Editor

Levine, P.H. (ed.) *Epstein-Barr Virus and Associated Diseases* (1985)
Becker, Y. (ed.) *Virus Infections and Diabetes Mellitus* (1987)
De Clercq, E. (ed.) *Clinical Use of Antiviral Drugs* (1988)
Revel, M. (ed.) *Clinical Aspects of Interferons* (1988)

DEVELOPMENTS IN MOLECULAR VIROLOGY

Yechiel Becker, Series Editor
Julia Hadar, Managing Editor

Becker, Y. (ed.) *Herpesvirus DNA* (1981)
Becker, Y. (ed.) *Replication of Viral and Cellular Genomes* (1983)
Becker, Y. (ed.) *Antiviral Drugs and Interferon: The Molecular Basis of Their Activity* (1983)
Kohn, A. and Fuchs, P. (eds.) *Mechanisms of Viral Pathogenesis from Gene to Pathogen* (1983)
Becker, Y. (ed.) *Recombinant DNA Research and Viruses. Cloning and Expression of Viral Genes* (1985)
Feitelson, M. *Molecular Components of Hepatitis B Virus* (1985)
Becker, Y. (ed.) *Viral Messenger RNA: Transcription, Processing, Splicing and Molecular Structure* (1985)
Doerfler, W. (ed.) *Adenovirus DNA: The Viral Genome and Its Expression* (1986)
Aloni, Y. (ed.) *Molecular Aspects of Papovaviruses* (1987)

DEVELOPMENTS IN VETERINARY VIROLOGY

Yechiel Becker, Series Editor
Julia Hadar, Managing Editor

Payne, L.N. (ed.) *Marek's Disease* (1985)
Burny, A. and Mammerickx, M. (eds.) *Enzootic Bovine Leukosis and Bovine Leukemia Virus* (1987)
Becker, Y. (ed.) *African Swine Fever* (1987)
De Boer, G.F. (ed.) *Avian Leukosis* (1987)
Liess, B. (ed.) *Classical Swine Fever and Related Viral Infections* (1987)
Darai, G. (ed.) *Virus Diseases in Laboratory and Captive Animals* (1988)

CLINICAL ASPECTS OF INTERFERONS

edited by

Professor Michel Revel
The Weizmann Institute of Science
Israel

Kluwer Academic Publishers
Boston • Dordrecht • London

Distributors

for the United States and Canada: Kluwer Academic Publishers, 101 Philip Drive, Assinippi Park, Norwell, MA 02061

for the UK and Ireland: Kluwer Academic Publishers, Falcon House, Queen Square, Lancaster LA1 1RN, UK

for all other countries: Kluwer Academic Publishers Group, Distribution Centre, P.O. Box 322, 3300 AH Dordrecht, The Netherlands

Library of Congress Cataloging-in-Publication Data

Clinical aspects of interferons / edited by Michel Revel.
p. cm. — (Developments in medical virology)
Includes bibliographies and index.
ISBN 0-89838-371-4
1. Interferon—Therapeutic use. 2. Cancer—Immunotherapy.
3. Virus diseases—Immunotherapy. I. Revel, Michel. II. Series.
[DNLM: 1. Interferons—therapeutic use. QV 268.5 C6407]
RC271.I46C58 1988
615′.37—dc19
DNLM/DLC 88-1248
for Library of Congress CIP

PRINTED IN THE UNITED STATES.

CONTENTS

II CLINICAL USES IN CANCER

CONTRIBUTING AUTHORS

Oded Abramsky
Department of Neurology
Hadassah University Hospital
Hebrew University
Hadassah Medical School
Jerusalem , 91120 , Israel

Sefik S. Alkan
Pharmaceutical Research Div.
Department of Inflammation
and Immunology
Ciba-Geigy Ltd.
CH-4002, Basel, Switzerland

Frances. R. Balkwill
Biological Therapy Laboratory
Imperial Cancer Research Fund
Lincoln's Inn Fields
London, WC2A 3PX, UK

Henrik C.F. Bauer
Department of Orthopedic Surgery
Karolinska Hospital
S-104 Stockholm, Sweden

P. Beuzeboc
Service de Medecine Oncologique
Institut Curie
26 Rue d'Ulm
75231 Paris V,France

Velio Bocci
Istituto di Fisiologia Generale
Universita degli Studi di Siena
Via Laterina 8
53100 Siena, Italy

Ernest C. Borden
Department of Human Oncology
The University of Wisconsin
Clinical Cancer Center
Madison, WIS 53792, USA

Otte Brosjo
Department of Orthopedic Surgery
Karolinska Hospital
S-104 Stockholm, Sweden

L.-A. Brostrom
Department of Orthopedic Surgery
Karolinska Hospital
S-104 Stockholm, Sweden

Giuseppe De Palo
Istituto Nazionale per lo Studio
e la Cura dei Tumori
Via Venezian 1
Milan 20133, Italy

T. Dorval
Service de Medecine Oncologique
Institut Curie
26 Rue d'Ulm
75231 Paris V,France

Stephen Einhorn
Department of Oncology
Radiumhemmet
Karolinska Hospital
S-104 Stockholm, Sweden

B. Erikson
Ludwig Institute for Cancer Research
Uppsala Branch
Akademiska Sjukhuset
S-751 85 Uppsala, Sweden

Ernesto Falcoff
Section de Biologie
Institut Curie
Pavillon Pasteur
26, rue d'Ulm
75231 Paris V, France

R. Feinman
Department of Microbiology
New York University Medical Center
550 First Avenue
New York, NY 10016, USA

Kenneth A. Foon
Department of Medicine
Division of Hematology and Oncology
University of Michigan
Ann Arbor, Michigan 48109, USA

Robert P. Gale
Department of Medicine
Division of Hematology and Oncology
Transplantation Biology Unit
University of California Los Angeles
Los angeles, Cal 90024, USA

E. Garcia-Giralt
Service de Medecine Oncologique
Institut Curie
26 Rue d'Ulm
75231 Paris V,France

Mario Ghione
Istituto di Microbiologia Medica
University of Milan
Via Mangiagalli 31,
Milan 20133, Italy

Jordan U. Gutterman
Department of Clinical Immunology
M.D. Anderson Hospital
and Tumor Institute
1515 Holcombe Bd
Houston, Texas, 77030, USA

Frederick G. Hayden
Departments of Internal Medicine
and Pathology
University of Virginia
School of Medicine, Box 473
Charlottesville, Virginia 22908, USA

Robert Herndon
Center for Brain Research
University of Rochester Medical Center
Rochester, NY, 14263 USA

Heinz-Kurt Hochkeppel
Department of Immunology
Ciba-Geigy Ltd
CH-4002, Basel, Switzerland

Lawrence D. Jacobs
Dent Neurologic Institute
State University of New York
School of Medicine
Buffalo, NY 14263 USA

M. Jouve
Service de Medecine Oncologique
Institut Curie
26 Rue d'Ulm
75231 Paris V, France

Hagop Kantarjian
Department of Hematology
M.D. Anderson Hospital and
Tumor Institute
1515 Holcombe Bd
Houston, Texas, 77030, USA

Susan E. Krown
Memorial Sloan Kettering Cancer Center
1275 York Avenue
New York, NY 10021, USA

Razelle Kurzrock
Departments of Clinical Immunology
and Hematology
M.D. Anderson Hospital
and Tumor Institute
1515 Holcombe Bd
Houston, Texas, 77030, USA

Eugene Leibowitz
Department of Pediatrics and
Pediatric Research Institute
Kaplan Hospital
Rehovot, 76100, Israel

Stanley Levin
Department of Pediatrics and
Pediatric Research Institute
Kaplan Hospital
Rehovot, 76100, Israel

Hilton B. Levy
Molecular Virology Section
National Institute of Allergy and
Infectious Diseases
N.I.H.
Bethesda, Md 20982, USA

Masakatsu Nagai
Department of Neurosurgery
Dokkyo University School of Medicine
Mibu
Tochigi 321-02, Japan

U. Nilsonne
Department of Orthopedic Surgery
Karolinska Hospital
S-104 Stockholm, Sweden

Olle S. Nilsson
Department of Orthopedic Surgery
Karolinska Hospital
S-104 Stockholm, Sweden

K. Nooter
Primate Center TNO
P.O. Box 5815
2280 HV Rijswijk, The Netherlands

I. Nordheim
Department of Internal Medicine
University Hospital
S-751 85 Uppsala, Sweden

Kjell Oberg
Ludwig Institute for Cancer Research,
Uppsala Branch
Akademiska Sjukhuset
S-751 85 Uppsala, Sweden

Robert K. Oldham
Biological Therapy Institute
Hospital Drive
Franklin, TN 37064 , USA

Judith A. O'Malley
Department of Biological Resources
Roswell Park Memorial Institute
and Microbiology Department
State University of New York
Buffalo, NY 14263, USA

T. Palangie
Service de Medecine Oncologique
Institut Curie
26 Rue d'Ulm
75231 Paris V, France

Pierre Pouillart
Service de Medecine Oncologique
Institut Curie
26 Rue d'Ulm
75231 Paris V, France

Jorge R. Quesada
Department of Clinical Immunology
and Biological Therapy
M.D. Anderson hospital
and Tumor Institute
6723 Berner Avenue
Houston, Texas 77030, USA

Peter A. Reese
Computer Science
Department of Biomathematics
Roswell park Memorial Institute
Buffalo, NY 14263, USA

Michel Revel
Department of Virology
Weizmann Institute of Science
Rehovot 76100, Israel

Andres M. Salazar
Walter Reed Army Medical Center
Neurology Department
Uniformed Services University of
the Health Sciences
Bethesda, Md 20814, USA

Amichai Schattner
Department of Internal Medicine and
Immunology Institute
Kaplan Hospital
Rehovot 76100, Israel

Huub Schellekens
Primate Center TNO
P.O. Box 5815
2280 HV Rijswijk, The Netherlands

S.M. Schmid
Department of Human Oncology
The University of Wisconsin
Clinical Cancer Center
Madison, WIS 53792, USA

Geoff M.S. Scott
Clinical Microbiology
University College Hospital
Grafton Way
London WC1E 6AU, UK

Jacob Shoham
Department of Life Sciences
Bar-Ilan University
52100 Ramat-Gan, Israel

Gerald Sonnenfeld
Department of Microbiology
and Immunology
Schools of Medicine and Dentistry
University of Louisville
Louisville, KY 40292, USA

Bernadina Stefanon
Istituto Nazionale per lo Studio
e la Cura dei Tumori
Via Venezian 1
Milan 20133, Italy

Israel Steiner
Department of Neurology
Hadassah University Hospital
Hebrew University
Hadassah Medical School
Jerusalem , 91120 , Israel

Hans Strander
Department of Oncology
Radiumhemmet
Karolinska Hospital
S-104 Stockholm, Sweden

Moshe Talpaz
Departments of Clinical Immunology
and Hematology
The University of Texas
M.D. Anderson Hospital
and Tumor Institute
1515 Holcombe Bd
Houston, Texas, 77030, USA

P.H. Van Der Meide
Primate Center TNO
P.O. Box 5815
2280 HV Rijswijk, The Netherlands

Jan Vilcek
Department of MIcrobiology
New York University Medical Center
550 First Avenue
New York, NY 10016, USA

INTRODUCTION: An overview of Interferons action.

Interferons (IFNs) are one of the first example of cellular hormones, or cytokines, brought by molecular biology to medical applications. They will certainely not be the last as interleukins, tumor necrosis factor, growth factors, hemopoietic colony stimulating factors, tissue plasminogen activator and many others, are on their way to medical usage. But the registration by the U.S. FDA of recombinant IFN-α2 in June 1986, marks the dawn of a new phase in biotherapy. As expounded by Dr Robert K. Oldham (chapter 2.1), IFNs can serve as models of biotherapeutic drugs for cancer and viral diseases, coming to complement, if not to eventually replace chemotherapy which has dominated the scene in the last 20 years. Produced naturally in only small amounts and difficult to purify in sufficient quantities from human cells, IFNs were ideal candidates for demonstrating the power of recombinant DNA technology to manufacture the large supply of pure protein needed for clinical trials. Genetic engineering also revealed the unexpected complexity of the IFN system and the number of human genes producing molecules with IFN-type activity (1,2). Understanding this complexity, the mode of action of the various IFNs, and their interactions with other cytokines (see chapter 3.9) will require many more years of research. A brief summary of the state of the art is, nevertheless, necessary to allow a better grasp of the Clinical Aspects of Interferons. The reader is refered to reviews for more detailed information and citations (3-6).

Functions of IFNs

The first IFN was discovered by Isaac and Lindenman in 1957 by its ability to inhibit growth of viruses and to protect infected cell cultures against the viral cytopathic effects (antiviral effect). The subsequently discovered cell growth inhibitory action (antigrowth effect), as well as multiple cell regulatory effects (affecting cell structure, activity and differentiation) are probably as important for IFN clinical applications. Particularly important are the immunoregulatory functions of IFN (7) as increase in natural killer (NK) lymphocyte activity, increase in Histocompatibility antigens of class I (all IFNs) or class II (IFN-γ), monocyte/macrophage activation (IFN-γ) and B-cell functions (all IFNs) (7,8).

Classification of IFNs

The classification was originally based on immunological criterias: the IFNs made by virus-infected leukocytes, which could be separated in several protein molecules, were all neutralized by the same antibodies and designated as IFN-α, while the IFN made by fibroblasts and other solid tissues was neutralized by different antibodies and designated IFN-β. Immune IFN, produced by T-cell mitogen and antigen stimulation was yet immunologically distinct and called IFN-γ (9). Cloning of the cDNAs and genes for these molecules and sequencing of the cDNAs and part of the proteins, allowed to establish a more sophisticated nomenclature. IFN-α activity was found to be due to more than 20 protein molecules produced by genes clustered on human chromosome 9 (1). These IFN-α proteins are now designated by numbers, e.g. IFN-α2 (itself subdivided in 2a, 2b or 2c) is the species mostly used in clinical trials. In their secreted form, IFN-αs are 165-6 aminoacid long (23 residues being removed from the N-terminus) and share 70-90% sequence homology. A more remote group has been named IFN-αII. The major species induced by double stranded (ds) RNA in fibroblasts, IFN-β1, has also 166 aminoacids but is a glycoprotein with 30-40% homology to the various IFN-αs; it is encoded by a more distant gene on chromosone 9 which like the IFN-α genes has no introns. The IFN-γ glycoprotein has no significant homology to the others, and its gene on chromosome 12 has introns. Additional IFN species are known to exist: cells undergoing differentiation produce low titers of autocrine IFN, neutralized by antibodies to IFN-β, but apparently different and which participate in growth arrest during terminal differentiation (2,10); IFN-β-2 (BSF-2/ HGF/ HSF), a glycoprotein with 15-20% homology to other IFN-αs and β, and which stimulates differentiation and growth of B-lymphocytes, and induces acute phase proteins in liver cells (10); autoimmune IFN-α circulating in patients with autoimmune diseases and AIDS, which in contrast to the other species is acid labile (destroyed at pH2) (12); epithelial cell IFN which may again be different, and the list is probably not closed (1,2).

IFN measurements

Standard measurements of IFN activity is based on the antiviral effect, one unit being the dilution per ml providing at least 50% protection to an infected cell monolayer. Most purified IFNs have typically around $5x10^8$ units per mg protein on human cells, with some species showing 10-100 fold lower specific activities, as IFN-α1 or IFN-γ) or a few fold higher as IFN-αC. The activity may vary with the cells and virus used and must be standardized by an international reference IFN solution. Typically the action of IFN is species specific. However, human IFN-αs are active on bovine and porcine cells, and modification of the molecules (as formation of hybrids between IFN-α1 and α2) can lead to high activity on murine cells (1). It is likely that the binding affinity of the protein to a human or animal cell surface receptor is the main factor determining species specificity.

Mechanism of action of IFNs

The IFN-αs and β species (type I IFNs) compete for binding to the same human cell surface receptor which is encoded by a gene on chromosome 21 (12). In contrast, IFN-γ (type II IFN) binds to a separate receptor mapped on chromosome 6 (13). The type I receptors are sparse (2-5000 per cell) but affinity is high (Kd about 10^{-11}M); the type II IFN receptors appear more abundant. The receptors are large proteins (110 and 90 Kd respectively) which appear to be present on most cell types, although cell-specific differences may exist. The immediate events following IFN binding are not entirely clear: changes may be observed affecting lipid membranes, cytoskeleton, and cGMP levels. What is clear is that after a few minutes, the cell responds by activating the transcription of a number of specific genes (6). In this respect, IFNs are similar to most protein hormones and growth factors. The IFN-receptor complex is slowly transported into the cell, but it is still debated if some molecules reach the nucleus fast enough to trigger IFN response or if a second messenger molecule is involved.

It is estimated that at least 20 genes are activated in cells exposed to IFNs (6). A number of these genes have already been cloned, and their promoters appear to contain a common short sequence which determines activation by IFN. The products of these genes can be classified in several groups:

i) the translation regulatory enzymes which were first isolated from extracts of IFN-treated cells by their ability to inhibit translation in untreated cell extracts. The two main enzymes are the ds RNA activated protein kinase which phosphorylates initiation factor eIF-2, thereby impairing protein synthesis initiation, and the (2'-5') oligo A synthetase which polymerizes ATP into $ppp(A2'p)_nA$ oligonucleotides, whose only known function is to activate a latent Ribonuclease, leading to ribosomal RNA nicking and to the arrest of translation (3-5). This Ribonuclease and a 2'-phosphodiesterase degrading the (2'-5') oligoadenylate are also IFN-inducible in some cell systems. The (2'-5') oligo A synthetase cDNAs and gene were cloned and at least 4 forms of the enzyme are produced in IFN-treated cells (14). Assay of this enzyme in peripheral white blood cells can serve to monitor IFN action in vivo (see Chapter 3.3)

ii) surface proteins among which the HLA class I and $\beta_2$microglobulin gene products are probably best understood. These HLA molecules bind non-self antigens and are recognized by the cytotoxic T-lymphocytes; their increase on cells can favor destruction of infected cells and tumor cells. Per antiviral unit, IFN-γ is 100 times more efficient than type I IFNs to induce HLA class I. The HLA class II genes are only activated by IFN-γ and in some cells by IFN-β. The HLA class II antigens are essential to initiate an immune response and for presentation of foreign antigens by monocytes (4,6,7,16). Immunoglobulin Fc receptors are also induced by IFNs. Other cell surface and secreted proteins are increased in IFN-treated cells but their function is still unknown (6). A 17Kd protein membrane protein may have an antigrowth action (15) and one form of the (2'-5') oligo A synthetase (69Kd) is also membranal (14).

iii) Intracellular proteins, a few of which have known functions. One example is the Mx protein in mouse cells: mouse genetics revealed that the Mx gene responsible for the resistance of the animal to influenza virus lethal effect, encodes a type I IFN-induced 72 Kd nuclear protein that inhibits flu virus RNA transcription (17). In man, this protein exists in the cytoplasm but may have other functions. Another example is indol oxydase, induced by IFNs (mainly IFN-γ) and which leads to tryptophan deprivation. This may be the mechanism by which growth of the toxoplasma parasite is inhibited by IFN (see Chapter 1.7). Xanthine oxydase and an enzyme that increases degradation of guanine into neopterin (often excreted in the urine of patients treated by IFN) are further examples of IFN-induced catabolytic enzymes (18). Many IFN-induced proteins are known only as dots on two-dimensional gel electrophoresis, or as cDNAs found by screening libraries for IFN-induced sequences (e.g. C56, ISG-54, ISG-15, 1-8, 6-16) (see 6 for review). Such DNA screens also indicated that Metallothionein (a detoxification enzyme) and thymosin β4 (a lymphocyte differentiation factor) gene expression is increased by IFN.

The relative induction of these genes by type I and II IFNs varies. Some, like the Mx, ISG-54, ISG-15 and 6-16 proteins, are induced only by type I IFN. Conversely, a group of proteins induced by IFN-γ only was defined, it includes HLA-DR, Fc receptor α chains and a protein related to Platelet Factor 4 with a possible role in chemotaxis (see 6 for review). The basic working hypothesis on IFN action is that these proteins mediate the various biological effects of IFNs. Recombinant DNA technology allows now to test this hypothesis by introducing IFN-activated genes into cells in such a way as to obtain constitutive production of the proteins. Thus, the Mx protein was shown to confer to mouse cells a specific resistance to influenza virus (17). Similarly, we could show that the (2'-5') oligo A synthetase mediates by itself the resistance to picornavirus infections and reduces the clonogenic potential of some tumor cells (19). Despite these beginnings, it is much too early to attempt to explain the antiviral, antigrowth and cell-regulatory functions of IFNs by the induction of one or another of the induced proteins. Interesting correlations can be attempted such as the study of tumor cells having acquired resistance to the antigrowth effect of IFNs: e.g. Burkitt lymphoma Daudi cells which are very sensitive to the antigrowth effect can readily form Daudi^R growth-resistant clones which have lost the ability to induce some of the genes in response to IFN while still inducing others. In this respect, the 17Kd membrane protein which is no more induced in Daudi^R, and inhibits growth when added to other cells, is particularly interesting (15). The (2'-5') A oligonucleotide product of the IFN-induced synthetase has also antimitogenic effects and suggestive correlations betwen the level of the enzyme and growth have been observed. This enzyme appears also induced periodically at the end of the S-phase of the cell cycle (6). A particularly usefull approach has been the study of IFN effects on oncogene expression. Inhibition of c-myc, c-ras and other oncogenes, have been observed in correlation with cell growth arrest induced by IFN (20,21). However, many tumor cells appear to have mutated to resistance against this IFN effect. Both antiviral action and antigrowth effect are probably mediated by several mechanisms, making correlations with single molecular events impossible. Multiple mechanisms of action may account for the fact that no IFN-resistant virus mutant has yet been found.

Direct and indirect IFN effects

The in vivo tumor inhibitory action of IFN observed in animal models cannot be explained only by the direct antigrowth action on tumor cells (22). Study of xenografts (see Chapter 3.5) have shown that there are indirect antitumor effects mediated by the host, and that even cells resistant to the antigrowth effect of IFNs can be eliminated by the animal. Some of the effects observed are host mediated catabolic alterations in phospholipids, and may involve interplay with other cytokines such as Tumor Necrosis Factor (TNF) (23). IFNs stimulate the secretion of TNF by monocytes (24). IFN-γ potentiates the killing of tumor cells by TNF and TNF was shown to kill virus-infected cells but not normal cells 24). Indirect mechanisms mediated by the host immune system are probably important also for the antiviral action of IFN . The role of immune cells in IFN's antiviral and antitumor effects appear complex (see Chapter 3.8 and 3.9), involving secretion of TNF by monocytes, of lymphotoxin by T-lymphocytes and other cytokines, changes in HLA-A,B,C antigen ratios which make the cells more vulnerable to killer lymphocytes, as well as activation of killer cells themselves (e.g. NK cells) (7). Increased NK activity is itself not straightforward since IFN-γ has also a protective effect on cells against NK killing. Fever, leucocyte migration to the periphery with leucopenia (25), which characterize IFN action in vivo, may also result from complex cytokine interactions. Killing of the virus-infected cells as an end result of IFN's indirect actions, and further helped by the HLA increase, is probably crucial for recovery from acute and chronic viral infections where just reducing viral proliferation would be insufficient without the active participation of the immune system. IFNs may have evolved to orchestrate these different defense mechanisms; an interesting case is IFN-β-2 (BSF-2/HGF/HSF) which combines antiviral and antigrowth actions with stimulation of antibody secretion by B-lymphocytes, plasma cell proliferation and induction of acute phase proteins by liver cells, regulating many components of inflammation (reviewed in 10).

The scope of IFN actions is thus much broader than just its direct antiviral and antigrowth activities. The numerous actions of IFNs in vivo makes it difficult to pinpoint the mechanism at play when testing IFN in trials against specific infections or a particular tumor. Nevertheless, we know from elegant in vivo experiments in animals injected with anti-type I IFN antibodies (26) , that the IFN system is of real importance in the defence of the organism against viral diseases and cancer.

Clinical aspects of IFN: why this book ?

With the limited biological information available, it may be surprising that so much clinical data on the efficacy of IFN have been generated in the past few years to warrant the compilation attempted in this volume. Immediately after its discovery, there was much hope that IFN would be the answer to many viral diseases. This has not been so and many years passed untill it was recognized, mainly through the work of Thomas Merigan and his group, that a 100-1000 fold more IFN has to be administred for clinical results than what is needed in vitro with cultured cells. With the availability of Kari Cantell's leukocyte IFN, and since the 80's of IFNs produced by recombinant DNA technology (1,2), a period of new exaltation began, followed by disappointment for those who wished for the magic bullet that would cure viral diseases and cancer. Excellent clinical work on specific indications has now reverted this trend again, and through the realization of what IFN can do in particular diseases (e.g. Hairy cell leukemia, Chapter 2.2), hope is growing again. IFNs have often significant efficacy as single agents, which justifies their use, but it is obvious that the future must be combinations of IFNs with antiviral chemicals and antitumor irradiations and chemo-therapeutic drugs, as well exposed in Chapters 2.1 and 2.11. More important even, although much more remote in time, is the potential of combinations with other cytokines, hemopoietic colony stimulating factors and other biotherapeutic agents.

Ideally, our goal was to provide physicians with enough reference information on the applications of IFNs in specific clinical situations, to prompt the use of IFN and to further explore its clinical potentials. The chapters in the first two parts of this volume, written by experts who often reported the first clinical trials on a given application, give a state of the art review of what has been achieved in viral infections, immune diseases, leukemias and solid tumors. The third part discusses the important topics of toxic and side effects of IFNs, of pharmacokinetics and of methods to monitor IFN therapy in man. The last 6 chapters describe clinically oriented research on animal models, the use of IFN inducers, targetting of IFN, the complex interactions with the immune system, and should give the clinician an insight on what is to come.

The fast pace of IFN clinical research, and other unfortunate reasons, have led to omission of some of IFN's application as for example in ophtalmology (briefly summarized in chapter 1.3). Some of the omitted topics deserve to be mentionned. Many studies have been devoted to Chronic active hepatitis where IFN, sometime in combination with Adenosine arabinoside, improve significantly the biological parameters (27). A recent study indicates that IFN-γ leads to improvements in Rhumatoid arthritis (28). Local injections of IFN-α2b appear very efficient in the treatment of Basal cell carcinoma (29). Finally a preliminary report indicates that IFN-α injections leads to disappearance of HIV in patients with asymptomatic AIDS, suggesting that IFN therapy may be helpfull to reduce contagion (30). More is obviously to come.

It is a pleasure to thank here the colleagues who have contributed their knowledge and skills to the chapters of this book. We are all deeply indebted to them for their lucid and informative presentations.

Michel Revel
Weizmann Institute of Science,
Rehovot, Israel

References.

1. Weissmann, C. and Weber, H. In:Progr. Nucl. Acids Res. Vol 33 (Eds W.E. Cohn an K. Moldave), Academic Press, New York, 1986, pp 251-300.
2. Revel, M. In: Interferon 5 (Ed. I. Gresser), Academic Press, London, 1983, pp 205-239.
3. Lengyel, P. Ann. Rev. Biochem. 51:251-282. 1982.
4. Revel, M. In: Antiviral Drugs and Interferon: The Molecular Basis of Their Action (Ed. Y. Becker), Martinus Nijhoff Publ., Boston, 1984, pp 357-433.
5. Sen, G.C. Pharmac. Ther. 24:235-257, 1984.
6. Revel, M. and Chebath, J. Trends Biochem. Sci. 11:166-170, 1986.
7. De Maeyer-Guignard, J. and De Maeyer, E. In: Interferon 6 (Ed. I. Gresser), Academic Press, London, 1985, pp 69-91.
8. Siegel, D.S., Le, J. and Vilcek, J. Cell. Immunol. 101:380-390, 1986.
9. Stewart, W.E.II (Committee on Interferon Nomenclature) Nature 286:110, 1980.
10. Revel, M., Zilberstein, A., Ruggieri, R., Rubinstein M. and Chen, L..J. Interferon Res. 7:221-228, 1987
11. Vilcek, J. In: Interferon 4 (Ed. I. Gresser), Academic Press, London, 1982. pp 129-154.
12. Aguet, M. and Mogensen, K.E. In: Interferon 5 , (Ed. I. Gresser), Academic Press, London, 1983, pp 1-22.
13. Pestka, S., Rashidbaigi, A., Langer, J.A. et al. In: Interferons as Cell Growth Inhibitors and Antitumor Factors (Eds. R.M., Friedman, T., Merigan and T. Sreevalsan), Alan R. Liss, New York, 1986, pp 259-268.
14. Chebath, J., Benech, P., Hovanessian, A., Galabru, J. and Revel, M. J. Biol. Chem. 262:3852-3857, 1987.
15. Knight, E., Fahey, D., Kutny, R. and Blomstrom, D.C. In: The Biology of THe Interferon System 1986 (Eds. K. Cantell and H. Schellekens), Martinus Nijhoff Publ., Boston, 1987, pp 93-98.
16. Rosa, F.M., Cochet, M.M. and Fellous, M. In: Inteferon 7 (Ed. I. Gresser), Academic Press, London 1986, pp 47-87.
17. Staeheli, P., Haller, O., Boll, W., Lindenman, J. and Weissmann, C. Cell 44:147-158, 1986.
18. Nathan, C. F. In: Interferon 7 (Ed. I. Gresser), Academic Press, London, 1986, pp 125-143.
19. Chebath, J., Benech, P., Revel, M. and Vigneron, M. in press, 1987.
20. Einat, M., Resnitzky, D. and Kimchi, A. Nature 313: 597-600, 1985.

21. Samid, D. and Friedman, R.M. In: Interferons as Cell Growth Inhibitors and Antitumor Factors (Eds. R.M. Friedman, T., Merigan and T. Sreevalsan), Alan R. Liss, New York, 1986, pp 413-422.
22. Gresser, I. In: Interferon 6 (Ed. I. Gresser), Academic Press, London, 1985, pp 93-126.
23. Gresser, I., Belardelli, F., Tavernier, J. et al. Intern. J. Cancer 38:771-778, 1986.
24. Wallach, D. In: Interferon 7 (Ed. I. Gresser), Academic Press, London, 1986, pp 89-124.
25. Schattner, A., Meshorer, A. and Wallach, D. Cell. Immunol. 79:11-25, 1983.
26. Belardelli, F., Vignaux, F., Proietti, E. and Gresser I. Proc. Natl. Acad. Sci. USA 81:602-606, 1984.
27. Lever, A.M.., Chan, G., Scully, L.J. and Thomas, H.C. In: The Biology of The Interferon System 1986 (Eds. K. Cantell and H. Schellekens), Martinus Nijhoff Publ., Boston, 1987, pp 477-484.
28. Obert H.-J., ibidem, 1987, pp 511-518.
29. Smiles, K., Peets, E., Tanner, D. and Taylor E. ibidem, 1987, pp 493-501.
30. Kovacs, Lane, H.C., Masur, H., Herpin, B., Feinberg, J. and Fauci, A.S. Abstracts 3d International Conference on AIDS, Washington, 1987, p18 (MP228).

CLINICAL ASPECTS OF INTERFERONS

I

Clinical uses in infectious and immune diseases

1

INTRANASAL INTERFERONS FOR CONTROL OF RESPIRATORY VIRAL INFECTIONS

FREDERICK G. HAYDEN, M.D.
Departments of Internal Medicine and Pathology,
University of Virginia School of Medicine,
Box 473, Charlottesville, Virginia 22908 (U.S.A.)

INTRODUCTION

Shortly after the discovery of interferon, its broad in vitro antiviral spectrum and potential ability to be applied directly to the upper respiratory tract were recognized as desirable characteristics for use in respiratory viral infections (1). In 1973 Merigan et al showed that intranasal administration of human leukocyte interferon (HuIFN-α [Le]) in a total dosage of 14 Mu over 4 days reduced seroconversion and illness rates following experimental rhinovirus type 4 challenge (2). A subsequent study at the MRC Common Cold Unit by Scott et al utilizing a lower dosage ($\geq$ 0.6 Mu) of fibroblast-derived interferon (IFN-β) found no evidence of protection (3). Scott et al later showed that high dosages of purified IFN-α(Le) (90 Mu over 4 days) protected against experimental rhinovirus type 9-induced illness (4). Studies conducted at Baylor College of Medicine identified rapid nasal clearance and insufficient contact time with the nasal mucosa, concentration dependency of antiviral action, and perhaps inactivation of IFN-β by nasal secretions as factors contributing to the relatively high interferon amounts needed to achieve antiviral effects in the nasal mucosa (5-9).

During the past five years the availability of highly purified, recombinant DNA-produced alpha (rIFN-α), and more recently beta (rIFN-βser) and gamma (rIFN-γ), interferons has stimulated considerable investigation. This review summarizes the results of recent clinical studies using intranasal interferons for the prevention and treatment of respiratory viral infections. It discusses the toxicity and virus-specific efficacy observed in studies of experimentally induced and naturally occurring infections.

EFFICACY IN EXPERIMENTAL INFECTIONS

Rhinovirus

A number of randomized, double-blind, placebo-controlled studies have tested the prophylactic efficacy of intranasal interferons in experimental rhinovirus infection (reviewed in ref. 10). These trials have conclusively shown that two closely related species, recombinant interferon-alfa2b (rIFN-α2b) and leukocyte A interferon (rIFN-α2a) provide protection against experimental colds (11-17). Protection has been documented against five different rhinovirus serotypes, a finding which correlates with *in vitro* observations indicating that interferons have broad spectrum although variable activity against rhinoviruses (18).

Dose-related efficacy. The level of protection appears to depend on both the interferon dosage and the duration of administration prior to virus challenge. Scott *et al* found that a high rIFN-α2b dosage (22.5 Mu/day in 9 divided doses) begun 1 day prior to challenge was associated with protection against rhinovirus infection (51% efficacy) and against definite colds (100% efficacy) (11). Hayden *et al* found that an rIFN-α2b dosage of 46 Mu/day in 4 divided doses was 78% effective in preventing infection and 100% effective in preventing colds due to rhinovirus type 39 (12). Dose-response studies by Samo *et al* found that rIFN-α2a 10.0 Mu/day in 2 divided doses prevented illness (89% efficacy) and had discernible effects on rhinovirus type 13 shedding, but allowed subclinical infection to take place (13). A dosage of 2.4 Mu/day was also protective against illness (60% efficacy) but not infection, whereas a dosage of 0.7 Mu/day provided no protection against illness (14). Treanor *et al* found that even relatively low doses of rIFN-α2b (2.0 Mu/day) were protective against both infection (63% efficacy) and illness (84% efficacy) when begun one week prior to virus challenge (15). Turner *et al* also observed protection with rIFN-α2b 2.0 Mu/day begun 7 days prior to virus exposure (16). The mechanism by which interferon reduces symptoms without preventing infection is undefined, but this pattern of protection could have the advantage of preventing illness but not the development of immunity to the infecting virus.

Several studies have assessed the durability of interferon's *in vivo* effects. In studies employing biopsies of nasal mucosal epithelium from uninfected volunteers, Greenberg *et al* found in an *in vitro* assay that significant antiviral effects lasted 18 hours but not 24 hours after _in vivo_ interferon exposure (9). In contrast, Harmon _et al_ found that human nasal epithelial cells exposed to HuIFN-α(Le) _in vitro_ exhibited a prolonged (48-72 hours) antiviral state (8). In volunteer challenge studies, Hayden _et al_ determined that a high rIFN-α2b dose (43 Mu) given once daily was associated with significant antiviral effects and provided protection (75% efficacy) against rhinovirus-induced illness (12). Phillpotts _et al_ used varying dose regimens and intervals between interferon administration and virus challenge to conclude that administration 3 times per day provided optimal protection against experimental colds (17). However, as discussed below, field trials have established that dosing once daily is protective against natural rhinovirus colds. The available evidence suggests that prolonged protection against rhinovirus infection occurs after interaction of interferon with the respiratory mucosa.

Type of interferon. No study has directly compared the relative activities of different interferons. The effects seen with rIFN-α2b or rIFN-α2a appear similar and comparable to those observed with HuIFN-α(Le) (4,10). Using HuIFN-α(Ly) 8.1 Mu/day in 3 divided doses Phillpotts _et al_ observed protection against illness (100% efficacy) but not infection (26% efficacy) after rhinovirus type 9 and 14 challenge (19). Higgins _et al_ found that rIFN-βser (6.5 Mu/day in 3 divided doses) also provided protection against rhinovirus-induced illness (67% efficacy) but not infection (24% efficacy) (20).

Method of application. The method of applying interferon to the nasal mucosa appears to be an important factor in determining its intranasal distribution and antiviral effects. Using different methods to increase contact time with the nasal mucosa, Greenberg _et al_ found that application of HuIFN-α(Le) by saturated cotton pledgets or by drops in volunteers pretreated with oral antihistamines, but not by drops alone, induced an antiviral effect in nasal epithelial cells (9). Relatively low doses of HuIFN-α(Le) (1-4 Mu), administered by saturated cotten pledgets or by aerosol to antihistamine-pretreated volunteers, were associated with modest reductions (40% efficacy) in illness frequency compared to placebo (21). When delivered by intranasal spray or drops, higher interferon dosages have been necessary to provide solid protection against experimental rhinovirus colds.

Using solutions of radiolabelled human serum albumin, Aoki and Crawley (22) and more recently Hardy et al (23) found greater coverage of the nasal passages following administration by drops than by spray. In a study of rIFN-α2b treatment of experimental rhinovirus infection, Hayden et al indirectly compared these methods of administration and found that dosing by drops may have been associated with greater antiviral and clinical effects than by spray (24). Further studies are needed to determine the optimal methods for delivering interferons to the upper respiratory tract.

Therapeutic activity. Little information has been published on the possible therapeutic activity of intranasal interferon given after experimental rhinovirus infection. Hayden et al found that administration of rIFN-α2b (27.0 Mu/day in 3 divided doses for 5 days) beginning 28 hours after rhinovirus type 39 challenge was associated with significant decreases in the duration of viral shedding and titers of virus recovered in nasal washings, but only modest clinical benefit (24). In this study subjects treated with interferon by nasal drops experienced 40-50% reductions in peak nasal symptoms scores, mucus weights, and nasal tissue use compared to placebo recipients. A reduced duration of virus shedding could potentially confer a reduced risk of transmitting infection.

Coronavirus

At the MRC Common Cold Unit, Higgins et al found that a relatively high rIFN-α2a dosage (8.8 Mu/day in 3 divided doses for 4 days) provided protection against both infection (60% efficacy) and illness (85% efficacy) following coronavirus 229E challenge (25). The proportion of subjects excreting virus was significantly reduced on all post-challenge days in the interferon group. Turner et al used longer exposure to a lower dosage of rIFN-α2b (2.0 Mu/day for 7 days before and after coronavirus 229E challenge) and found significant but incomplete protection (44% efficacy) against coronavirus-induced illness (26). No reductions in infection rates or viral shedding were observed. These findings suggest that interferon efficacy is also dose-dependent in experimental coronavirus infection.

Influenza Virus

Studies involving different types of interferon and influenza viruses have found no significant protection against infection and variable protection against influenza virus-induced illness. Merigan et al found that low dosages of HuIFN-α(Le) (0.8Mu/day) were not protective against influenza B virus challenge (2). Dolin et al found that intranasal administration of rIFN-α2b (10.0 Mu/day in 2 divided doses) starting two days before intranasal challenge with H1N1 subtype influenza A virus was associated with significant reductions in the number of days of virus shedding (36% fewer) and the frequency (66% efficacy) and severity of respiratory illness (27). Schiff used two lower dosages of rIFN-α2a (3.6 or 7.2 Mu/day) beginning on the same day as virus challenge with a H3N2 subtype influenza A virus and found no important effects on illness rates or the severity of illness (28). A trial by Phillpotts et al using HuIFN-α (Ly) 8.1 Mu/day found partial protection against illness (48% efficacy) following influenza A/H3N2 subtype challenge (19).

EFFICACY IN NATURAL INFECTIONS

The efficacy of intranasal rIFN-α2b in preventing naturally occurring respiratory viral infection and illness has been examined in randomized, double-blind, placebo-controlled field studies. As discussed below, there is considerable evidence that this interferon is effective for prophylaxis of natural rhinovirus infections. However, despite the broad antiviral spectrum of interferons in vitro and the protection observed in experimental coronavirus infections, no clear efficacy has been found against natural infection with viruses other than rhinovirus at the dosages tested (Table 1). For several viruses, insufficient information is

Table 1. Virus-Specific Prophylactic Efficacy of Intranasal rIFN-α2b

Respiratory Virus	Experimental Infection	Natural Infection
Rhinovirus	++	++
Coronavirus	++	0
Influenza A Virus	+	0
Parainfluenza Virus	N.D.	0
Respiratory Syncytial Virus	N.D.	N.D.
Adenovirus	N.D.	N.D.

ABBREVIATIONS: ++, definite protection; +, partial or variable protection; 0, no protection at dosages studied to date; N.D., not determined.

available to make determinations about specific efficacy. Possible explanations for the lack of broad prophylactic activity against all of the respiratory viruses include a lower susceptibility of these viruses to the _in vivo_ antiviral effects of interferon compared to rhinovirus, and the inability to deliver interferon to sites, such as the lower respiratory tract, where infection may be initiated or continue during the course of natural infection. Whether higher dosages of intranasal interferon will provide protection against these other viruses remains to be determined.

Rhinovirus

Seasonal prophylaxis. In 1982 Farr _et al_ conducted a 3-week prophylaxis study in Charlottesville, Virginia, during which adults self-administered nasal sprays of rIFN-α2b (10.0 Mu/day) or placebo once daily (29). Interferon was associated with complete protection (100% efficacy) against laboratory proven rhinovirus infection compared, to a 9% attack rate in the placebo group. In a similar study in Rochester, New York, Betts _et al_ found that the same total interferon dosage given in 2 divided doses was also associated with protection (93% efficacy) against natural rhinovirus infections (30). However, these studies found discrepancies between virologic and clinical efficacy, and both were terminated one week earlier than planned because of the occurrence of nasal irritation in interferon recipients. Although interferon was associated with reductions in the occurrence of cough, the total number of episodes of respiratory illness which included nasal symptoms were significantly higher in interferon than placebo recipients due to the nasal side effects of the interferon.

Studies using lower interferon dosages have been conducted to identify clinically acceptable dose regimens for long-term use. Hayden _et al_ found that rIFN-α2b (2.5 Mu/day in 2 divided doses) was associated with a high rate of minor nasal side effects (44% with blood in mucus) (31). This study was terminated after 12 days without evidence of protection, although it found no evidence of sensitization in those who had been previously exposed to the same interferon. During a 4-week study in Adelaide, Australia in 1983, Douglas _et al_ found that rIFN-α2b (2.0 Mu/day in 2 divided doses) was associated with 87% protective efficacy against rhinovirus infection, compared to the 3.9% infection rate in the placebo group (32). However, 8% of interferon recipients withdrew because of nasal side effects and 20% experienced blood-tinged mucus or bleeding for much or all of the study (32). No reductions in episodes or specific symptoms of respiratory illness were found. In 1983 Monto et al conducted two parallel 4-week studies at the University of Michigan (33). Among the 400 students assigned to rIFN-α2b 3.0 Mu/day in 2 divided doses or placebo, the protective efficacy against rhinovirus infection in the interferon group (3.5% infection rate) was 76% compared to placebo (14.6%). Among the 150 students given rIFN-α2b 2.5 Mu once daily or placebo, the efficacy against rhinovirus infection was 59% in the interferon group (6.7%) compared to placebo (16.2%). However, neither group of interferon recipients had reductions in numbers of respiratory illness episodes, although the severity of illness was significantly less in symptomatic interferon than placebo recipients. Both interferon groups had approximately 3-fold higher frequencies of blood-tinged nasal mucus during the study compared to placebo.

Although these studies have not directly determined dose-response in regard to efficacy, one study using a low dosage (1.7 Mu/day) found marginal protection against natural rhinovirus infection (34). The results of the different trials suggests that the minimally effective interferon dosage for prophylaxis of natural rhinovirus infections is approximately 2-3 Mu/day. However, long-term administration of rIFN-α2b at this dose level is not feasible for prophylaxis of rhinoviral infections in healthy adults because of the high incidence of nasal adverse effects.

Studies in high risk populations, such as those with asthma or chronic obstructive pulmonary disease, have found patterns of local nasal toxicity similar to those observed in healthy adults, but no evidence of lower respiratory tract intolerance determined clinically and by spirometric measurements (35). One 4-week study in asthmatic children and adults by Michael et al found no significant overall clinical benefit or protection against rhinovirus colds at a rIFN-α2b dosage of 2.0 Mu/day (36). A study in asthmatic children at the same dosage given over three 4-week periods showed no differences in the frequency of colds or asthma symptoms between interferon and placebo recipients (37).

Postexposure prophylaxis. Investigators at several institutions have completed studies to determine the usefulness of short-term intranasal rIFN-α for preventing illness in household contacts exposed to family members with respiratory illness. In a study involving 147 families in Basel, Switzerland, Herzog et al administered low dosages of rIFN-α2a (0.3 or 1.5 Mu/day in 2 divided doses for 5 days) to both the index case and the contacts (38). Interferon did not prevent transmission of colds to family contacts, in whom secondary attack rates were 24% (placebo), 20% (IFN 0.3 Mu/day), and 17% (IFN 1.5 Mu/day), but its use appeared to reduce the severity and duration of illnesses that occurred in the contacts (38). In an 8-month study involving 60 familes in Charlottesville, Virginia, Hayden et al found that higher doses of rIFN-α2b (5.0 Mu once daily for 7 days) reduced the occurrence of total respiratory illness (14% secondary attack rate) in healthy family contacts by 39% compared to placebo (23%) (39). The protective effect appeared limited to rhinovirus infections, which were reduced by 88% in interferon recipients during spray use. In family members on interferon who were contacts of an index case with a laboratory documented rhinovirus cold, the risk of developing a respiratory illness or a proven rhinovirus cold was reduced by 79% compared to the placebo group. Minor mucosal bleeding occurred more often in interferon recipients (13.6% of spray uses) than placebo (7.7%), but no evidence of cumulative toxicity was observed (39).

Two subsequent studies utilizing a similar design have been conducted at other centers. Douglas et al found comparable results to the Charlottesville trial in 97 families studied over 6 months in Adelaide, Australia (40). This trial found a 41% reduction in episodes of definite respiratory illness and a 33% reduction in secondary days of nasal symptoms in interferon recipients compared to placebo. The protective efficacy against proven or suspected rhinovirus related illness was 62% overall and 78% in instances when rhinovirus was recovered from the index case. In a study involving 53 families in Seattle, Foy et al found a trend toward fewer illnesses (52% efficacy) during interferon (15% secondary attack rate) compared to placebo (32%) spray use, when rhinovirus was isolated from the index case or other family member (41).

These trials established that it is possible to interrupt the transmission of rhinovirus colds in the family setting by post-contact prophylaxis. This is the first strategy for using intranasal interferon to clearly show prevention of respiratory illness under natural conditions, primarily because rIFN-α2b is generally well-tolerated when used in this manner. The efficacy and safety of postexposure prophylaxis needs to be studied in groups at increased risk for the complications of rhinovirus infection.

Therapeutic activity. Herzog et al reported that low dosages of rIFN-α 2a (0.3 or 1.5 Mu/day) had no therapeutic effect in natural colds (42). In a field trial involving adults with colds of diverse viral etiologies, Just et al recently found that rIFN-α2a (12.0 Mu/day in 4 divided doses for 5 days) provided no symptom benefit compared to placebo (43). A placebo-contolled study of rIFN-α2b treatment (10.0 or 20.0 Mu/day in four divided doses for 5 days) of natural rhinovirus colds has been recently completed at the University of Virginia (44). Preliminary analysis indicates that interferon treatment did not provide clinical benefit. These negative findings could relate to inability to initiate treatment early after the onset of symptoms, difficulty in effectively delivering interferon to the nasal mucosa in symptomatic individuals, or to the possibility that ongoing viral replication is not central to the pathogenesis of symptoms in rhinovirus infections.

Coronavirus

Very limited information is available about efficacy against natural coronavirus infections. Douglas et al found no evidence of protection against coronavirus infection in their seasonal prophylaxis study during which approximately 2% of interferon or placebo recipients had serologic evidence of coronavirus infection (32). This group also found no reduction in coronavirus infection in interferon recipients during the family-based prophylaxis study described above (40).

Parainfluenza Virus

During seasonal prophylaxis studies utilizing relatively low dosages of rIFN-α2b (2.0-3.0 Mu/day), no protection has been found against parainfluenza virus infection (31-33). Depending on the dose schedule, Monto et al found similar infection rates in those given interferon once (5.3% of subjects) or twice (7.1%) daily and in the corresponding placebo group (6.8% or 6.0%) (33). In those with parainfluenza virus-related infection, the severity but not the duration of illness appeared to be less in interferon recipients compared to placebo. In a family-based contact prophylaxis study, Hayden et al also found no evidence of protection against parainfluenza virus infection or illness, although the number of infections for analysis was small (39).

Influenza Virus

One field trial conducted by Isomura _et al_ found that small doses of HuIFN-α(Le) (0.01 Mu/day for 8 weeks) did not effect the frequency of serologically documented influenza or febrile episodes in young children, compared to placebo, but appeared to shorten the duration of episodes of fever and respiratory illness (45). A subsequent 3-month study by Saito _et al_ used 0.05 Mu/day in adults and found no reduction in influenza A virus-related illness (38% attack rate in each group) (45). In a recent 4-week prophylaxis study in Newcastle, Australia, Tannock used rIFN-α2a dosages ranging from 1.5 to 6.0 Mu/day but did not observe protection against influenza A virus infection or illness (47). No evidence of protection against influenza virus transmission was observed in the family-based contact prophylaxis studies (39-41).

TOXICITY OF INTRANASAL INTERFERONS

Systemic Tolerance

In contrast to parenterally administered interferons, the only documented systemic toxicity of intranasal interferon has been transient leukopenia (Table 2). Reversible decreases in granulocyte counts have observed in some (12,24,29) but not all studies (4,11,30) utilizing higher interferon dosages. During a 3-week prophylaxis study, Farr _et al_ observed leukopenia (WBC <4,000/mm^3) in 11% of those given rIFN-α2b 10 Mu/day (29).

Table 2. Adverse Effects Related to Intranasal Interferon

Toxicity	Symptoms	Signs
Systemic	None	Leukopenia
		Anti-IFN antibody
Nasal	Dryness, stuffiness	Mucosal friability
	Blood in mucus	Punctate bleeding
	Epistaxis	Erosions, ulcerations
		Histopathologic changes

Nearly two-thirds of the leukopenia patients had symptoms and signs of nasal irritation. In rhinovirus-challenged volunteers, Hayden _et al_ found that 21% given rIFN-α2b 44 Mu/day developed mild leukopenia (12). The cause of the changes in leukocyte counts has not been determined, but high interferon dosage and associated nasal mucosal irritation appear to be risk factors for its occurrence.

Another concern is the immunogenicity of synthetic interferons after repetitive or prolonged exposure. Two of over 1,300 individuals given intranasal interferon have been reported to develop transient circulating antibody against rIFN-α2b without recognized consequences (48). Limited attempts have failed to detect antibody in nasal secretions after exposure (49).

Local Tolerance

Although intranasal application avoids most of the systemic toxicities observed with parenteral administration, nasal mucosal toxicity has been a significant clinical problem (Table 2). The occurrence of nasal symptoms and mucosal abnormalities depends on both the dosage and duration of exposure. High dosages of HuIFN-α(Le) and HuIFN-α2b have been associated with a discernible increase in nasal symptoms after four days in some healthy subjects (4,11). As discussed above, relatively low doses of rIFN-α2b (2-3 Mu/day) are associated with significant increases in nasal complaints relative to placebo within 2 weeks of initiating administration (31-33). Similar types of dose-dependent nasal irritation have been observed within several weeks of initiating treatment with intranasal rIFN-α2a (12,13) or purified HuIFN-α(Le) (50). Indirect evidence suggests that local intolerance occurs sooner and perhaps more frequently when the same dosage is given in divided doses rather than once daily (29,30,33).

Placebo-controlled tolerance studies have found that intranasal rIFN-αs are associated with reversible histopathologic changes in nasal biopsy specimens (49, 51). In one study by Hayden et al healthy adults were given rIFN-α2b 8.5 Mu or placebo once daily for 28 days (49). Fifty-eight per cent of interferon recipients but none of those in the placebo group developed moderate or severe degrees of subepithelial chronic inflammatory cell infiltration. A subsequent study of rIFN-α2a (9.0 Mu/day for 4 or 10 days) used immunohistochemical techniques to determine the nature of the inflammatory infiltrate (51). This study found that 56% of interferon recipients developed increased numbers of subepithelial lymphocytes by the fourth day of exposure, prior to any symptoms of nasal irritation, and 60% by the tenth day. T-helper cells were the principal type of lymphocytes observed in both the normal and the interferon-exposed nose. The histologic changes observed in the nasal mucosa appear to antedate the clinical intolerance observed and to be secondary to the immunologic activity of rIFN-α.

It is unclear whether it will be possible to identify an effective interferon regimen that is also well-tolerated during chronic administration. Other interferons that may offer the possibility of better therapeutic ratios include rIFN-βser (20, 52), IFN-αcon1, or possibly combinations of rIFN-α and rIFN-γ (53). In a recent tolerance study, Hayden et al found that rIFN-βser 3.0 or 12.0 Mu/day for 25 days was associated with less pronounced effects on nasal histopathology and better clinical tolerance, compared historically to studies of rIFN-α (52). Interventions which could modify the inflammatory but not the antiviral effects of intranasal interferon might improve the therapeutic ratios of available interferons.

SUMMARY

Studies have established that intranasal interferon is effective in preventing both experimental and naturally occurring rhinovirus infections. The results of studies with the current formulations of recombinant α-interferons do not suggest that long-term or seasonal prophylaxis is a feasible strategy for preventing respiratory viral infections in healthy adults. Postexposure prophylaxis with intranasal rIFN-α2b is clinically useful in preventing transmission of rhinovirus colds in the family setting. Treatment studies in natural colds have yielded discouraging results. Trials are needed to define the mechanisms of local toxicity and to assess alternate dose schedules, modes of delivery, and other interferon preparations.

REFERENCES

1. Isaacs, A. Practitioner 183:601-605, 1959.
2. Merigan, T.C., Hall, T.S., Reed, S.E. and Tyrrell, D.A.J. Lancet 1:563-567, 1973.
3. Scott, G.M., Reed, S., Cartwright, T. and Tyrrell, D. Arch. Virol. 65:135-139, 1980.
4. Scott, G.M., Phillpotts, R.J., Wallace, J., Secher, D.S., Cantell, K. and Tyrrell, D.A.J. Br. Med. J. 284:1822-1825, 1982.
5. Johnson, P.E., Greenberg, S.B., Harmon, M.W., Alford, B.R. and Couch, R.B. J. Clin. Microbiol. 4:106-107, 1976.
6. Harmon, M.W., Greenberg, S.B. and Couch, R.B. Soc. Exp. Biol. Med.

152:598-602, 1976.
7. Harmon, M.W., Greenberg, S.B., Johnson, P.E. and Couch, R.B. Infect. Immun. 16:480-485, 1977.
8. Harmon, M.W., Greenberg, S.B. and Johnson, P.E. Pro. Soc. Exp. Biol. Med. 164:146-152, 1980.
9. Greenberg, S.B., Harmon, M.W., Johnson, P.E. and Couch, R.B. Antimicrob. Agents Chemother. 14:596-600, 1978.
10. Hayden, F.G. In:Antiviral Chemotherapy: New Directions for Clinical Application and Research (Eds. J. Mills and L. Corey), Elsevier Science Publishing Co., Inc., New York, 1986, pp. 28-40.
11. Scott, G.M., Wallace, J., Greiner, J., Phillpotts, R.J., Gauci, C.L., and Tyrrell, D.A.J. Lancet 2:186-188, 1982.
12. Hayden, F.G. and Gwaltney, J.M., Jr. J. Infect. Dis. 148:543-550, 1983.
13. Samo, T.C., Greenberg, S.B., Couch, R.B., Quarles, J., Johnson, P.E., Hook, S. and Harmon, M.W. J. Infect. Dis. 148:535-542, 1983.
14. Samo, T.C., Greenberg, S.B., Palmer, J.M., Couch, R.B., Harmon, M.W. and Johnson, P.E. J. Infect. Dis. 150:181-188, 1984.
15. Treanor, J.J., Betts, R.F., Reichman, R.C. and Dolin, R. Abstr. Twenty-third Interscience Conference on Antimicrobial Agents and Chemotherapy, no. 920, p. 251, 1984.
16. Turner, R. and Meschievitz, C. Abstr. Fourteenth International Congress of Chemotherapy, no. WS-19-8, p. 67, 1985.
17. Phillpotts, R.J., Scott, G.M., Higgins, P.G., Wallace, J., Tyrrell, D.A.J. and Gauci, C.L. Antiviral Res. 3:121-136, 1983.
18. Came, P.E., Schafer, T.W. and Silver, G.H. J. Infec. Dis. 133:A136-A139, 1976.
19. Phillpotts, R.J., Higgins, P.G., Willman, J.S., Tyrrell, D.A.J., Freestone, D.S. and Shepherd, W.M. J. Interferon Res. 4:535-541, 1984.
20. Higgins, P.G., Al-Nakib, W., Willman, J. and Tyrrell, D.A.J. J. Interferon Res. 6:153-159, 1986.
21. Greenberg, S.B., Harmon, M.W., Couch, R.B., Johnson, P.E., Wilson, S.Z., Dacso, C.C. Bloom, K. and Quarles, J. J. Infect. Dis. 145:542-546, 1982.
22. Aoki, F.Y. and Crawley, J.C.W. Br. J. Clin. Pharmac. 3:869-878, 1976.
23. Hardy, J.G., Lee, S.W. and Wilson, C.G. J. Pharm. Pharmacol. 37:294-297, 1984.
24. Hayden, F.G. and Gwaltney, J.M., Jr. J. Infect. Dis. 150:174-180, 1984.
25. Higgins, P.G., Phillpotts, R.J., Scott, G.M., Wallace, J., Bernhardt, L.L. and Tyrrell, D.A.J. Antimicrob. Agents Chemother. 24:713-715, 1983.
26. Turner, R.B., Felton, A., Kosak, K., Kelsey, D.K. and Meschievitz, C.K. J. Infect. Dis. 154:443-447, 1986.
27. Dolin, R., Bett, R.F., Treanor, J., Erb, S., Roth, F.K. and Reichman, R.C. Abstr. Thirteenth International Congress of Chemotherapy, no. SE4.7/1-7, p. 60/20-23, 1983.
28. Schiff, G., personnal communication.
29. Farr, B.M., Gwaltney, J.M., Jr., Adams, K.F. and Hayden, F.G. Antimicrob. Agents Chemother. 26:31-34, 1984.
30. Betts, R.F., Erb, S., Roth, F., Reichman, R.C., Treanor, J., Beutner, K. and Dolin, R. Abstr. Thirteenth International Congress of Chemotherapy, no. SE4.7/1-5, p. 60/13-15, 1983.
31. Hayden, F.G., Gwaltney, J.M., Jr., and Johns, M.E. Antiviral Res.

5:111-116, 1985.
32. Douglas, R.M., Albrecht, J.K., Miles, H.B., Moore, B.W., Read, R., Worswick, D.A. and Woodward, A.J. J. Infect. Dis. 151:731-736, 1985.
33. Monto, A.S., Shope, T.C., Schwartz, S.A. and Albrecht, J.K. J. Infect. Dis. 154:128-133, 1986.
34. Monto, A., personnel communication.
35. Ishikawa, S., Linzmayer, I., Riley, W.A., Falliers, C.J., Herron, J.M., Webb, R.D., Ohman, J.L., Haddad, Z.H., Wright, G.E. and MacDonnell, K.F. abstr. Fourteenth International Congress of Chemotherapy, no. WS-19-6, p.67, 1985.
36. Michaels, D., Foy, H., Krouse, H., Meschievitz, D and Albrecht, J. J. Allergy Clin. Immunol. 77(Supplement):160A, abstr. no. 158, 1986.
37. Geha, R., Maguire, J., McIntosh, K. and Meschievitz, C. J. Allergy Clin. Immunol. 77(Supplement):160A, abstr. no. 157, 1986.
38. Herzog, C., Berger, R., Fernex, M., Friesecke, K., Havas, L., Just, M. and Dubach, U.D. Antiv. Res. 6:171-176, 1986.
39. Hayden, F.G., Albrecht, J.K., Kaiser, D.L. and Gwaltney, J.M., Jr. N. Eng. J. Med. 314:71-75, 1986.
40. Douglas, R.M., Moore, B.W., Miles, H.B., Davies, L.M., Graham, N.M.H., Ryan, P., Worswick, D.A. and Albrecht, J.K. N. Eng. J. Med. 314:65-70, 1986.
41. Foy, H.M., Fox, J.P. and Cooney, M.K. N. Eng. J. Med. 315:513-514, 1986.
42. Herzog. C., Berger, R., Fernex, M. Friesecke, K., Havas, L., Just, M. and Dubach, U.D. Lancet 1:1089-1090, 1986.
43. Just, M., Berger, R., Ruuskanen, O., Ludin, M. and Linder, S. J. IFN Res. 6(Supplement 1):32, 1986.
44. Hayden, F.G., unpublished observations.
45. Isomura, S., Ichikawa, T., Miyazu, M., Naruse, H., Shibata, M., Imanishi, J., Matsuo, A., Kishida, T. and Karaki, T. Biken J. 25:131-137, 1982.
46. Saito, H, Takenaka, H., Yoshida, S., Tsubokawa, T., Ogata, A., Imanishi, F. and Imanishi, J. Rhinology 23:291-295, 1985.
47. Tannok, G., personal communication.
48. Spiegel, R.J., Spicehandler, J.R., Jacobs, S.L. and Oden, E. Am. J. Med. 80:223-228, 1986.
49. Hayden, F.G., Mills, S.E. and Johns, M.E. J. Infect. Dis. 148:914-921, 1983.
50. Scott, G.M., Onwubalili, J.K., Robinson, J.A., Dore, C., Secher, D.S. and Cantell, K. J. Med. Virology 17:99-106, 1985.
51. Hayden, F.G., Winther, B., Donowitz, G., Mills, S. and Innes, D. J. Infect. Dis., in press.
52. Hayden, F.G., Innes, D.J., Mills, S.E. and Levine, P.A. J. IFN Res. 6(Supplement 1):31, 1986.
53. Ahmad, A.L.M. and Tyrrell, D.A.J. Antiviral Res. 6:241-252, 1986.

2

INTERFERON THERAPY OF HERPES SIMPLEX DERMAL INFECTIONS AND PAPILLOMAVIRUS CONDYLOMATA

J. SHOHAM AND M. REVEL

Department of Life Sciences, Bar-Ilan University, 52100 Ramat-Gan and Department of Virology, Weizmann Institute of Science, Rehovot, Israel

INTRODUCTION

Herpes and papilloma virus infections will probably be among the first indications for which interferon (IFN) will find its place as a therapeutic agent. The data to be analysed below suggest that, when properly used, interferon offers the combination of efficacy and safety that has not been achieved thus far by any of the other currently available treatments for these two viral infections. However, in order to put this analysis in its proper framework, it will be appropriate to precede it with a short comparative description of the viruses and the diseases caused by them.

VIRUS AND DISEASE CHARACTERISTICS

Table 1 brings together some of the main features of the two viruses and the diseases caused by them. More details can be found in other texts and reviews (1-11). It is interesting that in spite of substantial differences in virus structure, genetic make up, growth characteristics and basic pathological lesions, there are several comparable features in disease presentation, complications and epidemiology. Accordingly, the comparative analysis of interferon effect on these 2 viral infections is a meaningful exercise. Some specific types of each of these viruses (HSV-2, HPV-6,10,11,16,18) cause genital lesions which comprise today a major medical and social problem. The prevalence

M. Revel (ed.), CLINICAL ASPECTS OF INTERFERONS. Copyright 1988. Kluwer Academic Publishers, Boston. All rights reserved.

Table 1. Virus and disease characteristics for herpes simplex virus (HSV) and human papilloma virus (HPV) infections.

	Herpes simplex virus (HSV)	Human Papilloma virus (HPV)
Virus		
Size	180 nm	55 nm
Genome	ds DNA, 154 kb.	ds DNA, 8 kb.
Types	1 & 2	25
Disease		
Basic lesion	Vesicle	Tumor (papilloma)
Body distribution	Genital (type 2) facial (type 1)	Genital (6,10,11,16,18) varia (other types)
Time course	Primary - latency - frequent recurrencies	Continuous; latent; recurrent
Symptoms	Prominent, maybe severe	Usually symptom free
Complications	Primary - encephalitis; neuritis	-
	Neonatal - systemic; encephalitis (early)	Neonatal - laryngeal papillomatosis (late)
	Cancer - Uterine cervix	Cancer - Uterine cervix
Epidemiology		
Transmission	Genital - sexual contact (60-80% infectivity)	Genital - sexual contact (60% infectivity)
	Labial - saliva	Other sites - direct "
Virus shedding	During attack only	Continuous
Incidence (genital)	100-200/10^5 (USA)	20-40/10^5
Age distribution	Type 1 - childhood	Nongenital - any age
	Type 2 - sexual life	Genital - sexual life
Diagnosis		
Clinical	Typical appearance	Typical appearance
Laboratory	Virus culture	No in vitro growth
	Cytology, immunocytology	Cytology, immunocytology
	DNA hybridization	DNA hybridization
Differential	Other vesicular & ulcerative lesions	Condylomata lata, Molluscum contagiosum, sebaceous cysts, tumors

of the genital form of both viral infections is increasing alarmingly. Their incidence has been increased by more than 2 fold during the last decade. They are frequently associated with each other, an association which is suspected to augment synergistically their oncogenic potential (12). They are also associated with other venereal infections. Moreover, the treatments that may be offered to these patients are distressingly unsatisfactory. This review chapter should be read in this context.

INTERFERON IN THE TREATMENT OF HERPES SIMPLEX INFECTIONS OF THE SKIN

Preclinical data

Early observations (13,14) suggested that herpes viruses are relatively resistant to the antiviral effects of IFN, *in vitro*. However, more recent studies clearly pointed out that herpes virus yields are reduced in interferon treated cells (15-17). Mean IFN titer needed to protect against HSV infection of cells was not significantly different from that needed for vesicular stomatitis virus (VSV), although dose response curve was shallower with HSV as compared to VSV (18). Different clinical isolates of HSV exhibit a similar degree of sensitivity to IFN, *in vitro* (18). In a comparative *in vitro* study IFNß was shown to have more consistent and reproducible effect on HSV replications than IFNα(19). It was found that IFN inhibits HSV replication at a very early stage, either prior to or during the synthesis of immediate early proteins of the virus (20).

The role of IFN in HSV infections was studied also *in vivo*. Resistance to HSV-1 infections in mice was found to be associated with IFN induction by the viral infection (21). IFN treatment may protect mice from lethal HSV-2 infection depending on the routes of the introduction of the virus and of the IFN (22). Recombinant hybrid human IFNαAD, active on mouse cells, was used in this study.

Clinical studies

The preclinical data suggest that IFN should be tested in herpes virus infections. However, only a few clinical studies followed this lead. These studies differ in the type of IFN used (α or ß), the route of its administration (systemic or topical) and the aim of the treatment (preventive or therapeutic). There is a remarkable discrepancy between the results of the studies: some (23,24) report complete failure and others (25-30) a prominent, significant success. This discrepancy may be related, in the first place, to differences in mode and route of treatment and probably, also, to the type of IFN used.

Systemic treatment with IFN was attempted by Ho and his colleagues in two consecutive studies (31,23). The studies were based on the observation that HSV was reactivated within 3-4 days in about 60% of patients who had a history of herpes labialis and had undergone microneurosurgical decompression of the trigeminal sensory root for treatment of trigeminal neuralgia (32). This observation provided a unique opportunity to study prophylactic and therapeutic effect of IFN in a defined group of patients. In the first study (31), IFN (7×10^4 units/Kg body weight, intramuscular, IM), given for 5 days beginning on the day before the operation, had an apparently significant effet in reducing virus reactivation (9/19 patients with reactivation in IFN and 15/18 in placebo treated groups, $p < 0.05$). However, follow up of patients few weeks later (33) indicated a similar recurrence rate in both groups. Furthermore, a second study (23) which addressed the question of prophylactic vs therapeutic effect of IFN, did not show an effect of post surgical treatment, whereas presurgical (prophylactic) use of IFN apparently precipitated reactivation and significantly increased its rate (10/11, 5/10 and 3/9 patients in the presurgical IFN, postsurgical IFN and placebo groups, respectively). The number of patients in each group was obviously small and consequently the results were of borderline significance.

A more recent double blind study (24) had made use of

recombinant IFNα_{2b} in an attempt to <u>prevent</u> frequently recurring genital herpes. Treatment plan was $3x10^6$ international units (IU) subcutaneously (SC), 3 times weekly for 12 weeks. Thirty seven patients were randomized to get either IFN (20 patients) or placebo (17 patients). No difference was observed between the 2 groups in all the parameters tested (mean time to and duration of first recurrence, mean number of recurrences during and after treatment). Similar negative results were observed by the same group when IFN was used systemically for the treatment, rather than the prevention, of genital herpes (34).

These negative results are in sharp contrast to the most favourable results observed with topical application of IFN cream or ointment to the diseased skin. It is interesting that although this mode of treatment constituted the first attempt to demonstrate a protective effect of IFN in man (35), this approach has not been followed by many investigators. The first report on the use of IFN ointment for the treatment of HSV infections came from Ikic and his colleagues (36-38). They used native, crude IFNα, prepared by Cantell's method. In herpetic gingivo-stomatitis (primary HSV-1 infection in children), 12 hospitalized patients received lyophilized IFNα (2000 units) into the mouth hourly (12 times a day), for 4-6 days. Other 7 children received placebo and 15 were on conventional treatment. There was a significant shortening of duration of fever and time to epithlization (38,25). In an open study, Ikic et al treated 69 and 68 patients with recurrent labial or genital herpes relapses, respectively, using IFNα ointment (2000 u/g applied on the lesions 5-6 times daily. Patients that were started on treatment during prodromal phase of attack (itching, pain, tension, swelling and redness at relapse site), <u>aborted</u> the attack completely (in 30% of patients) or had it in a much milder and shorter form (49%) as compared to those started later. The study seems to imply that the sooner local IFN treatment is started during an attack the better the effect.

A series of studies (open and double blind) were conducted during the last few years on the use of native IFNß (FRONE^R) cream or gel in HSV dermal infections (26-30, 39). FRONE^R is prepared by Inter-Yeda, Israel, for Serono-ARES, using normal diploid human foreskin fibroblasts induced by poly (rI)(rC) and is purified to $>10^7$ IU/mg protein, the preparation containing essentially the IFN/ß 1 subspecies (40). Lyophilized IFNß was mixed with either polyethylene glycol or carboxymethylcellulose (10^5 IU IFNß per gram of base) to get FRONE^R cream or gel, respectively.

In a preliminary trial (26), Isacsohn et al treated 15 patients with dermal HSV infections by FRONE^R (10^5 IU/gm cream, applied generously and massaged into the lesions 5-6 times a day). Twelve out of the 15 patients had a shorter and milder attack as compared to 5 placebo treated (non randomized) patients and to the past history of the IFN treated patients.

Three additional small studies with FRONE-cream lend further support to Isacsohn's study (26). Clerico et al (30) treated 12 patients with facial herpes and 8 patients with genital herpes in an open study. They reported good response without any side effects. Romano (39) treated 13 patients with periocular HSV infections (4 primary, 9 recurrent) with remarkable shortening of disease duration. Paldi and his colleagues treated 19 patients (17 females, 2 males) with recurrent genital herpes, using $0.5-1.5x10^6$ IU of FRONE cream in 1-3 daily applications for 7 days. There was a dramatic relief of pain and suffering in 18 out of 19 patients, about 2 fold reduction in duration of attack and more than 3 fold increase in interval between attacks (unpublished data).

However, the most conclusive data came from 3 consecutive studies, designed as phase II, III (double blind) and IV (post-marketing survey) studies (27-29, and Glezerman, Movshovitz, Doerner, Shoham and Revel, submitted for publication). In the first study, 31 patients with a long history of HSV infections (mean 18 ± 3.5 and 105 ± 18 months for genital and facial infections,

Table 2. Native FN-ß (FRONE[R]) cream in the treatment of genital and facial HSV infections. Results of 2 open studies.

	Phase II study		Phase IV study	
	Genital	Facial	Genital	Facial
Number of Patients	19	12	109	36
Duration of eruptions				
(days, mean ± SE)				
Before treatment	13.5±1.5	14.8±1.1	11.0±0.6	11.7±0.7
With treatment	5.5±1.5	6.9±0.7	4.4±0.4	6.0±0.8
p	<0.001	<0.001	<0.001	<0.001
Time to recurrence				
(days, mean ± SE)				
Before treatment	35±5.2	40±9.9	42±4	43±5.2
With treatment	151±258	239±67.3	160±15.9	151±23.9
p	<0.001	= 0.008	<0.001	<0.001
Frequency decrease				
(fold, mean ± SE)	5.2±0.9	6.2±1.8	5±0.6	4.4±0.7

respectively, were treated at the Sheba Medical Center, Tel Hashomer, Israel (27) using FRONE-cream applied to the affected skin area 4-6 times daily (about 2×10^4 IU IFNß in each time)during attack and twice daily between attacks. Diagnosis was verified by virus isolation in each case before initiation of treatment. Mean (± SE) follow up period: 15 (± 1.3) months. The results of this study (Table 2, Phase II) point out highly significant shortening of duration of attack (time to complete epithelization) and prolongation of the average remission time. This was accompanied by profound subjective improvement in symptoms (pain, itching) and in a reduction of HSV titer in skin vesicles by more than 2 orders of magnitude within 12-24 hr, while placebo cream had no efffect (27). Moreover, it was found that application of IFNß cream during the prodromal period could prevent eruption in almost half the patients.

Table 3. Native IFN-ß (FRONE[R]) gel in the treatment of recurrent labial and genital HSV infections. Results of a double blind placebo controlled trial.

	Placebo group	IFNß group
Number of patients		
Total	13	12
Labial	7	7
Genital	6	5
Duration of eruptions		
(days, mean $\pm$SE)		
Before treatment	8.3 ± 0.8	7.0 ± 0.8
With treatment	7.4 ± 0.8	4.5 ± 0.63
p	= 0.45	= 0.007
Eruptions per year		
(number, mean $\pm$SE)		
Before treatment	6.9 ± 1.6	6.6 ± 1.6
With treatment	6.7 ± 1.3	1.6 ± 0.2
p	= 0.9	= 0.006
Frequency decrease		
(fold, mean $\pm$SE)	1.09 ± 0.1	3.8 ± 0.8

Next we carried out a double blind placebo controlled trial at Soroka Medical Center, Beer Sheva, Israel. The 25 patients enrolled to this study were with prolonged (average 9.2 years, range 1-27 years) history of recurrent HSV dermal infections (Table 3). Patients were enrolled during acute attack, and diagnosis confirmed by virus isolation. Treatment was initiated on subsequent attack (during prodromal phase) using coded tubes containing either IFNß ($1x10^5$ IU/g) in gel or carrier placebo gel only. Follow-up period - 2 years. The response in terms of duration of attacks, and number of eruptions per year was highly significant with FRONE[R]-gel, whereas placebo had no effect

(Table 3). Symptoms were relieved in 92% of IFNß treated and only 8% of placebo treated patients. Facial and genital herpes responded equally well.

Finally, a phase IV study was carried out on 167 patients (16 primary, 151 secondary HSV infections to whom FRONE^R-cream had been prescribed for treatment of herpes dermal infections. Patients applied the cream six times daily during eruptions. The results obtained in this study were strikingly comparable to those of the Phase II study (Table 2). Here also treatment during the prodromal period could partially prevent eruption in 39% of all patients, 13% showing no eruption at all. Interestingly, about one third of all patients reported no recurrence for one year after the first episode treated. Decrease in frequency of attacks was greater in patients with frequent recurrences than in those with rarer ones. It should be emphasized that in the last 2 studies, treatment was applied only during acute attack (and not during the interval between attacks, as in the former study, 27) and nevertheless a significant, comparable reduction in recurrence rate was achieved in all these studies.

Herpes-zoster skin infections (shingles) also responded favourably to IFNß (FRONE) cream treatment (26; Dr. Romano, personal communication).

<u>Conclusions</u>

1. Topical application of IFN is far better than systemic (IM or SC) treatment. These two modes of treatment were not directly compared in one study. However, the complete failure of studies using systemic treatment (23,24,34) stands in sharp contrast to the highly significant success which was consistently achieved with topical IFN (25-30). These results make sense, since systemic treatment cannot reach the high local tissue concentration obtainable with topical application. Moreover, side effects are negligible with topical use of IFN as contrasted with the significant, frequent adverse reactions which accompany its systemic

use. Compliance of patients is much better when topical rather than injectable medication is used. Also, amount of IFN needed and consequently, treatment price should be much lower with the topical treatment.

2. IFN-ß is apparently better than IFNα for dermal HSV infections. This inference is indicated by some of the clinical (25 compared to 26-30) and preclinical (19, and see above) data but should be tested directly in a comparative study.
3. IFN topical treatment should be given as early as possible during the attack, preferably during prodromal phase.
4. There is no advantage in applying the medication during the interval between attacks.
5. Among alternative treatments, acyclovir is definitely the best (41-48). However, acyclovir applied topically is not effective against recurrences, although it reduces virus proliferation and may fasten healing (41-44). Acyclovir is effective in reducing the risk of recurrences, but only if taken systemically (e.g. orally) at high doses and continuously (daily) (45-48), raising problems of patients' compliance, long term toxicity and possible emergence of drug resistant HSV strains. All these drawbacks do not exist with topical IFN treatment, a treatment which is at least comparable to acyclovir in the treatment of the acute attacks. IFN delays recurrences without the need for continuous prophylactic treatment.
6. The data reviewed here suggest that IFN will turn out to be the treatment of choice for HSV infections. However, improvement of treatment plan can be envisaged in terms of better control of both the acute attack and recurrence rate. This may be reached by optimization of dose and dose schedule, (49), combination of two interferons (e.g. IFNß and IFNγ), combination with other immunomodulatory agents (50), thymic factors (51-53) and specific antibodies (54) or with chemical antiviral agents (55-56).

INTERFERON IN THE TREATMENT OF CONDYLOMATA ACUMINATA

In contrast to HSV, papilloma viruses cannot be grown *in vitro* thus precluding the possibility to assess IFN effect on HPV replication in culture. On the other hand much more clinical data were accumulated in regard to HPV infections, particularly the genital ones, than HSV infections. Since the clinical results were presented in a similar fashion by different investigators, it is possible to summarize the available clinical data on IFN treatment of condylomata acuminata patients in a tabular form (Table 4). The Table brings the results of 19 studies performed by 11 clinical research groups (25,57-69). These results can be summarized as follows:

1. *The overall response* to IFN treatment is consistently good, sometimes excellent. Indeed, all the double blind, placebo controlled studies, pointed out a significant therapeutic effect of IFN in patients suffering from condylomata acuminata. Obviously, the results in individual cases are not uniform. Although complete cure, without recurrence, was achieved in a considerable number of the patients (54,60,67), in many cases there was a gradual return of the lesions, even if the initial response was complete (64,68,69). In some other cases only partial response or even complete lack of response was noticed. The observed heterogenicity in clinical response can be analysed in terms of both disease and treatment parameters.

2. *Disease parameters*

   a. *Size of lesions.* Generally, the smaller the lesion the better the response (69). For small lesions topical application with relatively low doses of IFN may be sufficient (59,60). However, many cases of large widespread lesions responded well, even completely. Such lesions should be treated by IM or IL routes (59,60). In one study (64) topical high dose IFN treatment was found to be effective for large lesions.

Table 4. Summary of clinical studies on IFN in the treatment of condylomata acuminata

Investigator[a] (Reference)		Study design	No.[b] Eval. Pts.	IFN treatment type	route	dose Mu/d	schedule	duration	Response No. of cases CR	PR	NC	DP
Ikic	(25)	Open	40 F	α(n)	Top	0.01	x5/d	10d	36	-	4	-
	(57)	Open	32 M	α(n)	Top	0.01	x5/d	10d	14	-	18	-
Stefanon	(58)	Open	5	ß(n)	IL	2	qdx5	1-3 cyc.	3	1	1	-
Marcovici	(59)	Open	18 (S)	ß(n)	Top	0.5	x4/d	4w	15	-	3	-
		Open	7	ß(n)	IL	0.1		7 inj	7	-	-	-
Schonfeld	(60)	Open	16 (L)	ß(n)	Top	0.5	x5/d	5w	-	16	-	-
		Open	5	B(n)	IL	3	q2d	4 inj	5	-	-	-
		Open	5	ß(n)	IM	3	q2d	5 inj	5	-	-	-
		Open	5	ß(n)	IM	9	q2d	5 inj	5	-	-	-
		DB	IFN-11	ß(n)	IM	2	qd	10 inj	9*	-	2	-
			PLB-11	-	IM	-	qd	10 inj	2	-	9	-
Gross	(61)[c]	Open	11M	ß(n)	IM	3	q3d	6 inj	6	-	5	-
Geffen	(62)	Open	10	α(n)	IL	3 or 8.5	biw	9-28 inj	5	2	3	-
	(63)	DB	IFN-12	α(n)	IL	0.6	biw	8 w	10*	1	1	-
			PLB-13	-	IL	-	biw	8 w	1	1	11	-

Investigator[a] (Reference)	Study design	No.[b] Eval. Pts.	IFN treatment type	route	dose	schedule Mu/d	duration	Response No. of cases CR	PR	NC	DP
Vesterinan (64)	DB	IFN- 8	α(n)	Top	12	qd	2wx4	3	2	3	-
		PLB- 5						-	-	3	2
Marchionni (65)	Open	20	ß(n)	IM	2-3		6-10 inj	6	3	11	-
Gall (66,67)	Open	16	α(LB)	IM	5/m	qd-tiw	4w-2w	11	4	1	-
	Comp.	30	α(LB)	IM	3/m	qd-tiw	2w-4w	17	10	3	-
		14	α(LB)	IM	1/m	qd-tiw	2w-4w	6	5	3	-
Vance (68)	DB	IFN-HD-30	α-2(r)	IL	1	tiw	2 w	16*	NR	NR	NR
		IFN-LD-32	α-2(r)	IL	0.1	tiw	3 w	6	NR	NR	NR
		PLB-29	-	IL	-	tiw	3 w	4	NR	NR	NR

a Only name of first investigator is listed in the table

b Abbreviations: No. eval. pts., Number of evaluable patients; Mu, mega units; CR complete response; PR partial response; NC, no change; DP, disease progression; F, female; M, male; S, small lesions; L, large lesions; PLB, placebo; DB, double blind study; HD, high dose; LD, low dose; (n), native IFN ; ß(n), native IFNß; α(LB), lymphoblastoid IFN ; α-2(r), recombinant IFNα-2; Top, topical treatment; IM, intramuscular; IL, intralesional; d, day; w, week; qd, once a day; q2d, every other day; biw, twice a week; tiw, thrice a week; cyc., cycles; inj, injections; NR, not reported; * highly significant difference ($p<0.01$, at least) between IFN and PLB groups in DB studies.

c Data based on listed reference + personal communication

b. Site of lesion. In one study (68), and using the same treatment protocol, it was observed that the prominent success with the genital forms of HPV infection stands in sharp contrast to the complete lack of response in the non-genital forms. This may be related to tissue factors (mucusa vs multilayered skin, respectively), or to different sensitivity of various HPV subtypes to the IFN type and the treatment protocol used in that study. Even within the group of condylomata acuminata patients differences in results between some of the studies which employed similar IFN preparations and treatment schedule (60,61) may be attributed to different virus subtypes. However, in another study such differences were not observed (69).

c. HPV subtype was determined in 13 patients, in one study (67), prior to IFN treatment and using only one DNA probe (HPV-6) which was available to the investigators. No conclusions as to the correlation of virus subtype and response to IFN treatment could be drawn from that study. It is clear that this important correlation should be explored more systematically with DNA probes of various HPV subtypes and in connection with different IFN species and treatment protocols. The known differential sensitivity of different viruses to different IFN species (49,70) may be extended to different virus strains of the same virus, and thus may form the basis for specific, tailor-made treatment for each patient.

d. Disease duration. In general, the younger the lesion the better the response (69). However, since this is also true for spontaneous regression (69) a therapeutic effect of IFN in persistent long standing ($>$ 2 yrs) lesion is more significant. Indeed, IFN was repeatedly shown to be effective in such persistant cases (57-69).

e. Prior treatments by different means, have no effect on subsequent response to IFN treatment.

f. Age and sex of patients do not seem to modify response to IFN treatment.

g. Spontaneous regression of the lesions can be observed in a considerable number of patients, more so in early lesions as mentioned before. In early lesions spontaneous regression and cure may be observed in about one third of the patients (67). This should always be taken into consideration in the analysis of the results of the open studies. It may be the sole or main reason for placebo effect in double blind studies.

3. Treatment parameters

a. IFN type. IFNα from several sources, (native, lymphoblastoid or recombinant α -2) was tested. In contrast, IFNß tested thus far is from one source (native IFNß - FRONE[R] - of Inter-Yeda, Israel). There are no reports on the use of IFNγ for condylomata acuminata. Both IFNα and IFNß are effective in the 3 treatment forms tested (topical, IL, IM). No direct comparison of their relative efficacy has been made. However, by comparing the different studies, and taking into consideration the other treatment parameters (route, dose, schedule), the apparent impression is that patients treated by IFNß responded better. Obviously, this impression should be directly tested in a comparative trial.

b. Route of administration. (i) Topical use was shown to be effective with either IFNα (25,57,64) or IFNß (59,60). Our impression is that this mode of treatment is effective in small, early lesions. However, the study of Vesterinen et al (64) indicate that high dose IFN cream may be useful in the treatment of widespread vaginal flat condylomas. (ii) Intralesional treatment is most effective provided that a high enough dose is used. Thus 1 megaunit (Mu) but not 0.1 Mu recombinant IFNα- 2 were effective (68). The 3 studies based on IL treatment with IFN α (63,68,69) arrived at fairly similar results by using a comparable treatment protocol (Table 4). It was stressed, however, that (a) the IL treatment is essentially topical, i.e. other lesions in the vicinity of the treated lesion were not affected and that (b) the treatment is not curative; there is a gradual return of the lesions even in complete reponders (69). With IFNß (FRONE) - 4 (60) or 7 (59) IL injections were sufficent to achieve long lasting complete response in all the treated patients. (iii) Intramuscular treatment seems to comprise the best choice in terms of effectiveness and relative convenience and compliance. IM administration hits effectively all lesions, whereas with the IL route the effect is confined to the injected lesion only, even not its neighboring lesions. This apparent discrepancy is probably related to the relatively low dose used in IL treatment and probably also to consumption and destruction processes within the lesion. It should be mentioned that systemic IFN treatment (IM), was shown to be effective on other HPV infection - laryngeal papillomatosis (71-75). The dose was found to be important with lymphoblastoid IFNα (5 Mu/m^2 were much better than 3 or 1 Mu/m^2 (66,67). According to our experience with IFNß, dose dependency was less critical, and 2 Mu were essentially as effective as 3 or 9 Mu (60). This study also brings another important observation which establishes the potential of IFNß for systemic use. It was known that, in contrast to IFNα, IM (or SC) administration of IFNß, does not result in measurable serum levels of antiviral activity. Accordingly, it was assumed that IFNß is degraded locally or that there is a serum inhibitor to it. At any rate it has been generally accepted that IFNß does not have a systemic effect (76). Our study (60) suggests that this assumption is incorrect. Indeed, IFNß cannot be detected in serum after IM injection but it evidently has systemic biological effects as measured by induction of oligo A synthetase, antiviral state and natural killer cell activity in peripheral blood leucocytes.

c. Dose. Response was correlated with dose when IFN was used (IL, ref 68, or IM, ref 66,67) less so with IFNß (60), as mentioned before.

d. Schedule. Topical treatment seems to require daily application (probably several times a day). IL or IM treatment can be less closely spaced (2-3 times a week were sufficient, with both IFNα or ß).

e. Duration. The range tested is between 8-10 days to 6-8 weeks.

4. Side effects

No side effects were reported when IFN was used topically. IL or IM administration is associated with the well known "flu" like symptoms (60,62,66,68,69) with subsequent tachyphylaxis of the symptoms (62,69). Daily IM administration had more prominent side effects than the three times per week regimen (66). Hematological derangements were few, minimal and completely reversible (69). According to our experience IFNß causes much less frequent and less severe adverse reactions (60).

In summary, the place of IFN in the treatment of condyloma acuminata has been established beyond any doubt. Considering the limited effectivity of all the available alternative treatments (77-78), one can safely conclude that interferon should be the treatment of choice for this ailment. On the basis of the data reviewed here a tentative treatment plan can be formulated as follows:

1. Small, early lesions can be treated by applying cream containing 1×10^5 IU/g interferon, 3-4 times daily for 3 weeks.
2. Large and/or persistent lesions ($>$ 1 year old) should be treated by IM route. With IFNß, 2×10^6 IU, t.i.w. for 3

weeks may be sufficient (60). With IFNα, higher dose and a longer treatment period may be needed (e.g. 5 Mu/m^2 qd for 4 weeks followed by same dose t.i.w. for 2 weeks may be needed according to Gall et al, ref 66,67)

3. In case of failure of first treatment, alternative treatment protocols can be attempted, including, change to another IFN type, increased dose, IL administration or combination with other modes of therapy (e.g. cryosurgery, laser, etc.).

Further work has to be done in order to establish treatment protocols more conclusively, including individualised treatments based on HPV typing, combination of IFNs or combination with other modes of therapy, as mentioned before.

CONCLUDING REMARKS

Data on IFN effect on some HSV and HPV infections were reviewed and analysed. The results of the clinical studies taken together suggest that IFN has a strong beneficial, sometimes curative, effect in these infectious diseases and that accordingly it may well become the treatment of choice for them. The best alternative treatment for HSV infections is definitely acyclovir. However, by comparing preferred treatment plans for both (short term topical IFN vs long term systemic acyclovir treatment), use of acyclovir, but not of IFN, is associated with problems of patients' compliance, long term toxicity and possible emergence of drug resistant HSV strains. The alternative treatments for condylomata acuminata are primarily local destructive ones (i.e. surgery, podophyllin, laser, etc.). In addition to their being inconvenient they do not guarantee effective control or complete eradication of the disease process.

Formulation of optimal treatment plan has not yet been completed. However, on the basis of available data, one may suggest topical treatment for HSV dermal infections and IM treatment for condylomata accuminata. The difference in preferred route of treatment may be related to the differences in the two causative agents and to the organization of the skin lesions.

However, small condylomas may also be treated topically. IFNß is apparently better than IFNα for these two skin diseases. Generally, short term treatment is sufficient (1-2 weeks for HSV infections, during acute attack, and 3-6 weeks for condylomata) and there is no advantage in long term prophylactic treatment. Side effects are virtually non-existing for topical treatment and usually tolerable and entirely reversible for IL or systemic treatments.

Future studies should be addressed to problems such as comparison of different IFN (including IFNγ) or combinations thereof, optimization of treatment schedules and combination with other treatment modalities such as immunomodulators or chemical agents in an effort to further improve the current results and hopefully achieve complete cure of these viral infections.

REFERENCES

1. Glaser, R. and Gottlieb-Stematsky T. (Eds.). Human Herpes Virus Infections: Clinical Aspects. Marcell Dekker, Inc. New York, 1982.
2. Shiota, H., Cheng, Y. and Perusoff W.H. (Eds). Herpes Virus: Clinical, Pharmacological and Basic Aspects. Excerpta Medica, Amsterdam, 1982.
3. Corey, L., Adams, H.G., Brown, Z.A. and Holmes K.K. Ann Intern. Med. 98: 958-972, 1983.
4. Corey, L. and Holmes, K.K. Ann. Intern. Med. 98: 973-983, 1983.
5. Poste, G., Hawkins, D.F. and Thomlinson, J. Obstet. Gynecol. 40: 871-890, 1972.
6. Oriel, J.D. Brit. J. Vener. Dis. 47: 1-13, 1971.
7. Lutzner, M.A. Arch. Dermatol. 119: 631-635, 1983.
8. Singer, A., Walker, P.G. and McCance, D.J. Brit. Med. J. 288: 735-737, 1984.
9. Feldman, Y.M. Cutis 33: 118-120, 1984.
10. Syrjanen, K.J. Obstet. Gynecol. Survey 39: 252-265, 1984.
11. Ferensy, A., Mitao, M., Magai, N., Silverstein, S.J. and Crum, C.P. New Eng. J. Med. 313: 784-788, 1985.
12. Zur-Hansen, H. Lancet 2: 1370-1372, 1982.
13. Ho, M. and Enders, J.F. Virology 9: 446-447, 1959.
14. Glasgow, L.A., Hanshaw, J.B., Merigan, T.C. and Petralli, K. Proc. Soc. Exp. Biol. Med. 125: 843-849, 1967.
15. Rasmussen, L. and Farley, L. Infect. Immunity 12: 104-108, 1975.
16. Lerner, M.A. and Baily, E.J. J. Infect. Dis. 134: 400-410, 1976.

17. Munoz, A. and Carrasco, L. J. Gen. Virol. 65: 1069-1078, 1984.
18. Lazar, R., Breinig, M.K., Armstrong, J.A. and Ho, M. Infect. Immun. 28: 708-712, 1980.
19. Rasmussen, L.E., Chen, P.T. and Merigan, T.C. Antimicrob. Agents Chemother. 26: 599-600, 1984.
20. Gloger, I. and Panet, A. J. Gen. Virol. 65: 1107-1111, 1984.
21. Zawatsky, R., Gresser, I., Demayer, E. and Kirchner, H. J. Infect. Dis. 146: 405-410, 1982.
22. Fish, E.N., Banerje, K and Stebling, N. J. Gen. Virol. 64: 2317-2321, 1983.
23. Ho, M., Pazin, G.J., Armstrong, J.A., Haverkos, H.S., Dummer, S. and Jannetta, P.J. J. Infect. Dis. 150: 867-872, 1984.
24. Eron, L.J., Harvey, L., Toy, C. and Santomauso, D. Antimicrob. Agents Chemoter. 30: 608-610, 1986.
25. Ikic, D., Trajer, D. Cupak, K., Petricevic, I., Prazic, M. Soldo, I. Jusic, D. Smerdel, S. and Soos, E. Yugoslavian J. Clin. Pharmacol. Ther. Toxicol. 19: 498-505, 1981.
26. Isacsohn, M. Berson, B. Sternberg, I. and Morag, A. Isr. J. Med. Sci. 19: 959-962, 1983.
27. Movshovitz, M. Shewach-Millet, M., Kriss-Leventon, S., Shoham, J., Doerner, T. and Revel, M. In: Herpes Viruses and Virus Chemotherapy (Ed. R. Kono) Elsevier Science Publishers, Amsterdam, 1985, pp. 285-292.
28. Doerner, T. Revel, M., Shoham, J. and Glezerman, M. In: The Interferon System. Proc., 1985, TNO-ISIR Meeting.
29. Glezerman, M., Lunenfeld, E., Doerner, T., Cohen, V., Sarov, I. and Revel, M. In: The Interferon System. Proc., 1985, TNO-ISIR Meeting.
30. Clerico, R., Ribuffo, M. and Pala, S. Dermatologia Clinica 3: 249-253, 1983.
31. Pazin, G.J., Armstrong, J.A., Tai Lam, M., Tarr, G.C., Jannetta, P.J. and Ho, M. New Eng. J. Med. 301: 225-230, 1979.
32. Jannetta, P.J. Prog. Neurol. Surg. 7: 180-200, 1976.
33. Haverkos, H.W., Pazin, G.J., Armstrong, J.A. and Ho, M. New Eng. J. Med. 303: 699-670.
34. Eron, L.J., Toy, C. and Santomauro, D. In: Herpes Viruses and Virus Chemotherapy (Ed. R. Kono), Elsevier Science Publisher, Amsterdam, 1985, pp. 277-278.
35. Finter, N.B. In: Interferons and Interferon Inducers (Ed. N.B. Finter), American Elsevier, New York, 1973, pp. 363-378.
36. Ikic, D. Smerdel, S., Soos, Ê. and Jusic, E. Rodovi Immun., Zavoda 17: 69-72, 1974.
37. Ikic, D., Smerdel, S. Rajniger-Mihotic, M., Soos, E. and Jusic, D. In: Proc. 9th Internat. Immunobiol. Symp. Yug. Acad. Sci. Arts, Zagreb, 1975. pp. 195-202.
38. Ikic, D., Prazic, M., Smerdel, S., Jusic, D., Delinar, N. and Soos, E. In:Proc. 9th Internal. Immunobiol. Symp. Yug. Acad. Sci. Arts, Zagreb, 1975, pp. 203-205.

39. Romano, A. Revel, M. and Doerner, T. Proc. Int. Symp. on Interferon (Katholike Universitat Leuven, Belgium, 1983) (Eds. McDougel, P.C. and Missollen, L.), Dr. W. Jund Publisher, Belgium, 1985, pp. 349-355.
40. Knight, E. and Faley, D., J. Biol. Chem. 256: 3609-3611, 1981.
41. Corey, L., Nahmias, A.J., Guinam, M.E., Benedetti, J.K., Critchlow, C.W. and Holmes, K.K. New Eng. J. Med. 306: 1313-1319, 1982.
42. Kinghorn, G.R., Turner, E.B., Barton, I.G., Potter, C.W., Burke, C.A. and Fiddien, A.P. Antiviral Res. 3: 291-301, 1983.
43. Barton, I.G., Kinghorn, G.R., Rowland, M. Jeavons, M., Al-Omer, L.S. and Potter, C.W. Antiviral Res. 4: 293-300, 1984.
44. Shaw, M., King, M., Best, J.M., Banatvala, L., Gibson, J.R. and Klaber, M.R. Brit. Med. J. 291: 7-9, 1985.
45. Strauss, S.E., Takiff, H.E., Seidlin, M., Bachrach, S., Liningen, L., DiGiovana, J.J., Wester, K.A., Smith, H.A., Lehrman, S.N., Creach-Kirk, T. and Alling, D.W. New Eng. J. Med. 310: 1545-1550, 1984.
46. Douglas, J.M., Critchlow, C., Benedetti, J., Mertz, G.J., Conner, J.D., Hintz, M.A., Fahnlander, A., Remington, M., Winter, C. and Corey, L. New Eng. J. Med. 310: 1551-1556, 1984.
47. Mindel, A., Faherty, A., Hindley, D., Weller, I.V.D., Sutherland, S., Fiddian, A.P. and Adler, N.W. Lancet 2: 57-59, 1984.
48. Strauss, S.E., Seidlin, M., Takiff, H.E., Roney, J.F., Lehrman, S.N., Bachrach, S., Felser, J.M., DiGiovana, J.J., Grimes, G.J., Krakawer, H., Hallaha, C. and Alling, D. Antiviral Res. 6: 151-159, 1986.
49. Shoham, J. Advan. Biotechnol. Processes 2: 209-269.
50. Cerrutti, I. and Chany, C. Infect. Immuno. 42: 728-732, 1983.
51. Klein, A.S., Fixler, R. and Shoham, J. J. Immunol. 132: 3159-3163, 1984.
52. Tovo, P.A., Colleselli, P., Miniero, R., Palomba, E., Martino, S., Roncardo, M.G., Pastore, G. and Pugliese, A. In: Thymic Factor Therapy (Eds. Hobbs, J.R. and Byron, N.A.) Raven Press, New York, 1984, p. 199.
53. Shoham, J. In: Thymic Hormones: Immunologic Mechanisms and Therapeutic Actions (Eds. Kauttab, N.N. and Marcel Dekker, New York, 1987. In Press.
54. Langford, M.P., Villarreal, A.L. and Stanton, G.J. Infect. Immun. 41: 214-218, 1983.
55. Masonari, B., Ito, M. Shigeta, S. and DeClercq E. Antimicrob. Agent Chemother. 25: 515-517, 1984.
56. Crane, L.R., Milne, D.A., Sunstrum, J.C. and Lerner, A.M. Antimicrob. Agent Chemother. 26: 557-562, 1984.
57. Ikic, D., Brnobic, A., Jurkovic-Vukelic, V., Smerdel, S., Jusic, D. and Soos, E. D. In: Proc. 9th Internat. Immunobiol. Symp. on Clinical Use of Interferon, Yug. Acad. Sci. Arts, Zagreb, 1975, pp.235-238.

58. Stefanon, D. The Cervix 1: 23-26, 1983.
59. Marcovici, B., Peretz, B.A. and Paldi, E. Isr. J. Med. Sci. 19: 104, 1983.
60. Schonfeld, A., Nitke, S., Schattner, A. Wallach, D., Crespi, M., Hahn, I., Levavi, H., Yarden, O., Shoham, J., Doerner, T and Revel, M. Lancet 1: 1038-1042, 1984.
61. Gross, G. Ikenberg, H. and Gissman, L. Lancet 1: 1467-1468, 1984.
62. Geffen, J.R., Klein, R.J. and Friedman-Kien, A.E. J. Infect. Dis. 150: 612-615, 1984.
63. Friedman-Kien, A.E., Plasse, T.F., Cremin, P., Geffen, J.R., Castro, B., Badiak, H., Fedorczyk, D. and Trout, J.R. Workshop Proceedings on Papilloma Viruses, Molecular and Pathogenetic Mechanisms (Eds. Syrjanen, K. and Syrjanen, S.), Finland, 1985, p. 17.
64. Vesterinen, E., Meyer, B., Cantell, K. and Purola, E. Obstet. Gynecol. 64: 535-538, 1984.
65. Marchioni, M. Vallecchi, C., Zuccuti, G., Novelli, A. and Periti, P. Farmacia & Terapia 2: 1-5, 1985.
66. Gall, S.A., Hughes, C.E. and Trofatter, K. Am. J. Obstet. Gynecol. 153: 157-163, 1985.
67. Gall, S.A., Hughes, C.E., Mounts, P. Sagniti, A., Weck, P.K. and Whisnant, J.K. Obstet. Gynecol. 67: 643-651, 1986.
68. Vance, J.C., Bart, B.J., Hansen, R.C., Reichman, R.C., McEwen, C., Hatch, K.D., Berman, B. and Tanner, D.J. Arch. Dermatol. 122: 272-277, 1986.
69. Eron, L.J., Judson, F., Tucker, S., Prawer, S., Mills, J., Murphy, K., Hickey, M., Rogers, M., Flannigan, S., Hien, N., Katz, H.I., Goldman, S., Gottlieb, A., Adams, K., Burton, P., Tanner, D., Taylor, E. and Peets, E. New Eng. J. Med. 315: 1059-1064, 1986.
70. Rubin, B.Y. and Grupta, S.L. Proc. Nat. Acad. Sci. 77: 5928-2933, 1980.
71. Haglund, S., Lundquist, P-G, Cantell, K. and Strander, H. Arch. Otolaryngol. 107: 327-332, 1981.
72. McCabe, B.F. and Clark, K.F. Ann. Otol. Rhinol. Laryngol. 92: 2-7, 1983.
73. White, E.C. J. Am. Med. Assoc. 250: 1815, 1983.
74. Bomholt, A. Arch. Otolaryngol. 109: 550-552, 1983.
75. Naiman, H.B., Ruben, R.J., Doyle, A.T. and Kadish, A.S. Ann. Otol. Rhinol. Laryngol. 93: 483-487, 1984.
76. Grunberg, S.B., Harmon, M.W. and Couch, R.B. In: Interferon and Interferon Inducers (Ed. Stringfellow O.A.) Marcel Dekker, New York, 1980, pp. 57-88.
77. Margolis, S. Rev. Infect. Dis. 4: S829-S836, 1982.
78. Adler, M.W. Brit. Med. J. 288: 213-215, 1984.

3

INTERFERON THERAPY IN ACUTE VIRAL ILLNESSES

STANLEY LEVIN, MD. and EUGENE LEIBOWITZ, MD.

Department of Pediatrics and Pediatric Research Institute, Kaplan Hospital, Rehovot, affiliated to the Hebrew University-Hadassah Medical School, Jerusalem, Israel.

INTRODUCTION

Thirty years have elapsed since the discovery of interferon (IFN) as a naturally occurring biological substance with antiviral activity against a broad spectrum of RNA and DNA viruses, but reports of its use in clinical trials of acute viral infections have been surprisingly meager. There are several reasons for this, the major one being that for the first 20 years after its discovery, stocks were in very short supply, and when available, very expensive.

Additionally, when the subsequently discovered properties of the IFNs to suppress cell growth and modulate the immune response became known, clinicians were excited with the possibility of finding a new approach to the control and cure of cancer. At this point, the antiviral activity of IFN seemed to have become of secondary importance to physicians, for after all, unlike cancer, many of the cases of viral disease for which IFN was being used were not life-threatening. The recent appearance of several antiviral chemotherapeutic agents like acyclovir, active against many of the viruses which are responsive to IFN, may have also led to the paucity of clinical trials of IFN as an antiviral agent. With the advent of newer technology, increasing amounts of the different types of IFN were produced and during the last few years genetically engineered recombinant IFNs have become available cheaply and in vast amounts, allowing for their use in clinical trials of viral diseases as well as cancer.

Tissue culture and animal studies have shown that IFN has a remarkable antiviral effect on most viruses but the effect is not a direct one. IFN inhibits virus replication and spread in the cellular microenvironment by activating several intracellular antiviral enzymes whose functions are to prevent recognition of the virus, translation of viral mRNA and assembly of new viruses. Normally the presence of a virus leads to the protective production of IFN molecules by the cell and these are then secreted into the extracellular spaces and into the blood stream where they will interact with IFN receptors and other cells leading to the induction of an antiviral state. Under certain conditions, and for reasons not as yet apparent, this IFN system may not be activated and the virus will replicate and spread uncontrolled until overwhelming infection leads to the demise of the patient. This has been reported to occur in cases of fulminant hepatitis, herpes encephalitis and in the terminal stage of AIDS, and has been termed the "IFN deficiency syndrome" (1). The giving of exogenous IFN in these cases has led to activation of the IFN system and has proved to be life-saving.

The main stimulus for pursuing clinical trials of IFN in viral disease came from Merigan and his co-workers in Stanford, California. It is more than 10 years since they began their trials to prove the efficacy of IFN in the prevention and treatment of herpes infections, particularly in the immune-suppressed host. They persisted in their efforts to establish IFN as a promising therapeutic agent in chronic viral hepatitis, and encouraged other workers in the field. Although some dramatic effects were often seen in individual cases, clinical trials as a whole were not decisive but encouraging. A major problem was the absence in the early years of well controlled trials. It appeared that IFN would be effective in certain cases whereas in others, the response was poor, if at all. This was most evident in the conflicting reports of the beneficial, or otherwise, effect of IFN in reducing or eliminating the viral markers in chronic viral hepatitis. All this has led to a feeling that IFN has as yet to prove itself as a good and efficient antiviral agent.

The following is a brief description of our experience in the treatment of 80 cases of severe, acute, life-threatening viral infections with HuIFN- , and a short review of published reports of the use of IFN therapy in a variety of acute viral illnesses.

PATIENTS AND METHODS

Since 1980, 80 patients of all ages suffering from life-threatening viral diseases have been admitted to an open study evaluating the effectiveness of HuIFN- treatment. Most patients were extremely ill with over 50% being in deep coma when entered into the study, usually as a last resort when all other therapies had failed. The trial was conducted according to a predetermined protocol which included pharmakinetic studies of several parameters of the IFN system before, during and after IFN therapy.

HuIFN- prepared by Drs H. Rosenberg and T. Bino at the Israel Institute for Biological Research, Nes Ziona, was given intramuscularly once a day in a 1ml solution of $3x10^6$ international units ($5x10^6$ units/mgm of protein). In small infants the daily dose was 70,000 to 100,000 units/kgm body weight. The course of treatment varied according to the illness or the patient's condition and response to therapy. For example, in cases of herpes infection of the skin, 1 to 5 (average-3) injections proved sufficient to stop the progress of the infection and for healing to begin, whereas in fulminant hepatitis, the average course lasted about 10 days. In a case of laryngeal papillomatosis the treatment was continued for 10 months. Only patients receiving at least 3 daily injections of IFN- were included in the final evaluation.

The methods used in laboratory studies of the IFN system have been described in detail (2) and included the following:
1) Plasma IFN levels as an indication of in vivo IFN production in response to the viral infection.
2) Production of IFN- and IFN- by stimulated peripheral blood mononuclear cells (PMBC) in vitro, as an indication of the competency of cells to produce IFNs.
3) Evaluation of the antiviral state (AVS) of PBMC by testing whether they supported viral replication in culture or not.

RESULTS

One hundred and twenty patients ranging in age from 1 month to 70 years were treated with HuIFN- . Of these 80 were felt to fulfil the requirements of the protocol and are reported here (Table 1). The remaining patients were excluded because they were either proved later not to have an acute viral illness, or died before completing 3 days of therapy.

Acute progressive and fulminant hepatitis

There were 30 patients with acute hepatitis who developed acute hepatic failure and whose condition was serious enough to cause concern for their lives. Eighteen were confirmed as being caused by hepatitis A virus (11 of whom were in grade III+IV coma when treatment was begun and the rest grades I+II), 6 were due to hepatitis B virus infections (5 in grade IV coma), and 6 non A-non B hepatitis including two with herpesvirus infection, all of whom were in grade IV coma. Seventeen patients (57%) recovered of whom 10 were in grades III+IV coma. Our results indicated the best prognosis in adults and older children where 15/22 (77%) recovered whereas only 2/8 children below 4 years of age survived. In more than half these cases the IFN response was defective prior to IFN therapy and in almost every case the IFN system became activated within a few days of beginning IFN therapy (3).

Table 1. Outcome of patients with severe viral illnesses treated with HuIFN- .

Diagnosis	Treated	Recovered	Unchanged	Died
Progressive or fulminant hepatitis	30	17	(2)	13
Encaphalitis	14	9	(2)	5
Herpes with immune-deficiency	19	19	-	-
Cytomegalovirus	6	3	1	2
Measles	4	3	-	-
Subacute sclerosing panencephalitis	3	-	3	-
Poliomyelitis (bulbar)	2	2	-	-
Varicella myelitis	1	1	-	-
Laryngeal papillomatosis	1	1	-	-
Total	80	55	4 (4)	21

( ) No. in brackets - cases recovered, but no evidence that IFN had any effect.

Acute viral encephalitis

There were 14 patients who were diagnosed as having viral encephalitis of whom 12 were in states III or IV coma. Eight cases were confirmed by serology as having herpes encephalitis, two measles, one varicella and in three the virus was not identified. Nine patients recovered, including 8/12 comatose patients. Nine patients were aged 1 month to 9 years, seven of whom survived. In 2 survivors there was no apparent change in the patient's condition while receiving IFN (4).

Herpes in immune-compromised patients

Seventeen cases with herpes zoster and 2 with herpes simplex received an average of 3-5 daily injections of IFN- . One critically ill patient with extensive spreading pyoderma gangrenosum and leukemia was treated for 16 days before recovering (5). In most cases pain was alleviated within 1-2 days, no new lesions appeared after 2-3 days therapy and healing was noted within 3-5 days. In 3 cases it appeared that IFN- therapy had no effect on the infections after 3-5 injections. In general, the treatment was well tolerated, symptoms were alleviated and the usual course of the disease appeared to be shortened.

Cytomegalovirus infections

Six patients with CMV infections were treated. Of these three recovered: a 6-month-old with hepatomegally and severe immune-deficiency, another with severe hemolytic anemia and a 4-year-old with X-linked lymphoproliferative disease and pneumonia. Two others with lymphomas died of pneumonia, and the condition of a baby with chronic cytomegalovirus encephalitis remained unchanged.

Measles

Only very severe cases of measles with complication were accepted for IFN therapy. Of four cases treated, three had pneumonia and one encephalitis. Three children recovered, but a 1-year-old baby with IFN deficiency syndrome died with secondary infected pneumonitis.

Poliomyelitis

An adult and an infant, both with progressive ascending poliomyelitis affecting the bulbar area of the brain, were treated with IFN- . The immediate results were quite dramatic, with the disease progression stopping within 1-2 days and improvement beginning after 4-5 days of therapy. In the adult, who was severely paralyzed and receiving assisted ventilation, dramatic improvement occurred within 48 hours, followed later by gradual recovery (6).

Varicella transverse myelitis

An adult with varicella developed a clinical picture of progressive transverse myelitis. Following a few days of IFN- treatment recovery began and was complete within a few weeks.

Laryngeal papillomatosis

A 2-year-old child with severe recurrent laryngeal papillomatosis had undergone 18 operations for removal of recurrent papillomata over a period of 9 months. Following IFN- therapy no further growth occurred and the child recovered completely after 10 months of injections. A six-year follow-up showed no recurrence.

Subacute sclerosing panencephalitis (SSPE)

Three cases of SSPE showed no response to 2-4 weeks of IFN- therapy.

Interferon deficiency syndrome

Eighteen cases of acute progressive viral disease with seriously defective in vivo and in vitro IFN responses have been diagnosed. These include patients with acute fulminant hepatitis, progressive herpes encephalitis and terminal AIDS. Whereas these cases are invariably fatal without IFN therapy, treatment with IFN has led to reactivation of the IFN system and recovery in most cases (1).

DISCUSSION

Our clinical experience over a period of 7 years in which HuIFN- has been used in an open study for the treatment of 80 cases of acute life-threatening viral disease indicates that this therapy is safe and effective when given at a dosage level of $3x10^6$ U/day, and the final results are most encouraging. Individual cases have shown dramatic responses especially in patients with acute fulminant hepatitis. Well controlled clinical trials are now needed to confirm the efficacy of IFN treatment in patients who are severely ill with viral illnesses.

A review of the literature shows that IFN therapy in man has been studied in no more than twenty different viral illnesses, and in only a few have well controlled clinical trials been attempted. In most early studies HuIFN- or HuIFN- was used, but during the last few years different recombinant IFNs have been tried. The following is a brief summary of some of the more promising results, with final confirmation in most instances awaiting more extensive controlled clinical trials.

Herpesviruses

Of the approximately 80 herpesviruses which have been at least partially characterised, only 5 have been isolated from humans (7). Many of these herpesviruses have been shown to be sensitive to IFN in tissue culture and animal studies, and infections in humans with these viruses were among the first to be investigated in clinical trials. These viruses vary greatly in their biological properties. Some, like herpes simplex, have a wide host-cell range, multiply efficiently, and rapidly destroy the cells which they infect. Others, such as Epstein-Barr virus have a narrow host-cell range, and still others, such as cytomegalovirus seem to multiply slowly and are less destructive to cells. An ubiquitous property of herpesviruses is their capacity to remain latent in the host cell in which they multiply, and be triggered to reactivation by a variety of mechanisms. Many infections with these viruses are benign and self-limiting, but under certain circumstances, such as in the immune-compromised host, the infections may become life-threatening and lead to death of the host.

Herpes simplex virus infections (HSV): Two antigenic types of HSV have been described: HSV-1 associated chiefly with non-genital infections of the mouth, lips, eyes and central nervous system (although it may also cause genital disease), and HSV-2, most commonly associated with genital and neonatal (birth canal) infections, but may also cause oral and CNS infections. Persistent and recurrent infections may be characterized by "fever" blisters, genital herpes or dendritic corneal ulcers. Disseminated herpesvirus infections, meningoencephalitis and eczema herpeticum (Kaposi's varicelliform eruption) are usually life-threatening. Infections in immunosuppressed and cancer patients and those with immunodeficiency can be fatal and require specific antiviral treatment. It must be stressed that the earlier treatment is begun, the better the results.

Skin and mucous membrane herpetic infections: Several studies have shown that IFN given systemically or topically as an ointment was effective in shortening the course of the disease and ameliorating symptoms (8,9).

Systemic herpetic infections: These have responded to daily Im injections of $3x10^{6}$ U HuIFN- although no well-controlled trials have been reported. We have used it in some immune-compromised patients who recovered rapidly from the infection.

Herpetic eye infections: Recurrent herpetic keratitis and dendritic keratitis have been treated with IFN drops alone (10) or combined with other therapies such as acyclovir ointment (11, 12), trifluorothymidine drops (13, 14) and secretory IgA (15). All these studies, and many others, have shown the superiority of these over other treatments, including more rapid healing as well as rapid disappearance of photophobia and pain.

Herpes encephalitis: Recently, acyclovir has become the accepted treatment for herpes encephalitis although in some cases Ara-A is preferred. However, in a recent study reported by us, 9 patients aged 7 months to 60 years suffering from severe herpes encephalitis (7 of whom were in coma) were treated with HuIFN- $3x10^{6}$ U daily Im. Six patients recovered. In cases where IFN was started 6 days or more after onset of symptoms, only 3/6 survived, whereas all three who received IFN within 4 days of onset of symptoms recovered (4). A single case report of a newborn baby who received intrathecal IFN for 6 days indicated no beneficial effect from this therapy (16).

Other HSV infections: Used prophylactically in a double-blind study, IFN injected prior to microneurosurgery of the trigeminal nerve significantly diminished the occurrence of post-operative herpes labialis as well as HSV viral shedding from the oropharynx (17). On the other hand, prophylactic IFN did not seem to have any significant effect on the development of HSV infections in patients undergoing renal transplantation (18).

Genital herpes infection: This is usually due to HSV-2, and is probably the commonest cause of genital ulceration in females and males. Of great clinical importance is the fact that a large group of these patients tend to have recurrent or chronic infections which often give rise to severe clinical and psychological problems.

Acyclovir ointment has been used successfully in treating the acute episodes. However the infections tend to recur, and in some cases are markedly disabling. HuIFN- and have been used topically as an ointment in uncontrolled studies and the results reported as highly significant (9).

Varicella-zoster (VZ) infections: VZ belongs to the herpes group of viruses and is a common, usually benign skin infection of children and adults. Zoster, which is common in adults, is a reactivation of latent VZ virus and is characterised by localized crops of varicella-like lesions along the course of the sensory nerve distribution of dorsal root ganglia or extramedullary ganglia of cranial nerves. In general, both chickenpox and herpes zoster (HZ) are characterised by negligible mortality but are relatively irritating, particularly with regards to itching, pain and loss of time from school or work. HZ in particular may be followed by neurologic pains lasting weeks to years, and this may cause severe disability. However, in the immunosuppressed patient, whether due to cancer or following therapy, HZ and varicella viruses tend to multiply in the viscera and disseminate and lead to more severe illness which may end fatally in 10% of cases. In these patients, IFN therapy has proved useful in well-controlled studies (19,20) as well as in our own open study. Pain is rapidly reduced in most cases, and the appearance of new vesicles suppressed. Post-neuralgic pain is less common in IFN-treated patients (19). In our experience a dose of $3x10^6$ U HuIFN- Im daily for 3-5 days, has proved sufficient in most cases that we have treated. Larger doses may be more effective, but increases the chances of toxicity. IFN-based ointments seem to be useful for controlling localised pain and pruritis. Ara-A and acyclovir also have beneficial effects in VZ infections, the latter being somewhat better (21).

The Epstein-Barr virus (EBV): The EBV was discovered in 1960 and has been shown to cause a wide spectrum of illnesses in children and adults, from a mildly contagious, self-limited febrile illness with lymphoid hyperplasia called infectious mononucleosis, to prolonged chronic illness. It is also associated with certain neoplastic diseases such as Burkitt's lymphoma and naso-pharyngeal carcinoma. Like other viruses of the herpes group, EBV can be fatal if it occurs in the immune-compromised host. Very limited experience has been reported on the use of IFN in EBV infections, where in general, the mildness of the illness does not warrant systemic therapy. In one study of EBV infections occurring in renal transplant recipients, IFN appeared to decrease virus excretion when compared to patients receiving antithymocyte globulin or placebo (22). A trial of its use in chronic EBV infections seems warranted particularly in the light of several reports suggesting some beneficial effects and even cure in nasopharyngeal carcinoma (23).

Cytomegalovirus (CMV): CMV is an ubiquitous virus with host interactions ranging over a large spectrum of health and illness. Virus shedding from the genital tract is common in asymptomatic women, and infection of the newborn at birth is not unusual. Clinical symptomology occurs only in a small proportion of these infected infants. However fulminant CMV infections can be alarming and life-threatening. Postnatal infections also occur, apparently by contact with infected secretions or even from contaminated breast milk. In older individuals, infection can occur from blood transfusions and bone marrow and organ transplants from infected donors (24).

Although experimental studies have indicated that IFN may diminish virus excretion in chronic excretors and delay CMV reactivation in transplant recipients (25), its use therapeutically in patients with generalised CMV infection or immunosuppressed patients (such as post bone-marrow transplantation) has been generally unsatisfactory. High doses of IFN, as well as combination therapy with acyclovir or Ara-A, has given rise to unacceptable toxic sequelae. One encouraging report (25) showed that $3x10^6$ U IFN- given prophylactically before and after renal transplantation ($1x10^8$ U over a period of about 4 months), markedly reduced clinical signs of CMV infection compared to a placebo control group. Opportunistic infections with aspergillus and pneumocystis carinii were only seen in the placebo group. Further experience using combination therapies is required in order to validate whether there is a place for IFN in the treatment of CMV infections, although a combination study with acyclovir appeared to be ineffective and quite toxic (26).

IFN has been used in chronic CMV infections, again with little success. Viral shedding can be diminished, but in most cases there is no apparent effect on the clinical illness. In one of our cases with severe congenital CMV infection and immunodeficiency treated with HuIFN- for 1 month, all clinical symptoms regressed, the grossly enlarged liver returned to normal, immunological abnormalities disappeared and the child was well and normal for 1.5 years, when he suddenly went into unexplained coma and died within 2 days. Autopsy was refused.

Vaccinia

Vaccinia, a poxvirus, has been used to virtually eradicate smallpox from the face of the earth. For this reason, since 1977 smallpox vaccination has been stopped and vaccine is no longer being produced. Serious complications of vaccinia such as eczema vaccinatum, progressive vaccinia in the immune-compromised host and vaccinia gangrenosum are all conditions that require treatment. Early clinical and animal studies showed that IFN had a beneficial effect in vaccinia infection when given systemically or locally (eye drops or ointment), but well-controlled trials are lacking. More recently a live recombinant vaccinia virus has been used in an attempt to prevent this infection in patients at risk.

Respiratory virus infections

Upper respiratory illnesses caused by viruses have the highest morbidity of all infections. Discomfort and sometimes "misery" lasting several days is the classic clinical pattern seen in the majority of patients with the common cold which is caused by a broad spectrum of viruses of which the rhinoviruses and coronaviruses are probably the most common. Influenza, parainfluenza and respiratory syncytial virus (RSV) may cause more severe illness when they involve the middle or lower respiratory tract, and RSV is the commonest cause of a severe, sometimes fatal illness of the bronchi and bronchioli affecting infants below the age of 1 year.

Numerous well-controlled studies of IFN- (natural and recombinant) used mainly as an aerosol spray of the upper respiratory tract or as nasal drops, have shown that in general this treatment is effective in partially preventing the spread of colds in households, schools or places of work if used according to a specific schedule. In a recently reported study in which household contacts of an infected individual were given INF as a spray once a day for 7 days beginning within 48 hours of onset of the illness, effectiveness in preventing colds was only 39%. However, in laboratory-exposed rhinovirus-infected volunteers the efficacy of IFN sprays was 88% as compared to placebo (27). Other studies showed that this treatment significantly reduced the duration and quantity of viral shedding, although neither spray (3x daily)or drops prevented rhinovirus infection or colds when given 28 hours after rhinovirus inocculation (28). Similar results have been reported from a double-blind placebo-controlled study in Australia (29). Another study suggested that intranasal lymphoblastoid IFN was less effective as prophylaxis against influenza A virus infections than against rhinovirus (30). However, one disconcerting effect of nasal spray or drops was the frequent occurrence of nasal irritation sometimes manifested by blood-tinged nasal mucus and superficial erosions of the nasal mucosa. This was also seen in a smaller percentage of patients receiving placebo sprays or drops.

It appears from a review of the large number of reported studies that IFN aerosol therapy will reduce remarkably the incidence of the common cold when used prophylactically. The best results were seen if the IFN was given early, although varying results were seen depending on the type and dosage of IFN used. However, aerosol sprays or drops used as symptomatic therapy for colds were not as effective although duration of viral shedding and nasal mucus production were often significantly reduced. Best results were seen in rhinovirus infections and less so with other virus infections. Further studies using modified recombinant IFNs appear to be promising in that nasal irritation is less common.

Papilloma viruses

Human papilloma viruses (HPV) are among a small group of viruses known to cause tumors in man. In general they cause benign papillomas or warts of the skin or mucous membranes including those of the genitalia and respiratory tract because of a fastidious selectivity for squamous epithelial cells. Under certain circumstances malignancies may develop, and HPV has been identified in cervical dysplasia and in invasive cancers of the cervix.

Laryngeal papillomatosis: In this condition papillomata growing on the larynx tend to block the airway to the lungs leading to a life-threatening condition. Accepted therapy has been surgical removal or resection by lasers. However the tumors tend to regrow, sometimes rapidly, necessitating repeated operations, as occurred in our case who required 18 operations in 9 months. In older patients the growths may stop after a while leading only to voice changes. Systemic IFN has been shown to be very effective in preventing regrowth of papillomata after removal. However treatment needs to be carried on for 6-12 months or longer, and in at least 50% of patients recurrent growth occurs when treatment is stopped. Retreatment will again inhibit growth and complete remission has been seen in about 30% of juvenile-onset laryngeal papillomatosis treated with IFN- (but not with IFN-), and in almost all cases of adult-onset type (31). Another 46% had a decrease in lesion size, with responses generally occurring within 2 months, and with prolonged therapy 90% could be maintained relatively symptom-free with minimum intervention. Recommended dosage is $3x10^6$ U IFN- Im daily for 1-2 weeks (or 70,000 U/kgm for small infants), then every other day or 3 times a week for 6-12 months. With this dose side effects are minimal.

Respiratory papillomatosis: Recurrent papillomatosis of the respiratory tract is commoner in adults than in children. IFN- has been shown to lead to a moderate or better response in the majority of patients receiving therapy over a period of at least 8 months (32).

Condyloma acuminata (genital warts): A common manifestation of sexually-transmitted papillomavirus involving the vulva, vagina or perianal regions (and the penis in the male) is a typical raised condyloma which can grow to the size of a fist. Many cases will get better with the simplest of treatment. However some cases remain resistant to all therapies including surgical and laser treatments, and these may benefit from either local IFN ointment treatment, systemic

IFN therapy or injection of IFN into the base of the condyloma (33). In one study using lymphoblastoid IFN, 33% of podophyllin-resistant cases were completely cured with 1-3x10^6 U 3x weekly for 6 weeks, and an additional 30% showed partial clearance, a total response rate of 88%. (34). Intramuscular IFN- was effective in clearing up mild primary lesions in 80% of cases treated (35). Intralesional injections of IFN- cured 5 or 8 patients, male and female (33). IFN-based ointments used 4-5x daily have also proved effective in some cases.

Cutaneous warts: These have been treated either with local injection into the base of the lesion or by systemic injections of IFN. Reports, mainly from Japan, indicate an 80% or better cure rate following intralesional injection of IFN. For multiple generalised cutaneous warts, systemic IFN has been used with limited success although improvement has been noted in some cases. It should be noted that spontaneous regression of warts may occur in as many as 2/3 of children within 2 years of onset. Therefore intralesional IFN treatment may be indicated in only the severest of cases, usually in cases with a single lesion or a few annoying ones. Dosage should be at least 1-10^5 U per lesion once or twice a week for several weeks. In rare cases, a single injection may suffice.

Enteroviruses

Very little has been published on the possible benefit of IFN therapy on infections caused by this group of viruses which include poliomyelitis, coxsackie viruses A+B (herpangina, pleurodynia and foot and mouth disease) and the large group of echoviruses. Most illnesses are transient, mild and non-fatal. However, myocarditis due to coxsackie B, enteroviral meningoencephalitis and poliovirus infections may end fatally. There is no specific antiviral treatment available for these conditions and the only report of the use of IFN for treating illnesses due to any of these viruses is the one we published in 1984 in which two cases of progressively ascending bulbar poliomyelitis received HuIFN- . Both cases responded remarkably in that progress of the disease was halted dramatically and regression began within 2-3 days of onset of therapy (6). It is felt that early treatment of severe progressive enteroviral infections with IFN may have beneficial results.

Measles

Measles is due to an infection with a paramyxovirus and is characterised by infection of the upper respiratory tract which is highly contagious. Serious complications involving the lungs and central nervous system with occasional fatalities occur in a small minority of cases. Immunization has all but eliminated this disease in developed countries, but in some underdeveloped countries measles takes a large toll especially in malnourished infants. Measles in the immune-suppressed patient can be very severe and even fatal. We have treated four very severe cases of measles including one with meningo-encephalitis with IFN- , three of whom recovered rapidly. One infant died on the 6th day with overwhelming pneumonia. The recommended dose of IFN- is 70,000-100,000 U/Kg Im daily.

Subacute sclerosing panencephalitis (SSPE): SSPE can be considered a late complication of measles with the features of a slow-growing viral infection. Electron-microscopic and serological studies suggest that the cause of the brain pathology is a defective variant of the virus. Several reports describing the use of intravenous, intramuscular and intrathecal IFN therapy in at least 14 patients showed no evidence of any therapeutic effect in any of the cases (36, 37). However in one other single case there was remarkable improvement with IFN therapy (38). Spontaneous improvement of clinical signs and symptoms is well known in this disease, so this single anecdotal report should be considered with caution.

Japanese encephalitis (JE):

JE is an endemic disease in southeast Asia affecting mainly children and is one of the more serious anthropod-borne viral infections caused by a flavivirus and characterised clinically by fever and CNS involvement leading to convulsions, paralysis, coma and death in about 20-30% of cases. Up to 50% of those that survive have permanent cerebral sequelae. Laboratory studies in Thailand showed that flavivirus causing JE and dengue hemorrhagic fever were sensitive to IFN- and IFN- . In a preliminary report, 2/4 comatose children with

JE who were treated with IFN- recovered (Prof. C. Harinasuta, personal communication). The authors suggested further studies be undertaken with both IFN- and IFN- given Iv and intrathecally.

Rabies

In animal studies IFN has been shown to provide protection against challenge by rabies virus only when it is administered before or shortly after virus challenge. Therefore the early use of IFN in rabies seems justifiable. Once clinical symptoms are present the disease is uniformly fatal and no treatment has proved of any use. In a report of 5 cases treated with IFN after symptoms appeared in another single-case report, none survived (39). It would seem appropriate to try IFN early in the incubation or prodromal stage of suspected rabies together with anti-rabies vaccine and immunoglobulin.

Acquired immune deficiency syndrome (AIDS)

This recently described disease has been shown to be due to a cytopathic retrovirus called the Human Immunodeficiency virus (HIV). Infection with this and related viruses is at present spreading at an alarming rate throughout the world by sexual dissemination in homosexuals and to a lesser degree in heterosexuals, as well as in intravenous drug users. Congenital infections of infants of infected mothers is being more commonly seen. At the time of writing this report, several agents appear to have promising anti-HIV activity in vitro. However, no particular one has been shown to have therapeutic value in well controlled trials. IFN has been used in numerous clinical studies, but as yet no definite conclusions have been drawn as to its efficacy in AIDS. In certain terminal cases of AIDS the body's IFN system has been shown to be completely defective (40), and at this stage IFN may prove useful in helping to combat secondary viral infections and prolonging life.

An unusual skin tumor called Kaposi's sarcoma commonly develops in AIDS patients. This may be accompanied by systemic symptomatology, and intercurrent and opportunistic infections are common due to the accompanying immunodeficiency. IFN- in high doses, which often leads to signs of toxicity, has been shown to lead to complete recovery in some cases and partial remissions in others. Low dosage schedules were usually ineffective as was treatment with IFN- . In general, some patients responded well to therapy, others only partially, and some failed to derive any benefit whatsoever. Intercurrent CMV and EBV infections do not appear to be affected by the IFN therapy, although opportunistic infections occurred less frequently in treated cases. Many studies are now in progress to determine the most appropriate type of IFN to use and the best dosage schedule. The use of IFN together with an immune-stimulant has also been suggested.

Rubella virus infection

Postnatally-acquired rubella infections are usually very mild, induce immunity and do not require any treatment. Rubella rarely may infect the brain giving rise to encephalitis, and even here complete recovery is the usual course. Of more importance is congenital rubella acquired by the fetus during pregnancy, and which leads to a high incidence of congenital malformations. A rubella syndrome is sometimes seen in newborns with involvement of the brain, liver, lungs and other organs. Petechiae are common and the clinical picture is similar to that seen in congenital CMV infections. A 14-month-old baby with this syndrome complicated by vasculitis was treated for 2 weeks with $3x10^6$ U IFN- /day, and all signs and symptoms cleared up rapidly although virus secretion in the urine continued and anti-rubella IgM antibodies persisted (41).

Juvenile diabetes mellitus

For many years there has been speculation that a virus may be involved in the onset of insulin-dependent diabetes mellitus in children, and a coxsackie B4 virus has been isolated from the pancreas of a patient who died from diabetic ketoacidosis (42). Our studies showed that at the time of onset of newly-diagnosed cases of juvenile diabetes the IFN system is often activated as is seen with viral infections (43). However, IFN treatment in 2 newly diagnosed cases failed to arrest the diabetes (44).

Accidental laboratory infections - ebola virus

A laboratory worker was accidentally infected whilst processing material from African patients with hemorrhagic fever. The patient developed a disease resembling Marburg disease, and a virus similar to but serologically distinct from Marburg virus was isolated. The patient received 14 days therapy with 6x10^{6} IFN- daily as well as convalescent serum and recovered following a relatively mild course. The role of IFN in this case is difficult to judge. This report suggests that IFN may be useful in early prophylactic treatment of accidentally infected persons with unknown or unusual viruses (45).

<u>Adenovirus</u>

Adenovirus infections of the respiratory tract are usually mild and do not require any specific therapy. However, rare cases of fatal pneumonia or bronchiolitis obliterans as well as meningo-encephalitis have been ascribed to adenovirus infections, and may prove responsive to IFN therapy if given early in the course of the disease. Adenovirus infections of the eyes may lead to keratoconjunctivitis which may be very irritating and even disabling. This infection often occurs in epidemics and the pain and irritation may last for several weeks. IFN- drops, 2-5x10^{5} U daily divided into 8-10 drops cured the disease within an average of 6 days compared to 22 days in a control group, and prevented the appearance of subepithelial keratitis which occurred in 57% of the controls (46). This and other studies indicated that early treatment with high dosage IFN drops is a very useful treatment.

<u>Viral hepatitis</u>

Viral hepatitis refers to a primary infection of the liver most commonly caused by at least 5 etiologically and immunologically distinct viruses: hepatitis A (HAV), hepatitis B (HBV), hepatitis D (HDV - probably a defective virus requiring HBV synthesis for its expression), and 2 or more non A-non B viruses. Hepatitis may also occur as a secondary infection during systemic herpes viral disease due to CMV, EBV, HS and VZ infections. Clinically, most patients with viral hepatitis have a mild to moderate illness from which they recover within a few weeks with no obvious sequelae. About 10% of cases develop evidence of chronic disease, either chronic persistent hepatitis or chronic active hepatitis which may be symptomatic or even debilitating, and in whom the viral markers may persist for years. There is a high correlation between the prevalence of HBsAG carriers and the incidence of primary hepatocellular carcinoma. Rarely, cases of acute hepatitis may progress uninhibited with the development of an acute fulminant picture of liver failure and encephalopathy. We have shown that in acute progressive and/or fulminant hepatitis, the interferon system is not as adequately activated by the viremia as it is in acute hepatitis or in most other viral infections, and in some cases these may be an absolute deficiency of the IFN response (1,3). This could possibly explain the unremitting progression of the disease in fulminant hepatitis. Treatment with exogenous IFN- reverses the process in many cases, and the IFN response usually returns to normal in patients that survive.

Acute hepatitis: Very few studies have been reported of the use of IFN in the treatment of acute viral hepatitis, particularly because of the benign nature of the majority of cases. However, because of the dangers of the development of chronic hepatitis with persistent antigenemia it seems reasonable to consider a controlled study of early treatment of acute viral hepatitis with IFN in order to prevent the complications mentioned above.

Once a hepatitis patient follows a downhill course with the development of encephalopathy the need for some specific treatment becomes urgent. As reported above, we have treated some 30 patients with progressive and fulminant hepatitis (HAV, HBV, non A-non B and herpes) with HuIFN- . Most patients were unfortunately referred for IFN treatment very late in the course of the disease when they were already in stage IV coma and when all other treatments, including acyclovir and Ara-A had proved of no use. Despite this, 59% survived. Of the patients in grade III-IV coma when IFN therapy was begun 10/21 (48%) recovered. The prognosis was poorest in patients under 4 years where only 2/8 (25%) recovered. Of the older patients 77% recovered.

These results, although based on an uncontrolled, highly biased (towards severity as the patients were usually entered into the study as a last resort) open trial are encouraging, especially as the survival rate in fulminant hepatitis in historical controls is between 20-30% in adults, and about 10% in small infants. The usual dosage of HuIFN- was $3x10^6$ U Im per day for an average of 10 days. When a positive response was seen, it usually began on the 4th or 5th day at the same time that an antiviral state developed in the peripheral blood mononuclear cells (3). It is likely that these results could be improved upon if patients with progressive disease would be treated earlier. Obviously, confirmation of these results are required from a well-controlled double-blind study. In a study from Italy where IFN- was used no effect on the prognosis of acute fulminant hepatitis was noted (47).

SUMMARY AND SCOPE

IFN has been shown to effectively help combat a variety of viral infections. It seems to have been more effective in controlled laboratory infections or in animal studies than in clinical human trials, possibly because we have not yet overcome the problems of treating the patient as a whole with the numerous interrelating immune mechanisms involved in viral infections. It appears that it may not be sufficient to prevent viral replication with IFN, but also necessary to stimulate effective defense and tissue repairing mechanisms which will allow the body to recover with a minimal of irreversible damage. For this reason, it is possible that although certain viruses are effected by IFN in vitro, they are less responsive in vivo.

The viruses which seem most responsive to IFN therapy are the herpes group of viruses, some respiratory viruses, papillomaviruses, hepatitis viruses, measles and possibly enteroviruses. Moderately to severly ill patients with infections due to these viruses should be offered treatment with IFN, preferably HuIFN- , particularly where no effective alternative therapy is available. At the dosages recommended ($3x10^6$ U/day Im) IFN- is well tolerated, toxicity is minimal and clinical response is often seen within a few days. However, at this stage it appears that many more well controlled clinical trials are necessary before IFN antiviral therapy can be placed on a sounder footing.

REFERENCES

1. Levin, S. and Hahn, T. Clin. exp. Immunol. 60:267-73, 1985.
2. Levin, S. and Hahn, T. Clin. exp. Immunol. 46:475-83, 1981.
3. Levin, S. and Hahn, T. Lancet i:592-4, 1982
4. Shanon, A. and Levin, S. Pediatric Rev. Commun. 1:-, 1987.
5. Shalev, Y., Berrebi, A., Green, L., Levin, S. et al. Arch. Dermatol. 120:922-929, 1984.
6. Levin, S. J. Infect. Dis. 151:745-6, 1985
7. Roisman, B. and Batterson, W. In: Virology (Ed. B.N. Fields), Raven Press. NY., 1985, p 497.
8. Isaacson, M., Berson, B., Steinberg, J. and Morag, A. Isr. J. Med. Sci. 19:959-962, 1983.
9. Ikic, D., Trajer, D., Cupak, K. et al. J. Clin. Pharmacol. Ther. and Toxicol. 19:498=505, 1981.
10. Jones, B.R. In: Herpetic eye diseases (Ed. R. Sundmacher), Bergman, Munich, 1981, pp. 395-400.
11. Mecere, P.J. and Van Bujsterveld, O.R. Antiviral Res. 1:225-228, 1985.
12. De Koning, E.W.J., Van Bijsterveld, O.R. and Cantell, K. Arch. Opthalmol. 101:1866-1868, 1983.
13. Sundmacher, R., Cantell, K. and Mattes, A. Arch-Opthalmol. 102:554-555, 1984.
14. De Koning, E.W.J., Van Bijsterveld, O.R. and Cantell, K. Brit. J. Opthalmol. 66:509-512, 1982.
15. De La Pena, N.C., Diaz, A., Darnel, A. et al. Biomedicine 28: 104-108, 1978.
16. Clercg, E.de, Edy, V.G. and Vlieger, H.de. J. Pediatr. 86:736-739, 1975.
17. Pazin, G.J., Armstrong, J.A. et al. N. Engl. J. Med. 301:225-230, 1979.
18. Cheesman, S.H., Rubin, R.H., Stewart, J.A. et al. N. Engl. J. Med. 300:1345-1349, 1979.
19. Merigan, T.C., Rand, R.H., Polard, R.B. et al. N. Engl. J. Med. 298:981-987, 1978.
20. Arum, A.M. Kushner, J.H., Feldman, S. et al. N. Engl. J. Med. 306:761-765, 1982.
21. Shepp, D.H., Dandliker, P.S. and Meyers, J.D. N. Engl. J. Med. 314:208-212, 1986.
22. Cheeseman, S.H., Henle, W., Rubin, R.H. et al. Ann. Intern. Med. 93:39-42, 1980.
23. Treuner, J. and Niethammer, D. In: Interferon 4: In vivo and clinical studies. (Eds. N.B. Finter and R.K. Oldham). Elsevier, Amsterdam, 1985, pp. 280-295.
24. Krugman, S. In: Infectious disease of children, 8th Edit. (Eds: S. Krugman, S. Katz, A.A. Gershon and C.M. Wilfert), Mosby Co, St. Louis, 1985, pp. 8-21.
25. Hirsch, M.S., Schooley, R.T. et al. N. Engl. J. Med. 308:1489-1493, 1983.

26. Wade, J.C., McGuffin, S.C. et al. Jnl. Infect. Dis. 148:557-562, 1983.
27. Hayden, F.G., Albrecht, J.K., Kaiser, D.L., Gwaltney, J.M. N. Engl. J. Med. 314:71-75, 1986.
28. Hayden, F.G., and Gwaltney, J.M. J. Infect. Dis. 150:174-188, 1984.
29. Douglas, R.M., Moore, B.W. et al. N. Engl. J. Med. 314:65-70, 1986.
30. Philpotts, R.J., Higgins, P.J. et al. Jnl. Interferon Res. 4: 535-541, 1984.
31. Leventhal, Kashima, H., Wishnant, J.K. and Tuttle, R. In: Interferon 4: In vivo and clinical studies. (Eds. NB Fintner and RK Oldham), Elsevier, Amsterdam, 1985, pp. 325-334.
32. Sessions, R.B., Dichtel, W.J. and Goepfert, H. ENT Journal, 63: 488-493, 1984.
33. Geffen, J.R., Klein, R.J., Friedman-Kien, A.E. J. Infect. Dis. 180:612-615, 1984.
34. Gall, S.A., Hughes, C.E. and Trofatter, R. Am.J. Obstet. Gynecol. 153:157-163, 1985.
35. Schonfeld, A., Schattner, A. et al. Lancet i:1038-1042, 1984.
36. Bye, A., Balkwill,, F., Brigden, D., Wilson, J. Develop. Med. Child. Neurol. 27:170-175, 1985.
37. Mizutani, N., Miyazu, M. et al. Tohoku J. Exp. Med. 146:277-284, 1985.
38. Nakagawa, M., Michihata, T. et al. Acta Paediatr. Scand. 74:309-310, 1985.
39. Merigan, T.C., Beer, G.M., Winkler, W.G. et al. Ann. Neurol. 16: 82-87, 1984.
40. Levin, S., Hahn, T., Handzel, Z.T. et al. Antiviral Res. 5:229-240, 1985.
41. Larsson, A., Forsgren, M. et al. ActaPaediatr. Scand. 65:105-110, 1976.
42. Yoon, J.W., Auskin, M., Onodera, T., Notkins, A.L. N. Engl. J. Med. 300:1173-1179, 1979.
43. Zadik, Z., Levin, S. and Handzel, Z.T. J. Pediatr. Endocrinol. 1:119-122, 1985.
44. Rand, K.H., Rosenblood, A.L. et al. Diabetologia. 21:16-19, 1982.
45. Emond, R.T.D., Evans, B., Boven, E.T.W., and Lloyd, G. BMJ. 2: 541-544, 1977.
46. Ben Moshe, H., Romano, A., Blumenthal, M. and Revel, M. Metabol. Pediatr. Ophthalmol. 5:191-193, 1981.
47. Milazzo,F., Galli, M., Fassio, P.G. Infections. 13:130-133, 1985.

4

KAPOSI'S SARCOMA AND AIDS: THE ROLE OF INTERFERONS IN TREATMENT

SUSAN E. KROWN

Memorial Sloan-Kettering Cancer Center, 1275 York Avenue, New York, NY 10021, U.S.A.

ABSTRACT

The term acquired immune deficiency syndrome (AIDS) refers to a disease complex encompassing a wide spectrum of clinical manifestations including characteristic infections and neoplasms. This manuscript describes the role of interferons in the treatment of the most common AIDS-associated tumor, Kaposi's sarcoma, and discusses the potential role of interferons in the treatment or prevention of some of the opportunisitc infections associated with the syndrome. The rationale for therapeutic trials of interferon alpha, alone or in combination with other antiretroviral agents, to evaluate efficacy in the treatment of the underlying infection with the human immunodeficiency virus is also discussed.

INTRODUCTION AND BACKGROUND

Recent reports indicating that interferon alpha (IFN-α), alone or in combination with other antiviral agents, may inhibit the growth of the human immunodeficiency virus (HIV) in vitro (1,2) have stimulated renewed interest in exploring the potential of IFNs as therapeutic agents in patients with AIDS. However, even before the virus responsible for AIDS was identified (3,4) and before the full spectrum of HIV-associated diseases was described, there were compelling reasons to consider evaluation of IFNs as treatment for various manifestations of the syndrome, in particular its most frequent neoplastic complication, Kaposi's sarcoma (KS).

Despite its relative rarity in the western world prior to the 1980's, sufficient information was already available about KS in

M. Revel (ed.), CLINICAL ASPECTS OF INTERFERONS. Copyright 1988. Kluwer Academic Publishers, Boston. All rights reserved.

1981 to suggest a rationale for therapeutic trials of an agent that combined immunomodulatory antiviral and antiproliferative activities. The increased incidence of KS in patients undergoing immunosuppressive therapy, and the well-documented occurrence of "spontaneous" tumor regression in some cases where such therapy had been withdrawn (5) supported trials of an agent with well-documented immunomodulatory activities (6) in a tumor appearing in the setting of immunosuppression of unknown cause. Although the agent responsible for AIDS had not yet been discovered, epidemiologic studies had already suggested that a virus with a transmission pattern similar to hepatitis B might be the etiologic agent (7,8). Prior seroepidemiologic studies of European, American and African patients with "nonepidemic" forms of Kaposi's sarcoma had also suggested a link between cytomegalovirus (CMV) infection and the development of this tumor (9,10) and it was reasoned that any antiviral activity connected with IFN treatment might affect tumor growth and/or underlying infection with a putative "AIDS virus". Finally, the direct antiproliferative activity of IFN-α *in vitro* (11), together with clinical trials documenting its ability to induce the regression of tumors unassociated with profound immunosuppression or virus infection (12,13), supported trials of IFN-α as a possible alternative to cytotoxic chemotherapy, as it was feared that the latter approach might further increase the susceptibility of patients with AIDS to infection. Like chemotherapy, IFN-α had been shown capable of inducing neutropenia in cancer patients, but infectious complications had been rare, bone marrow cellularity appeared preserved, and patients who developed bacterial infections during such treatment showed an appropriate leukocytosis.

Later studies to evaluate the integrity of the IFN system in AIDS patients provided additional support for trials of both IFN-α and IFN-γ. Lopez *et al*. (14), reasoning that the severe virus infections seen in some AIDS patients might be traced to deficient IFN production, studied the ability of peripheral blood mononuclear cells to generate IFN-α *in vitro* after stimulation with *Herpes simplex* virus and found a markedly reduced response in a large proportion of patients. The frequent occurrence of intracellular infections requiring activated macrophages for their control also prompted studies to evaluate the capacity of T cells from such patients to produce IFN-γ, a central lymphokine involved in macrophage activation [reviewed in (15)]. Indirect evidence for deficient IFN-γ production *in vivo* came from studies showing a decreased proportion of HLA-DR-positive monocytes in the circulation of AIDS patients, a defect which could be reversed by the addition of IFN-γ *in vitro* (16). More direct evidence for deficient IFN-γ production come from studies by Murray *et al.* (17), who cultured T cells from seropositive AIDS patients with *Toxoplasma gondii* or *Candida albicans* antigens and found subnormal proliferative responses as well as a marked reduction in IFN-γ generation. When separated mononuclear phagocytes from AIDS patients were cultured in the presence of IFN-γ, they were able to ingest and kill *T. gondii* as efficiently as similarly treated cells from normal controls (17). Although these findings supported trials of IFNs as treatment or prophylaxis for the various opportunistic infections associated with AIDS, the observation that many patients already had high levels of IFN in the circulation [usually characterized as an acid-labile IFN-α (18-20), but occasionally as IFN-γ (19)], led to some misgivings about the therapeutic potential of these agents and to much speculation about the role of endogenous IFN in inducing the constitutional symptoms associated with the syndrome (21).

THERAPY OF KAPOSI'S SARCOMA

The role of interferons in the treatment of AIDS-associated KS can be best defined by recognizing the marked variability in the clinical presentations of this neoplasm and the potential benefits and limitations of the other available treatment options. Prior to 1981, few medical oncologists in the U.S. were familar with KS, partly because of its infrequent occurrence, and partly because the classical form of the disease, seen most often in middle-aged or elderly men, generally pursued an indolent course and was usually controlled by local excision or radiation. The only substantial experience with systemic chemotherapy for the disease had been in equatorial Africa [reviewed in (22)], where the tumor had been seen in a localized, nodular form as well as in more aggressive systemic presentations, including a distinct "lymphadenopathic" form in children (23).

Some patients with AIDS-associated KS may have only a few, inconspicuous pink macules on the skin, while others may have the tumor diagnosed incidentally after biopsy of a lymph node enlarged by follicular hyperplasia. In such patients, given the lack of treatments proven to affect the underlying cause of the tumor or to alter the natural history of the underlying immunodeficiency, treatment directed specifically at the tumor may not be indicated. In other patients, skin lesions may be multiple, large and nodular, deep red or purple in color, and present a major cosmetic problem. Involvement of dermal lymphatics may produce debilitating edema of the extremities, the genitalia and the periorbital tissues, and may predispose to bacterial cellulitis, particularly in the legs. Lesions of the soles of the feet may produce pain and difficulty walking, and nodules of the hard and soft palate, the gingiva and the pharynx may produce local obstructive symptoms. Although asymptomatic lesions of the gastrointestinal tract are the rule, occasional patients may experience pain, diarrhea or bleeding. Finally, in a small proportion of patients, involvement of the bronchial tree and lung parenchyma, sometimes accompanied by pleural exudates, may cause severe respiratory compromise. Despite the significant morbidity associated with the presence of KS in some patients, in devising treatment strategies for this tumor it must be kept in mind that, with the exception of those with pulmonary involvement, few patients with AIDS-associated KS die as a direct consequence of their tumor. In addition, treatment must often take place in the context of concurrent infections and diminished bone marrow reserve.

The epidemic form of KS is indistinguishable histologically from the previously described "classical" and African forms of the disease, and there were no reasons to anticipate major differences in its response to treatment. However, for reasons that are not well understood, response rates have been lower and response durations shorter than in KS unassociated with HIV infection. Whereas in classical KS radiation therapy is often an effective and long-lasting treatment for localized disease, this approach is sometimes limited in those with epidemic KS by the widespread extent of skin lesions and the presence of visceral disease. Under certain circumstances, however (e.g. disfiguring facial lesions, bulky oropharyngeal disease, certain cases of tumor-associated edema or painful lesions) radiation therapy may provide good temporary palliation without systemic toxicity, but responses have sometimes been of short duration and normal skin or mucosal tolerance may be poor (24,25).

Preliminary trials of single-agent or combination chemotherapy have also pointed to differences in response patterns between the epidemic form of KS and KS arising in other clinical settings. For example, whereas the overall response rate to vinblastine has been estimated as 89% in patients without AIDS (22), a major response rate of only 26% has recently been reported in a group of 38 patients with AIDS-associated KS (26). Considerably higher response rates have been reported in some preliminary trials in which single drugs [vincristine (27), bleomycin (28), VP-16 (29)] or vinca alkaloid-containing drug combinations (28-32) were used. For at least some of the drug combinations, the frequency of unacceptable hematologic toxicity was high and opportunistic infections were frequent, even in patients showing objective tumor regression. In addition, responses to chemotherapy have usually been partial, and complete tumor regression, common in patients with other forms of KS, has been infrequent in the group with AIDS. In an early review of the subject, Longo (33) estimated that 30% of all AIDS patients reported on in preliminary studies of chemotherapy had already died and projected a 2-year survival rate of only 30%, with no disease-free remission lasting more than 12 months. Nonetheless, chemotherapy may provide effective palliation for KS in some patients, but it clearly offers nothing for treatment of the underlying immune deficiency or for HIV infection, which are ultimately responsible for the death of most patients.

In late 1981, when the first trials of IFN-α were initiated in patients with the newly described "epidemic" form of KS, there were no published reports of the efficacy of standard treatments and no experience with IFN in other forms of KS which could guide the choice of treatment regimens or patient selection. The initial observations of therapeutic activity were made in patients enrolled in a phase I trial of IFN-α2a, a study that was near completion at the time the first patients with AIDS-associated KS were seen. Thus, it was fortuitous that the patients with KS who were entered on this study received high doses of IFN, at or exceeding the maximum tolerated dose. The results of our initial trial, in which 5 of 12 patients treated with daily intramuscular doses of 36- to 54-million units (MU) showed complete or partial tumor regression (34) were later confirmed in larger trials by our group (35-37) and by others investigating IFN-α2b (38-46) and IFN-αN1 (41,42). Overall, when given in doses of $\geq$ 20 MU/m^2 of body surface area, the mean objective response rate has been approximately 30%. The route and schedule of IFN administration has been less important than dose, and similar responses have been seen in trials in which high doses of IFN were given by the intramuscular, intravenous or subcutaneous routes or on daily, alternate week or thrice weekly schedules.

By using gradual dosage escalation over 1 to 2 weeks, it has been possible to reduce some of the acute toxicity associated with high-dose IFN administration, namely high fever and chills (36,37), but dosage attenuation was later required in up to one-third of patients because of chronic fatigue, malaise and neutropenia. The chronic toxicity of high-dose IFN treatment, along with the wish to eventually combine IFN with other therapeutic agents and recognition that biological response modifiers did not always demonstrate clear-cut dose-response relationships, led to investigations of low-dose IFN treatment in some patients (36,38,41). Although lower IFN doses (2 to 15 MU/m^2) induced a low frequency of dose-limiting toxicity, the response rates were also low, and less than 10% of patients showed major objective responses in the studies cited. The difference in response rates between the high- and low-dose regimens has been interpreted by some as suggesting that IFN acts by a direct antiproliferative mechanism, although an alternative explanation for the dose-response relationship may be that higher IFN doses are required to affect host defense mechanisms or to inhibit virus replication.

Only preliminary studies have been conducted in patients with AIDS-associated KS to evaluate the role of IFN-α in combination with other agents. The combination of IFN-α with vinblastine was suggested by clinical trials demonstrating therapeutic activity of vinca alkaloids as single-agent therapy (26,27) and by studies showing that combinations of vinca alkaloids and IFN caused synergistic suppression of certain other tumor types in in vitro systems (32). Although two preliminary trials have suggested that the combination of IFN-αN1 and vinblastine has substantial anti-tumor activity (44,45), a randomized comparison of IFN-α2a alone or in combination with vinblastine showed no significant difference in response rates (24% vs. 30%), but the combination induced a significantly higher rate of severe constitutional and hematological toxicity (46). Similarly, a preliminary report suggested little activity for sequential or concurrent treatments with IFN-α2b and VP-16 (47), another drug with substantial single-agent activity (29).

In early studies of IFN-α as treatment for KS, patients were selected on the basis of standard criteria, i.e. performance status, blood counts, renal and hepatic function. More recently, retro-spective analyses have been performed to define more relevant pre-treatment characteristics that might identify individuals most likely to benefit from IFN-α treatment. Although the unusually high response rate reported in a preliminary trial of IFN-αN1 was tenta-tively attributed to the preponderance of patients with minimal disease confined to the skin (42), subsequent analyses by our group have shown no association between responsiveness to IFN-α2a and either tumor burden or the presence of gastrointestinal lesions (34-37). However, as few patients entered into these studies have had involvement of organs outside the lymph nodes or gastrointes-tinal tract (e.g. lung or endobronchial disease), it is not known whether patients with lesions in such locations would respond equally well.

The clinical features most consistently associated with a poor response to IFN-α have included a history of prior opportunistic infection (36,40,41), and the presence of systemic ("B") symptoms (weight loss, fever, night sweats). In our experience, no patient with prior opportunistic infection has shown a major response to IFN-α, an observation that has led us to exclude such patients from our more recent trials. Similarly, in our randomized trial comparing IFN-α2a alone and together with vinblastine, patients with more than one "B" symptom were excluded. Even with such stringent exclusion criteria, we found that patients with no "B" symptoms showed a significantly higher response rate than those with only a single symptom (46). Such observations suggest that the severity of systemic symptoms may directly reflect the severity of immunologic impairment in patients with epidemic KS. Anemia, another non-specific sign, was also correlated with a poor response to IFN in our randomized trial (46), and has been an independent variable associated with shorter survival in at least two reported retrospective analyses (48,49).

Although there are no convincing data to suggest that IFN-α treatment improves any of the standard measures of immunocompetence in patients with KS, there is considerable evidence that relative preservation of immune function at the time IFN treatment is started is associated with a higher frequency of therapeutic response. Various investigators have shown that the total lymphocyte count (41), the absolute number of CD4-positive cells in the circulation (41), the CD4 to CD8 ratio (46), delayed type hypersensitivity (50), or lymphoproliferative responses to microbial activators (50) were associated with subsequent responsiveness to IFN-α. In addition, several studies have indicated that the presence of acid-labile IFN-α in the circulation is associated with subsequent unresponsiveness to exogenous IFN administration (41,42,50). Although this association suggests that the presence of circulating, endogenous IFN renders the host's cells resistant to the effects of additional, exogenous IFN, our studies failed to show an association between the presence of endogenous IFN in the serum and the ability of IFN-α2a to augment natural killer cell cytotoxicity _in vitro_

(50).

In addition to providing a guide for identification of patients most likely to benefit from IFN treatment, these analyses may offer insights into the mechanism of IFN action in KS and into the validity of staging systems proposed for this tumor. The staging systems proposed thus far have been based, almost exclusively, on the location and extent of tumor nodules, i.e. cutaneous vs. visceral disease, localized vs. widespread cutaneous involvement (51,52). The fact that such anatomical distinctions have not correlated well with responsiveness to IFN-α is consistent with the hypothesis that KS is a tumor of multicentric origin (53,54), in which visceral disease does not have the same negative prognostic significance as in tumors metastasizing from a single primary site. The importance of measures of immune function, together with clinical features such as opportunistic infection, systemic symptoms and anemia that may directly reflect the severity of immune dysfunction, suggest that such factors may be important in staging this tumor and that IFN-α functions optimally in setting of a relatively preserved immune system.

The fact that patients with relatively preserved immune function have a higher frequency of response to IFN-α has direct bearing on another important question, i.e. does IFN treatment improve the quality or duration of survival? Certainly the quality of life has been improved in those individuals in whom cosmetically disfiguring lesions or those producing functional impairment regressed. In addition, patients showing tumor regression in response to IFN treatment have also shown a significantly lower frequency of opportunistic infection and longer survival than those whose tumors did not respond (36,50). However, since these responses occurred in patients with more intact immune function, it is not possible to determine whether the improved prognosis reflects an effect of IFN on the natural history of the disease or a higher response rate of patients whose disease would have taken a more favorable course regardless of therapy.

The Role of Interferon Gamma

The results of studies indicating deficient IFN-γ production in AIDS patients logically suggested attempts to use this cytokine as therapy. However, preliminary trials of both partially purified IFN-γ induced in human T cells (55) and highly purified recombinant IFN-γ (56,57) in patients with KS have failed to demonstrate significant antitumor activity. The apparent lack of antitumor efficacy at the relatively high doses used in these studies does not, however, exclude a potential role for IFN-γ in the treatment or prevention of some of the infectious complications of AIDS. Indeed, as recently demonstrated in cancer patients (58) and patients with lepromatous leprosy (59), very low, non-toxic doses of recombinant IFN-γ are capable of stimulating the oxidative metabolism of human mononuclear phagocytes, a function that is closely correlated with the ability of monocytes to kill intracellular pathogens. The findings reported in leprosy patients may have particular relevance to AIDS, since the lepromatous form of the disease is also characterized by deficient IFN-γ production in response to specific antigen stimulation *in vitro* (60).

Treatment of HIV Infection

As stated earlier, even before the discovery of an etiologic agent in AIDS a major reason for considering IFN-α as a potential therapeutic agent in KS was its antiviral activity. Although clinical trials of IFN-α have failed to demonstrate any significant activity against CMV, the virus tentatively implicated in the development of this tumor, *in vitro* studies have shown that IFN-α2a is capable of suppressing HIV replication in peripheral blood mononuclear cells at concentrations that are easily achieved *in vivo* (1). Preliminary studies have also suggested that administration of IFN-α2a to patients with KS may suppress HIV replication *in vivo* (61), but these early results require confirmation with more sensitive analytical methods and larger numbers of patients. Although the precise mechanism by which IFN-α inhibits HIV replication is not known, its effects on animal retroviruses appear to be directed primarily at late stages of virus assembly and release.

This has suggested the possibility of combining IFN-α with antiretroviral agents which act at earlier stages of virus replication. One such agent, azidothymidine (AZT), which directly inhibits HIV reverse transcriptase (62), has shown synergistic *in vitro* suppression of HIV replication when combined with IFN-α2a (2). Such observations raise the possibility of combination therapies that may affect both the underlying cause of AIDS as well as its major neoplastic complication. However, as these agents have their own significant and overlapping toxicities, the use of such drug combinations must be approached cautiously.

REFERENCES

1. Ho, D.D., Hartshorn, K.L, Rota, T.R., et al. Lancet 1: 602-604, 1985.
2. Hartshorn, K.L, Vogt, M.W., Chou, T-C, et al. Antimicrobial. Agents Chemother., in press.
3. Barré-Sinoussi, F., Chermann, J.C., Rey, F., et al. Science 220:868-871, 1983.
4. Gallo, R.C., Salahuddin, S.Z., Popovic, M., et al. Science 224:500-503, 1984.
5. Harwood, A.R., Osoba, D., Hofstader, S.L., et al. Am. J. Med. 67:759-765, 1979.
6. Cheesman, S.H., Rubin, R.H., Stewart, J.A., et al. N. Engl. J. Med. 300:1345-1349, 1979.
7. Curran, J.W. New Engl. J. Med. 309:609-611, 1983.
8. Centers for Disease Control Task Force on Kaposi's Sarcoma and Opportunistic Infections. N. Engl. J. Med. 306:248-252, 1982.
9. Safai, B., Miké, V., Giraldo, E., et al. Cancer 45:1472-1479, 1980
10. Giraldo, G., Beth, E., (Eds)The Role of Viruses in Human Cancer. 1:57-73, 1980.
11. Borden, E.C., Hogan, T.F., Voelkel, J.G. Cancer Res. 42: 4948-4953, 1982.
12. Gutterman, J.U., Blumenshein, G.R., Alexanian, R., et al. Ann. Int. Med. 93:339-406, 1980.
13. Quesada, J.R., Swanson, D.A., Trindale, A., et al. Cancer Res. 43:940-947, 1983.
14. Lopez, C., Fitzgerald, P.A., Siegel, F.P. J. Inf. Dis. 148:962-966, 1983.
15. Krown, S.E. In: Interferon 7, 1986 (Ed. I Gresser), Academic Press, London, 1986, pp. 185-211.
16. Heagy, W., Kelley, V.E., Strom, T.B., et al. J. Clin. Invest. 74:2089-2096, 1984.
17. Murray, H.W., Rubin, B.Y., Masur, H., et al. N. Engl. J. Med. 310:883-889, 1984.
18. DeStefano, E., Friedman, R.M., Friedman-Kein, A.E., et al. J. Infect. Dis. 146:451-455, 1982.

19. Buimovici-Klein, E., Lange, M., Klein, R.J., et al. Lancet 2:344, 1983.
20. Eyster, M.E., Goedert, J.J., Poon, M-C, et al. N. Engl. J. Med. 309:583-586, 1983.
21. Vilček, J. Prog. Med. Virol. 30:62-77, 1984.
22. Volberding, P.A., Conant, M.A., Stricker, R.B., et al. Amer. J. Med. 74:652-656, 1983.
23. Taylor, J.F., Templeton, A.C., Vogel, C.L., et al. Int. J. Cancer. 8:122-135, 1971.
24. Harris, J.W., and Reed, T.A. In: Cancer and AIDS. Frontiers of Radiation Therapy and Oncology, Vol. 19 (Ed. J.M. Vaeth), Basel, S. Karger, 1985, pp. 126-132.
25. Nisce, L.Z., and Safai, B. In: op cit, pp. 133-137.
26. Volberding, P.A., Abrams, D.I., Conant, M., et al. Ann. Int. Med. 103:335-338, 1985.
27. Mintzer, D., Real, F.X., Jovino, L., et al. Ann. Int. Med. 102:200-202, 1985.
28. Wernz, J., Laubenstein, L, Hymes, K., Walsh, C., and Muggia, F. Proc. Amer. Soc. Clin. Oncol. 5:4, 1986.
29. Laubenstein, L.J., Krigel, R.L., Odajnyk, C.M., et al. J. Clin. Oncol. 2:1115-1120, 1984.
30. Kaplan, L.D., Volberding, P.A., and Abrams, D.A. Proc. Amer. Soc. Clin. Oncol. 5:7, 1986
31. Minor, D.R., and Brayer, T. Proc. Amer. Soc. Clin. Oncol. 5:1, 1986.
32. Glaspy, J., Miles, S., McCarthy, S., Carden, J., and Mitsuyasu, R. Proc. Amer. Soc. Clin. Onco. 5:3, 1986.
33. Longo, D.L. Ann. Int. Med. 100:96-98, 1984.
34. Krown, S.E., Real, F.X., Cunningham-Rundles, S., et al. N. Engl. J. Med. 308:1071-1076, 1983.
35. Krown, S.E., Real, F.X., Krim, M., et al. Ann. NY Acad. Sci. 437:431-438, 1985.
36. Real, F.X., Oettgen, H.F., Krown, S.E. J. Clin. Oncol. 4:544-551, 1986.
37. Krown, S.E., Real, F.X., Gold, J.W.M., et al. Proceedings, International Conference on AIDS, Paris, France, June 23-25, 1986 Communication 88:S14F, p. 35.
38. Groopman, J.E., Gottlieb, M.S., Goodman, J., et al: Ann. Intern. Med. 100:671-677, 1984.
39. Volberding, P.A., Valero, R., Rothman, J., et al. Ann. NY Acad. Sci. 437:439-455, 1984.
40. Mitsuyasu, R., Volberding, P., Jacobs, A., et al. Proc. Amer. Soc. Clin. Oncol. 3:51, 1984.
41. Gelmann, E.P., Preble, O.T., Steis, R., et al. Amer. J. Med. 78:737-741, 1985.
42. Rios, A., Mansell, P.W.A., Newell, G.A., et al. J. Clin. Oncol. 3:506-512, 1985.
43. Aapro, M.S., Alberts, D.S., Salmon, S.E. Cancer Chemother. Pharmacol. 10:161-166, 1983.
44. Rios, A. Mansell, P., Newell, G., et al. Proc. Amer. Soc. Clin. Oncol. 4:6, 1985.
45. Fischl, M., Lucas, S., Gorowski, E., et al. J. Interferon Res. 6 (Suppl. 1): 4, 1986.

46. Krown, S.E., Gold, J.W.M., Real, F.X., et al. J. Interferon Res. 6(Suppl. 1): 3, 1986.
47. Lonberg, M., Odajnyk, C., Krigel, R., et al. Proc. Amer. Soc. Clin. Oncol. 4:2, 1985.
48, Vadhan-Raj, S., Wong, G., Gnecco, C., et al. Proc. Amer. Soc. Clin. Oncol. 5:6, 1986
49. Volberding, P., Kaslow, K., Bilk, M., et al. Proc. Amer. Soc. Clin. Oncol. 3:51, 1984.
50. Vadhan-Raj, S., Wong, G., Gnecco, C., et al. Cancer Res. 46:417-425, 1986.
51. Krigel, R.L., Laubenstein, L.J., Muggia, F.M. Cancer Treat. Rep. 67:531-534, 1983.
52. Mitsuyasu, R., Groopman, J.E. Semin. Oncol. 11:53-59, 1984.
53. Costa, J., Rabson, A.S. Lancet 1:58, 1983.
54. Templeton, A.C. Pathol. Ann. 315-336, 1981.
55. Odajnyk, C., Laubenstein, L, Friedman-Kien, A., et al. Proc. Amer. Soc. Clin. Onco. 3:61, 1984
56. Vadhan-Raj, S., Al-Katib, A., Pelus, L., et al. J. Clin. Oncol. 4:137-146, 1986.
57. NIH Workshop, Washington, May 3-4, 1984. (Unpublished) cited in Lotze MT: Treatment of immunologic disorders in AIDS. In: AIDS: ETiology, Diagnosis, Treatment and Prevention (Eds DeVita VT Jr, Hellman S, Rosenberg SA). Lippincott Company, New York, 1985.
58. Nathan, C.F., Horowitz, C.R., de la Harpe, J., et al. Proc. Natl. Acad. Sci. USA 82:8686-8690, 1985.
59. Nathan, C.F., Kaplan, G., Levis, W.R., et al. N. Engl. J. Med. 315:6-15, 1986.
60. Nogueira, N., Kaplan, G., Levy, E., et al. J. Exp. Med. 158:2165-2170, 1983.
61. Krown, S.E. Presented at the Second International Conference on AIDS, Paris, France, June 24, 1986.
62. Yarchoan, R., Klecker, R.W., Weinhold, K.J., et al. Lancet 1:575-580, 1986.

5

USE OF INTERFERON IN THE TREATMENT OF SUBACUTE SCLEROSING PANENCEPHALITIES

Israel Steiner, M.D., and Oded Abramsky, M.D., Ph.D.

Department of Neurology, Hadassah University Hospital and Hebrew University-Hadassah Medical School, Jerusalem 91 120, Israel.

Subacute sclerosing panencephalitis (SSPE) is a progressive fatal central nervous system (CNS) disease affecting children and young adults. Since first described by Dawson in 1933 (1) much knowledge regarding its etiology and pathogenesis, with relation to measles virus, has accumulated, yet there is still no effective treatment.

Onset of the disease is generally between 5 and 15 years of age (2,3). The initial clinical symptoms are subtle: disturbances of affect, forgetfulness, indifference, irritability and lethargy are followed by drooling, regressive speech and other cerebral and mental abnormalities (stage I). Within several weeks involuntary movements, incoordination of the trunk and limbs and myoclonic attacks also appear and intensify (stage II). The disease slowly but unremittingly progresses and the child sags into decerebrate rigidity without any response to external stimuli, has a startle response to noise and occasional generalized myoclonic jerk with lack of any cerebral cortex function (stage IV). SSPE leads to death in 55% of the patients within one year of onset of the neurological symptoms. Only 10-28% of patients survive more than 4 years after diagnosis (3,4).

Clinical diagnosis is fortified by several laboratory findings: abnormal electroencephalographic recording consisting of periodic bursts of brain electric activity at regular intervals; increased level of gamma globulin in the cerebrospinal fluid (CSF); presence of oligoclonal bands on CSF electrophoresis. The detection of measles antibody in serum and CSF of patients in the absence of acute measles infection is pathognomonic of SSPE (5,6).

The relationship between measles infection and SSPE is supported

M. Revel (ed.), CLINICAL ASPECTS OF INTERFERONS. Copyright 1988. Kluwer Academic Publishers, Boston. All rights reserved.

by studies from several disciplines. Epidemiologically, in the analysis of 375 confirmed SSPE cases recorded in the United States between 1969 and 1974,a history of previous measles infection could be obtained in 79% (7,8). The infection generally took place before two years of age and the mean time interval between measles infection and the onset of SSPE symptoms was seven years. The histopathological lesions observed in material obtained by brain biopsy or in post-mortem are widespread throughout the white and grey matter of the CNS. They consist of lymphocytes and plasma cells, perivascular cuffing of small blood vessels with proliferation of microglial cells and astrocytes. Intranuclear and intracytoplasmatic inclusion bodies like those present in measles virus infected cells are found. These bodies react positively with reagents designated to demonstrate measles antigens and electron microscopy shows that they contain tubular structures similar to the nucleocapsid of the paramyxovirus (9). The infectious virus can be recovered from SSPE brain cells by long term cocultivation and/or fusion with cells known to support measles virus replication (10). It is possible to produce a similar disease in experimental animal model by measles virus inoculation (11, 12). In experimental animals as in man, infection of this type can only be accomplished during a critical period in early life.

After the isolation of the virus, the agent was studied extensively to determine its structural and biological characteristics. The conclusion was that the SSPE agent present in patients' brain cells is a measles virus but for the fact that it does not contain the measles-virus M protein (13). This protein is required for nucleocapsid alignment beneath the cytoplasmic membrane before budding.

Several factors are associated with and contribute to the pathogenesis of SSPE. These include measles infection prior to two years of age, rural environment, larger family size and ethnic predisposition (14). Still, the pathogenetic mechanism responsible for SSPE is not fully understood. On the basis of present knowledge it is assumed that measles virus enters the brain during primary infection but encounters tissue that is relatively non-permissive for viral replication. The age of the patient, the inoculom of virus and other factors such as transient immune suppression at the time of primary infection may be important determinants. Patients remain well during the prolonged interval from primary infection till a significant enough amount of CNS tissue destruction takes place. During this period, it is thought that viral nucleocapsides, RNA and other gene products accumulate and spread from cell to cell through the CNS slowly destroying them. Lymphocytes and plasma cells are stimulated by viral antigens producing antibodies against them, but the lack of the M protein (thoughtto result from brain cell restriction of its synthesis) interferes with virus budding. Thus, the measles virus is unable to reach the extracellular space and is somewhat protected from the extracellular host defence reactions (15). However, this explanation does not clarify why an otherwise immunocompetent host does not eliminate the virus initially or why it fails to mount an effective cell mediated response against virus accumulating in the nervous system throughout years of persistant infection.

In conclusion: in SSPE CNS tissue is progressively destroyed by measles-like virus while the immune system is unable to clear the infection.

Treatments in SSPE have been aimed in two directions: boosting the hypothetical defective immune response or attacking the putative viral pathogen (Table 1). The assessment of any therapy influence upon SSPE course is difficult because of the small number of treated patients (disease incidence being one per million), the variable natural history (ranging from death within several months to survival beyond 4 years) and the inability to perform a double blind controlled study in some modalities. Most of the different therapies have little or no beneficial effect upon the course of SSPE (26). Two antiviral agents produce some promising but contraversial results and are currently being investigated. Amantadine is an anti-RNA agent that blocks the penetration of sensitive strains of virus into host cells (4, 26). Inosiplex has some inhibitory action on the process of viral replication, has pharmacological effects on the cellular and humoral immune response, and was also reported to augment interferon action (23-26).

Table 1.

TREATMENT OF SSPE

Treatment	Reference
A. Immune mediated	
1.Corticosteroids	16
2.Exchange transfusion	17
3.Transfer factor	18
4.Thymectomy	19
5.Measles vaccination	20
B. Antiviral	
1.Amantadine	21
2.Rifampin	4
3.Ether	4
4.5 bromo, 2 deoxyuridine & pyrancopolymer	22
5.Inosiplex	23-25

Interferon is a logical candidate for treatment of cell associated viral infection such as SSPE. It acts directly on cells to prevent viral replication by interfering with synthesis of viral messenger RNA and translation of RNA into viral proteins. It is synthetized in cells infected by a virus, released into the extracellular fluid and can initiate inhibition of viral multiplication in other cells (27). It has been tried in several conditions of neurological disease such as multiple sclerosis (28), and amyotrophic lateral sclerosis (29) as well as CNS viral infections (30,31). Modes of administration include intravenous, subacutaneous and intramuscular. However, an intact blood-CSF barrier effectively prevents large amounts of interferon from entering the CSF following systemic administration. Intravenous injection in monkeys of $3X10^7$ units of leukocyte interferon produced a CSF level of 20 units/ml for 6 hours (32). In patients with amyotrophic lateral sclerosis,subcutaneous administration of $6X10^6$ units resulted in CSF levels of 5 units/ml between 8 and 24 hours thereafter (33). Systemic administration has also been associated with dose related side effects: general malaise, hyperpyrexia, bone marrow suppression and disturbances in liver functions (34). On the other hand,intrathecal administration produced increase and substained levels of interferon in the CSF: 50 to 400 units/ml 48 hours following injection of $1x10^6$ units in one study (33) and 21 to 66 units/ml 72 hours following administration of a similar amount in another study (29).

Table 2.

INTERFERON TREATMENT IN SSPE

Mode of administration	Duration	No.of Patients	Outcome	Reference
Intramuscular	25,90 days	2	no change	36
Intramuscular & intrathecal	23 days	1	no change	36
Intravenus and intraventricular	21 days	1	transient improvement	37
Intravenus and intraventricular	18 days	1	improved	38
Intravenus and intrathecal	1-6 days	6	no change	39
Intrathecal	30 days	5	1 improved 4 unchanged	35
Intraventricular	6 months	3	improved	40
Intraventricular	21 months	3	2 improved slightly 1 stable	present study,41

Presently, there are several completed and ongoing studies on the effect of interferon treatment in SSPE (Table 2.) Systemic (I.M.) treatment had no effect (36). Intrathecal administration (with or without I.M. injections) was short term in 4 studies (36-39) lasting no more than one month and two of them reported some beneficial effect (37,38). Intraventricular interferon administration over a six months period was associated with clinical remissions(40). Side effects of intrathecal or intraventricular treatment in the various studies include fever, chemical meningitis and transient hemiparesis.

We have treated 3 SSPE patients with interferon for up to 21 months (Tabel 3.) Protocol of experiment is detailed elsewhere (41). Briefly: diagnosis of SSPE was based upon the clinical course, elevated serum and CSF antimeasles antibody and presence of oligoclonal immunoglobulins in the CSF, and was confirmed by brain biopsy. All patients were in stage II of SSPE. Following informed consent from patients' parents, ventricular reservoir (Ommaya) was placed into the anterior horn of the right lateral ventricle under

Table 3.

CLINICAL AND LABORATORY FINDINGS IN 3 SSPE PATIENTS UNDER INTRAVENTRICULAR INTERFERON TREATMENT

Patient	1	2	3
Sex/Age (years)	F/12	M/10	M/9
SSPE duration before treatment	6 weeks	3 weeks	3 months
Clinical Stage	II	II	II
Duration of treatment	20 months	17 months	21 months
Treatment Protocol (per week)	5 mo.$5X10^5$iu	5 mo.$5X10^5$iu	2 mo.$5X10^5$iu
	7 mo.$5X10^5$iu x2	12 mo.10^6iuX2	10 mo.10^6iuX2
	8 mo.10^6Xiu x2		
Complications	iatrogenic ventriculitis	Twice iatrogenic ventriculitis	
Clinical follow-up	slightly improved	stable	slightly improved
EEG	markedly improved	mild improvement	mild improvement
Antimeasles antibodies (CSF) before treatment	1:32,800	1:32,800	1:10,200
Antimeasles antibodies (CSF) during treatment	1:8,200	1:32,800	1:2,000

general anesthesia. Alpha interferon produced by a recombinant DNA method (42) was administered via the reservoir following complete recovery from the surgical procedure. The protocol of the interferon injection was modified with time and was increased during the period of therapy from a weekly injection of $5X10^5$ units to $1X10^6$ units twice a week. Baseline assessment and follow-up in all patients included monthly clinical evaluation and electroencephalographic recording, brain computerized axial scan and CSF levels of anti-measles antibodies performed every 3 months. Antimeasles antibodies were determined by the ELISA system. Preliminary results are given in Table 3. In the first patient improvement in speech and ambulation was accompanied by her ability to converse, recognize friends and care for bodily needs. The third patient, who prior to treatment was unable to communicate and was incontinent, is now able to answer questions, read simple sentences, and is continent. Electroencephalographic recordings improved in all patients and in two a decrease in CSF measles antibody titers was noted. In addition,to the long term influence we observed transient improvement following every injection. This lasted about 24 hours and included reduction in myoclonic jerks, increased motor activity and improved communication. The mechanism responsible for the acute effect is unknown. Recently, it has been suggested that alpha interferon may influence CNS activity as assessed by electrophysiological and behavioral experiments (43). Side effects of treatment were febrile reactions following each injection which were sometimes associated with an episode of general convulsion. Treatment was complicated by iatrogenic ventriculitis on 3 occasions. Infection responded to systemic antibiotic treatment and did not neccesitate removal of the intraventricular reservoir. In case 1, treatment was discontinued because of technical difficulties after 20 months of therapy. Within the 4 months following cessation of treatment, her condition rapidly deteriorated and she lapsed into stage IV becoming completely bed-ridden,severely demented and non-communicative. The possibility that therapy had no effect upon disease course cannot be ruled out. Nevertheless, all patients had a rapid deteriorating course prior to treatment, responded similarly to the interferon therapy and the disease's arrest was associated in all with marked improvement of their EEG recordings, and in two with decrease of antimeasles anti-bodies titers in the CSF.

Thus, it seems plausible to carefully assume that long term intraventricular interferon treatment can modify the natural history of SSPE. Therefore, it should be tried on a larger number of patients on a long term basis in a controlled multicenter study. At present, neither interferon nor any other therapy can cure SSPE victims. Further understanding of the causative mechanism underlying the disease, followed by adequate preventive measures will eventually eradicate this fatal disorder.

REFERENCES

1. Dawson, J.R. Am. J. Pathol. 9:7-16, 1933.
2. Jabbour, J.T., Garcia, J.H., Lemmi, H., Ragland, J., Duenas, D. A., Sever, J.L. JAMA 207:2248-2254, 1969.
3. Haddad, F.S., Risk, W.S., Jabbour, J.T. Ann. Neurol. 1:211-217, 1977.
4. Robertson, W.C., Clark, D.B., Markesbery, W.R. Ann. Neurol. 8: 422-425, 1980.
5. Sever, J.L. Rev. Infect. Dis. 5:467-473, 1983.
6. Johnson, R.T. Viral Infections of the Nervous system. Raven Press, New York, 1982, pp 244-252.
7. Modlin, J.F., Jabbour, J.T., Whitte, J.J. Pediatrics. 59: 505-512, 1977.
8. Modlin, J.F., Halsey, N.A., Eddins, D.L. J. Pediatr. 94: 221-236, 1979.
9. Kipps, A., Dick, G., Moodie, J.W. Lancet II: 1406-1410, 1983.
10. Payne, F.E., Baublis, J.V., Itabashi, H.H. N. Engl. J. Med. 281:585-589, 1969.
11. Raine, S.C., Byington, D.P., Johnson, K.P. Lab, Invest. 33: 108-116, 1975.
12. Johnson, K.P., Norrby, E., Swoveland, P., Carrigan, D.R.J. Infect. Dis. 144: 161-169, 1981.
13. Hall, W.W., Choppin, P.W. N. Engl. J. Med. 304: 1152-1155, 1981.
14. Soffer, D., Rannon, L., Alter, M., Kahana, E., Feldman, S. Am. J. Epidemiol. 103:67-74, 1976.
15. Griffith, J.F. N. Engl. J. Med. 313: 952-954, 1985.
16. Parker, J.C., Klintworth, G.K., Graham, D.G., Griffith, J.F. Am. J. Pathol. 61:275-291, 1970.
17. Sell, K., Ahmed, A., Bailey, D.W. Arch. Neurol. 32:502-503, 1975.
18. Valdimarsson, H., Angarsdottir, G., Lachmann, P.J. Arch. Neurol. 32:503, 1975.
19. Kolar, O., Obrucnik, M., Behounkova, L., Musil, J., Penickova, V. Brit. Med. J. 3:22-24, 1967.
20. Jabbour, J.T., Duenas, D.A., Modlin, J. Arch. Neurol. 32: 496, 1975.
21. Haslam, R.H.A., McQuillen, M.P., Clark, D.B. Neurology 19: 1080-1085, 1969.
22. Freeman, J.M. Neurology 18: 176-180, 1968.

23. Jones, C.E., Dyken, P.R.,Huttenlocher, P.R., Jabbour, J.T., Maxwell, K.W. Lancet I: 1034-1037, 1982.
24. Silverberg, R., Brenner, T., Abramsky, O. Arch. Neurol. 36: 374-375, 1979.
25. Haddad, F.S., Risk, W.S. Ann. Neurol. 7:185-188, 1980.
26. Taylor, W.J., DuRant, R.H., Dyken, P.R. Drug. Intell. Clin. Pharm. 18:375-381, 1984.
27. Friedman, R.M. Interferons. Academic Press, New York, 1981, pp. 47-71.
28. Jacobs, L., O'Malley, J., Freeman, A., Ekes, R. Science 214: 1026, 1981.
29. Mora, J.S., Munsat, T.L., Kao, K., Finson, L.J., Hedlund, W., Bradley, G.A., Schiefe, R., Georgiades, A. Neurology 36: 1137-1140, 1986.
30. Prange, H., Wismann, H. N. Engl. J. Med. 305:1283-1284, 1981.
31. Merigan, T.C., Baer, G.M., Winkler, W.G. Ann. Neurol. 16: 82-87, 1984.
32. Habif, D.V., Lipton, R., Cantell, K. Proc. Exp. Biol. Med. 149: 287-289, 1975.
33. Jablecki, C., Cantell, K., Connor, J., Kingsbury, D., Hamburger, R., Westall, F., Smith, R. Ann. Neurol. 10:82, 1981.
34. Priestman, T.J. Lancet II: 113-118, 1980.
35. Huttenlocher, P.R., Picchietti, D.L., Roos, R.P., Cashman, N.R., Horowitz, B., Horowitz, M.S. Ann. Neurol. 19:303-305, 1986.
36. Behan, P.O. Lancet I: 1059-1060, 1981.
37. Bartram, C.R., Henke, J., Treuner, J., Basler, M., Esch, A., Mortier, W. Eur. J. Pediatr. 138: 187-190, 1982.
38. Nakagawa, M., Michihata, T., Yoshioka, H., Sawada, T., Kunsunoki, T., Kishida, T. Acta. Pediatr. Scand. 74:309-310, 1985.
39. Bye, A., Balkwill, F., Brigden, D., Wilson, J. Develop. Med. Child.Neurol. 27:170-175, 1985.
40. Panitch, H.S., Gomez-Plascencia, J., Norris, H., Cantell, K., Smith, R.A. Neurology 36: 562-566, 1986.
41. Smith, R.A., Abramsky, O., Steiner, I., Panitch, H., Cantell, K. In: The Biology of the Interferon System 1985 (Eds. W.E.Stewart and H. Schellekens) Elsevier Science Publishers, 1986 pp. 505-510.
42. Cantell, K., Hirvonen, S. J. Gen. Virol. 39:541-543, 1978.
43. Dafny, N., Pieto-Gomez, B., Reyes-Vazquez, C. J. Neuroimmunol. 9:1-12, 1985.

6

MULTIPLE SCLEROSIS: CLINICAL TRIALS WITH INTERFERONS

JUDITH A. O'MALLEY[1], LAWRENCE D. JACOBS[2], PETER REESE[1], ANDRES SALAZAR[3], AND ROBERT HERNDON[4]

Roswell Park Memorial Institute, Buffalo, New York[1], Dent Neurologic Institute, Millard Fillmore Hospital, Buffalo, New York[2], Walter Reed Army Medical Center, Washington, D.C.[3], and University of Rochester Medical Center, Rochester, New York[4]

The rationale for the use of the interferons as treatment for multiple sclerosis (MS) is based upon a large body of evidence which documents a variety of immunological aberrations as well as various types of viral association in patients with this disease. Among the immunological abnormalities in MS are decreased numbers of suppressor T lymphocytes during exacerbations (1, 2), defective NK cell activity and defective in vitro interferon production (3). A possible viral etiology of MS has been proposed (4). Elevated serum and cerebrospinal fluid (CSF) levels of antibodies against viruses, particularly measles virus (5, 6), have been reported, as well as the isolation of a viral agent, termed the IM virus, from the CSF of several patients with MS (7, 8). Recently, evidence of a human T-lymphotropic virus type I (HTLV-I) type of virus has been found in patients with MS (9). A T-cell clone with some morphologic and histochemical properties of HTLV-I-infected cells has been established from an MS patient's CSF. This clone grew independently of exogenous interleukin 2 (IL-2). It was postulated that this lymphotropic virus or its products play a role in the pathophysiology of MS by altering immune regulation (10). The known immunomodulatory and antiviral effects of the interferons therefore might be utilized to provide regulatory functions which could alleviate the symptoms of MS. Several clinical trials of the human interferons alpha (α), beta (β), and gamma (γ) and an interferon inducer as treatment for MS will be reviewed here.

M. Revel (ed.), CLINICAL ASPECTS OF INTERFERONS.

INTRATHECAL INTERFERON TREATMENT

In early 1980, a study of intrathecal (i.t.) β interferon treatment was begun by Jacobs et al. (11). Twenty patients with MS were randomized to either receive interferon or to serve as controls. The study was not blinded. The interferon was produced at Roswell Park Memorial Institute in human fibroblasts (12) and was purified to a specific activity of approximately 1×10^7 units/mg protein (13). The interferon was administered via lumbar puncture at doses of 1×10^6 units/m^2 semi-weekly for the first 4 weeks and once per month for the next 5 months. After 2 years on study, the controls were crossed over to i.t. β interferon treatment at doses of 1×10^6 units/injection, following the same schedule as used for the first 10 patients. After 1.5 years on study, it was reported (11) that the 10 patients who received interferon for 6 months had a significantly reduced mean rate of exacerbations/year, 0.25, compared to their mean exacerbation rate prior to treatment, 1.8 ($p < .01$). At that time, there was no significant change in the mean exacerbation rate of the 10 control patients. Clinically, the conditions of 5 interferon recipients and 2 controls improved, those of 3 recipients and 4 controls were unchanged, and those of 2 recipients and 4 controls worsened. Toxic side effects consisted mainly of headache, low grade fevers, and weakness, occurring in 10, 6, and 6 patients respectively. Malaise occurred in 5 patients and myalgia in 4. Rarely occurring symptoms were nausea, vomiting, diarrhea, chills, and rash. Transient pleocytosis and increases in CSF total protein occurred in all recipients but there was no apparent relationship with systemic toxic effects or clinical neurologic profiles. A more extensive report (14) described in detail the clinical courses of the patients. Because the side effects of the i.t. interferon treatments were relatively mild and usually of short duration, and because of the significant reduction in exacerbation rates, the patients who served as controls were crossed over to interferon treatment.

Crossover of the 10 control MS patients to β interferon took place after they had been on study for a mean of 2.1 years, during which time their mean exacerbation rate had not significantly changed. Two years after the 6 months of interferon treatment began, the initial results of the crossover study were reported (15). The patients' (i.e. initial controls) mean pre-study exacerbation rate, 0.68/year, was reduced to 0.30/year ($p < .04$). This finding is similar to that obtained in the 10 patients who originally received i.t. β interferon. However, the level of significance was not as great in the cross-over group as in the original recipients, perhaps due to the fact that the initial pre-study exacerbation rate of the controls was lower than that of the interferon recipients. It was postulated that with time, the same level of significance might be achieved.

Follow-up studies of original interferon recipients and the crossed-over controls indicated that treatment resulted in a sustained reduction in exacerbation rates (16). After 5.3 years, the mean exacerbation rate of the original 10 recipients had decreased from the pre-study rate of 1.8/year to 0.20/year ($p < .001$). The controls' pre-study rate of 0.68/year had decreased to 0.30/year ($p < .03$) 2.9 years after crossover to interferon treatment.

Early results from the above studies indicated a need for a larger, multicenter, definitive study of the efficacy of i.t. β interferon in MS. Such a study was designed (17) to include a homogeneous population of MS patients whose exacerbation rates were relatively high, and a population large enough to ascertain within a 2-year period whether this form of treatment would be beneficial. An important feature of this study would be that it would be double-blinded. The study was carried out in 69 MS patients who had exacerbating/remitting disease and exacerbation rates of at least 0.6/year (18). The patients were randomly assigned to either an interferon recipient group or a control group. These groups had similar pre-study mean exacerbation rates, 1.79 and 1.98, respectively. The β interferon used was produced at Roswell Park as previously described (12, 13). Treatment consisted of doses of 1×10^6 units administered to recipients by lumbar punctures, weekly for 4 weeks, followed by 1/month for 5 months. The controls received placebo according to the same schedule but only the first and last treatments consisted of true lumbar punctures, in order to obtain CSF for analysis. The other treatments were false lumbar punctures, administered subcutaneously (s.c.) following a local anaesthetic. At each center, a treating physician administered the interferon or placebo, and different physicians, who were blinded regarding which treatment was given, served as examining physicians. Mean exacerbation rates for the patients before and 2 years after beginning the study, are shown in Fig. 1.

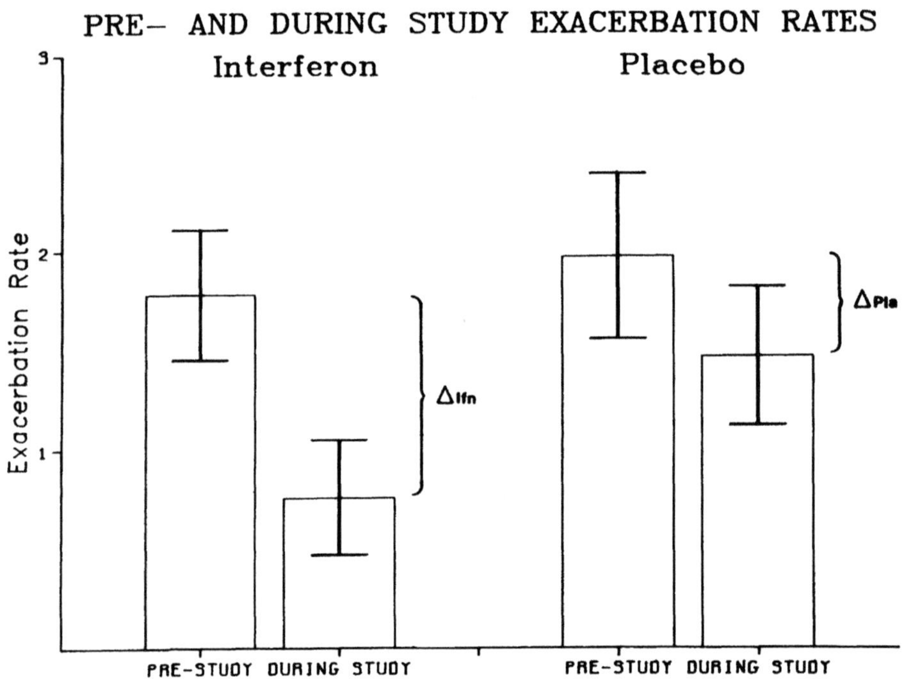

Fig. 1. Mean exacerbations/year. Vertical bars represent 2 SEM; Δ Ifn is reduction in β interferon recipient rates and Δ Pla is reduction in placebo controls during the study. (Reprinted with permission from ref. 18).

The recipients' mean pre-study rate of 1.79 fell to 0.76, significantly lower than the controls' rate which fell from 1.98 to 1.48 ($p < 0.001$). Although the exacerbation rates of both groups fell during the study, the reduction in the recipients' rate (1.02/year) was significantly greater than that of the controls (0.51/year) ($p < 0.04$). Clinically, 76.5% of the recipients and 60% of the controls were improved or unchanged, while 23.5% of the recipients and 40% of the controls were worse at the end of the study.

Headache, nausea, myalgia, and lethargy occurred in both groups, attributable in part to the successful blinding of the study, and in part to the use of indomethacin for the reduction of interferon's side effects. Although low-grade fevers occurred more frequently in recipients (75%) than in controls (31%), the blinding of the study was not broken. This was confirmed by questionnaires filled out by the patients which indicated that most patients in each group (79.3% of recipients and 63.6% of the controls) believed that they had received interferon treatment. Pleocytosis and rises in protein in CSF occurred in the interferon recipients, but the transient CSF abnormalities were less marked than in the preliminary study.

The mechanism by which the 57% reduction in mean exacerbation rate occurred in the recipient group is not known. Interferon injected i.t. flows upward over the surface of the cerebral convexities and comes into direct contact with brain parenchyma (19, 20). It is conceivable that this indirectly stabilized suppressor T lymphocyte fluctuations which may be an integral part of the MS exacerbation/remission cycle (1, 2). It is also possible that the interferon treatment eradicated a persistent virus infection in the CNS, thereby eliminating a viral trigger of exacerbations.

The mechanism by which the 25% reduction in mean exacerbation rate occurred in the control group is also not known but was probably due to a placebo effect of their false lumbar puncture treatments. Because of the blinded nature of the study, and the fact that most of the control patients indicated that they thought they were receiving interferon, the placebo effect seems most likely. The preliminary study described above, which was not blinded, showed no difference in the mean exacerbation rate of controls pre-study and after 2 years on study, prior to crossover. It is becoming more and more widely known that interrelations exist between stress and impaired immune competence (21). If stress was decreased by the belief that one was receiving effective therapy, then psychoneuroendocrine factors might be capable of restoring some degree of immune competence. It has been shown in animals that a conditioned increase of an immune response could be achieved without restimulation by its inducer, namely, the interferon inducer, poly I·poly C (22). That some form of neuroimmunomodulation contributed to the beneficial effect of placebo in the MS patients is a possibility. The same factors may have also contributed to a portion of the beneficial effects observed in the interferon recipients.

SYSTEMIC INTERFERON TREATMENT

An early pilot study of systemically administered natural α interferon treatment in a small number of MS patients did not show any therapeutic effect (23). A larger study of systemic treatment of MS with α interferon was undertaken at 2 centers (24). Twenty-four MS patients were randomized to receive either natural α interferon (25) or placebo which they self-administered by intramuscular (i.m.) injection. Interferon (5×10^6 units) or placebo was given daily for 6 months. A 6-month washout period followed, and then the patients were crossed-over to the alternate treatment for 6 months. This was followed by a second washout period. The 12 patients who received interferon during the first 6-month period had a mean pre-study exacerbation rate of 0.92/6-month period for the 2 years preceding the study. The mean exacerbation rate fell to 0.58 during the interferon treatment phase, was 0.50 during the first washout period, rose to 0.82 during the placebo phase, and was 0.55 during the second washout period. The 12 patients who received placebo during the first 6-month period had mean pre-study exacerbation rates of 0.88 and 0.63/6-month periods during year 2 and year 1, respectively, preceding the study. During the study, their mean exacerbation rates became 0.42 for the placebo period, 0.42 for the first washout period, 0.33 for the interferon treatment phase, and 0.09 for the second washout period. Although both groups of patients had mean exacerbation rates in all periods of the study that were lower than their pre-study rates, the difference between the rates during the interferon phase and the placebo phase was not significant. However, when a subgroup of 15 patients, whose disease was categorized as strictly of the exacerbating/remitting type, was analyzed, fewer exacerbations occurred during interferon treatment than during the placebo phase (p 0.08). The other 9 patients, who had a progressive decline in their neurological rating scale score, had exacerbated during the chronic progressive course of their disease and continued to have exacerbations during all phases of the study. Neurologic status of the patients was not improved by the interferon treatment.

Side effects included fatigue, malaise, myalgia, fever, nausea, hair loss, and depression, occurring in at least half of the patients during the interferon phase of the study. Occurring in fewer patients were headache, infection, lymphadenopathy, granulocytopenia, and SGOT elevation. The authors claimed that the study was double-blinded, but it is difficult to reconcile this claim with these toxic side effects that were experienced by interferon recipients, but not controls. The decrease in exacerbation rates for both groups of patients during both the interferon phase and the placebo phase might be due to the crossover design of the study. It was postulated that a learning phenomenon contributed to the greater reduction in exacerbation rate that occurred when the interferon phase followed the placebo phase.

Long-term follow-up of the 12 patients treated at one of the centers at which the study was carried out showed that their mean pre-study exacerbation rate of 1.33/year fell to 1.08/year after the first year on study, and to 0.75 during the second year (26). During a 2.3 year follow-up period, the mean exacerbation rate fell to 0.47/year, a significant decrease from the pre-study rate ($p < .01$). Although disability scores indicated an increase in overall disability, the increase was not statistically significant ($p > .1$). In general, this study indicated that systemic natural α interferon treatment provided some degree of benefit to those MS patients whose treatment began when their disease was of relatively short duration (less than 5 years).

Another form of human interferon, recombinant gamma (γ) interferon (Biogen), was tested in MS patients (27). A phase I trial was conducted in which 18 patients with exacerbating-remitting MS received low (1 μg), intermediate (30 μg), or high (1000 μg) doses of interferon by intravenous (i.v.) infusion twice weekly. During one month of treatment, exacerbations had occurred in 6 of the patients. This rate of exacerbations was significantly higher ($p < 0.01$) than the group's mean rate prior to the study or after the study. These episodes were mild to moderate in severity, and were not dose-related. All patients who received the 1000 μg dose experienced chills, fever, headache, fatigue, and myalgias. These side effects were less prominent in the patients who received 30 μg, and absent in those treated with 1 μg of interferon. Although the side effects were dose-related and could apparently be avoided through dose reduction, the authors recommended no further trials of γ interferon in MS.

Another approach to interferon treatment of MS is the use of an interferon inducer. The rationale for treating with an inducer is based upon the possible advantages of: 1). endogenous interferon produced in the patient may avoid side effects encountered with preparations of the natural interferons, 2). production of various molecular forms of interferon theoretically could affect different aspects of the pathogenesis of MS, and 3). longer periods of elevated serum interferon levels occur when inducers are administered systemically than with exogenous interferon treatment. Conversely, possible disadvantages to be considered are: 1). side effects may be caused by the inducer, 2). it is not yet known what effects combinations of different forms of interferons might have in MS, and 3). the relationship of circulating interferon and drug efficacy has not been established.

An open Phase I trial of the interferon inducer, Poly ICLC, was conducted in 18 patients with chronic progressive MS (28, 29). Poly ICLC is polyinosinic acid·polycytidylic acid complexed with poly (lysine) and carboxymethylcellulose (30). The drug was administered as a 30-60 minute i.v. infusion starting with a dose of 20 μg/kg. This was gradually increased to 100 μg/kg. Weekly treatments were given for 6 to 12 weeks, followed by biweekly or monthly infusions for up to 18 months. Of the 18 patients treated, 9 had rapidly progressive MS. Of these, 5 improved, but 1 of the 5 deteriorated after 15 months of treatment, 2 remained stable while on treatment, 1 stabilized for 5 months but then worsened in spite of continued treatment, and 1 deteriorated. Of the 9 patients who had slowly progressive MS, 4 remained stable, 4 withdrew due to side effects of treatment, and 1 died of a cause unrelated to therapy.

Side effects of Poly ICLC treatments were fever, transient elevated SGOT and SGPT, nausea, headache, and weakness in more than half of the patients. Occurring in fewer patients were vomiting, transient recurrence of scotoma, back pain, transient quadriparesis, urinary retention, prolonged SGOT and SGPT elevation, and hives. Polymorphonuclear leukocytosis and lymphocytopenia occurred after each treatment. Peak serum interferon levels occurred 8-12 hours after treatment and were greater than 50 units/ml after 82% of the infusions and greater than 500 units/ml after 15%.

In summary, clinical trials of interferon treatment of MS that have been reported to date have included various types of human interferon, α, β, and γ, and an interferon inducer, Poly ICLC, which have been administered by various routes, i.v., i.m., and i.t. The studies with i.t. administered natural β interferon have thus far shown the greatest degree of drug efficacy (11, 14-18). However, it is likely that variations in dosage, scheduling, and perhaps the use of recombinant interferons and/or combinations of interferons or interferon plus inducer may provide the optimal effects in the treatment of this disease.

REFERENCES

1. Antel, J.P., Arnason, B.G.W. and Medof, M.E. Ann Neurol. 5:338-342, 1979.
2. Reinherz, E.L., Weiner, H.L., Hauser, S.L., Cohen, J.A., Distaso, J.A. and Schlossman, S.F. N. Engl. J. Med. 303: 125-129, 1980.
3. Neighbour, P.A. and Bloom, B.R. Proc. Natl. Acad. Sci. U.S.A. 76:476-480, 1979.
4. Johnson, R.T. Adv. Neurol. 13:1-46, 1975.
5. Adams, J.M. and Imagawa, D.T. Proc. Soc. Exp. Biol. Med. 111:562-566, 1962.
6. Brody, J.A., Sever, J.L., Edgar, A. and McNew, J. Neurol. 22:492-499, 1972.
7. Melnick, J.L., Seidel, E., Inoue, Y.K. and Nishibe, Y. Lancet i:830-833, 1982.
8. Melnick, J.L., Seidel, E., Wang, S.-S. and Muchinik, G. J. Clin. Microbiol. 21:698-701, 1985.

9. Koprowski, H., DeFreitas, E.C., Harper, M.E., Sandberg-Wollheim, M., Sheremata, W., Robert-Guroff, M., Saxinger, C., Wong-Staal, F., Feinberg, M. and Gallo, R.C. Nature (London) 318:154-160, 1985.
10. DeFreitas, E.C., Sandberg-Wollheim, M., Schonely, K., Boufal, M. and Koprowski, H. Proc. Natl. Acad. Sci., USA 83:2637-2641, 1986.
11. Jacobs, L., O'Malley, J., Freeman, A. and Ekes, R. Science 214:1026-1028, 1981.
12. Havell, E.A. and Vilcek, J. Antimicrob. Agents and Chemother. 2:476-484, 1972.
13. Mikulski, A.J., Heine, J.W., Le, H.V. and Sulkowski, E. Prep. Biochem. 10:103-119, 1980.
14. Jacobs, L., O'Malley, J., Freeman, A., Murawski, J. and Ekes, R. Arch. Neurol. 39:609-615, 1982.
15. Jacobs, L., O'Malley, J.A., Freeman, A., Ekes, R. and Reese, P.A. Arch. Neurol. 42:841-847, 1985.
16. Jacobs, L., O'Malley, J.A., Freeman, A. and Reese, P. In: The Biology of the Interferon System (Eds. W.E. Stewart II and H. Schellekens), Elsevier Science Publishers B.V., 1986, pp. 497-504.
17. Jacobs, L., O'Malley, J.A., Freeman, A. and Reese, P. Arch. Neurol. 40:683-686, 1983.
18. Jacobs, L., Salazar, A.M., Herndon, R., Reese, P.A., Freeman, A., Josefowicz, R., Cuetter, A., Husain, F., Smith, W.A., Ekes, R. and O'Malley, J.A. Lancet, 20/27 Dec., 1986, 1411-1413.
19. Billiau, A. Arch. Virol. 67:121-123, 1981.
20. Smith, R.A., Landel, C., Cornelius, C.E. and Revel, M. J. Interferon Res. 5(suppl. 1):140, 1986.
21. Riley, V. Science 212:1100-1109, 1981.
22. Ghanta, V.K., Hiramoto, R.N., Solvason, H.B. and Spector, N.H. J. Immunol. 135:848s-852s, 1985.
23. Fog, T. In: Search for the cause of multiple sclerosis and other chronic diseases of the central nervous system (Ed. A. Boese), Verlag Chemie, Weinheim, 1980, pp. 490-493.
24. Knobler, R.L., Panitch, H.S., Braheny, S.L., Sipe, J.C., Rice, G.P.A., Huddlestone, J.R., Francis, G.S., Hooper, C.J., Kamin-Lewis, R.M., Johnson, K.P., Oldstone, M.B.A. and Merigan, T.C. Neurol. 34:1273-1279, 1984.
25. Cantell, K. and Hirvonen, S. J. Gen. Virol. 39:541-543, 1978.
26. Panitch, H.S. Arch. Neurol. 44:61-63, 1987.
27. Panitch, H.S., Haley, A.S., Hirsch, R.L. and Johnson, K.P. Neurol. 36(suppl. 1):285, 1986.
28. McFarlin, D.E., Bever, C.T., Salazar, A.M. and Levy, H.B. J. Biol. Resp. Modif. 4:544-548, 1985.
29. Bever, C.T., Salazar, A.M., Neely, E., Ferraraccio, B.E., Rose, J.W., McFarland, H.F., Levy, H.B. and McFarlin, D.E. Neurol. 36:494-498, 1986.
30. Levy, H.B., Baer, G., Baron, S., Buckler, C.E., Gibbs, C.J., Iadarola, M.J., London, W.T. and Rice, J. J. Inf. Dis. 132:434-439, 1975.

7

THE PROMISE OF INTERFERON THERAPY FOR PARASITIC DISEASES

GERALD SONNENFELD

Department of Microbiology and Immunology and Department of Oral Biology, Schools of Medicine and Dentistry, University of Louisville, Louisville, KY 40292, USA

ABSTRACT

Interferons were originally described as antivirals. Over the years, it has become clear that interferons have additional activities, inclduing modulation of host immune responses. As a result of these additional activities, the possibility exists that interferons could play some role in regulation of host resistance to various non-viral infections. Evidence will be discussed suggesting that interferons are induced during the course of several different types of parasitic infections of mammalian hosts. In addition, evidence obtained from in vitro and in vivo experiments will be presented to indicate that interferons could contribute to resistance of hosts to various parasitic infections. The potential for use of interferons in therapy of parasitic diseases will also be discussed.

INTRODUCTION

The original description of interferons described them as antiviral agents (1). Additional evidence utilizing pure interferons accrued over the past several years has made it clear that interferons have several other biological activities in addition to their antiviral activity (2,3). Included among those is the regulation of immune responses (2,3). As a result of the discovery of these additonal activities of interferons, interest has developed in a possible role for interferons in host resistance to non-viral infections. The range of the infections studied has included fungal (4), bacterial (5) and protozoan (6,7) infections.

The focus of this monograph will be on the presentation of

M. Revel (ed.), CLINICAL ASPECTS OF INTERFERONS. Copyright 1988. Kluwer Academic Publishers, Boston.

evidence suggesting that interferons may play a role in host resistance to parasitic infections. The discussion will conclude with an exploration of the possible clinical potential use of interferons for the treatment of parasitic infections.

PRODUCTION OF INTERFERONS DURING PARASITIC INFECTIONS

There have been many reports involving a wide variety of parasites indicating that infection of animals or of appropriate cells in tissue culture could result in the induction of interferons. The induction of interferons by exposure of animals or cells to parasites or parasite antigens appears to, in general, escape the generalized immunosuppression that occurs during many protozoan parasite infections (8).

Very early reports suggested that interferon could be induced when animals were infected with Toxoplasma gondii, the causative agent of toxoplasmosis which is such a threat to fetuses (9). In this early study, the interferon produced was not characterized.

An additional parasite for which interferon induction was reported very early was Plasmodium berghei, one of the agents responsible for malaria. As early as 1968, Huang and associates were able to show that interferon was induced in animals infected with P. berghei (10). The induction of the interferon was apparently related to the blastogenic effects of the malarial antigens. This finding was later extended by other workers who showed that interferon was also induced in the natural course of infection of humans with P. falciparum (11,12), another causative agent of malaria.

In extending these studies, Rhodes-Feuillete and associates were able to demonstrate the production of interferon-γ during the acute phase of infection of patients with P. falciparum (13). However, in several of these patients, including those with cerebral malaria, interferon-α was detected even earlier (96 hr) in the course of infection (13). This would suggest that interferon-α preceded interferon-γ production in patients with malaria (13). This type of progression and change in the type of interferon produced will surface as a recurring event in several different parasitic infections to be discussed in this monograph. Additional work by this group confirmed the correlation of interferon-γ production with the mitogenic activity of P. falciparum (14).

One of the most studied protozoan parasite interactions with the interferon system has been that involving Trypanosoma cruzi. T. cruzi is the South American trypanosome that is responsible for Chagas' disease. It causes an intracellular infection. Very early reports suggested that an interferon-like antiviral substance was produced in the sera of mice infected with T. cruzi (15). This was confirmed by later studies, which showed that the production of the interferon-like substance occurred early (24-72 hr post-infection) in the course of infection and that the time of production was related to the rapidity of the spread of infection to cells of the reticulo-endothelial system (16). This antiviral substance was later confirmed to be interferon-α/β (17). Wietzerbin and associates have recently reported that interferon-γ production can also be detected in the sera of T. cruzi-infected mice (18).

Additional studies have been carried out to determine the nature of the interaction of African trypanosomes with the interferon system. These trypanosomes are extracellular parasites that are the cause of African sleeping sickness. Resistant animals that are infected with these parasites have the trypanosomes go through several cycles of increased and decreased parasitemia, and the variant specific surface glycoprotein antigen of the parasite changes in each cycle of parasitemia (19). This allows the trypanosome to escape immune defenses of the host and to eventually take the life of the host.

Several species of animal and African trypanosomes, including T. equiperdum and Trypanosoma brucei, have been shown to induce interferons during the course of experimental mouse infections (20,21). The interferons produced during infection of mice with T. brucei were identified as a mixture of interferon-α/β and interferon-γ by Bancroft and associates (21). Additional studies with Trypanosoma brucei rhodesiense indicated that strains of mice that were very susceptible to infection with the parasite produced little to no interferon, and all that was produced was of the α type (22). Resistant strains of mice produced interferon-α/β early in the course of infection, followed by interferon-γ production as the mice began to control the first wave of parasitemia (22). The induction of interferon appeared to be due to stimulation by a common antigen that was expressed during all of the peaks of parasitemia, and not due to the variant specific surface glycoprotein antigen (23). In addition the production appeared to be due to true antigenic stimulation and not to the blastogenic effects of the parasite antigens (23). Here again the biphasic production of interferon in a parasitic infection was observed. The parasite-induced interferon production evaded the generalized immunosuppression, including suppression of mitogen (concanavalin-A) induced interferon induction, that occurs during infections of mice with African trypanosomes (23).

Leishmanial infections may be unusual with regard to induction of interferons. Individuals who were infected with Leishmania and had an active, visceral form of the disease, appeared not to have the ability of their cells to produce both interferon-γ and interleukin-2 (24). The immunosuppression induced by the disease appeared to overcome any interferon-induction ability of the parasite. In an animal model, mice infected with Leishmania mexicana amazonensis produced some interferon very early (4 hr) post-infection (25). However, mice that were immunized with parasite antigens produced no interferon upon challange (25). The situation with interferon induction by these parasites requires further study and remains to be clarified.

In general, it appears that many different protozoan parasites can serve as interferon inducers. The production of these interferons appears to occur during the natural course of infection of hosts with these parasites.

EFFECTS OF INTERFERON INDUCERS AND EXOGENOUS INTERFERONS ON PARASITIC INFECTIONS

In addition to several studies that have shown induction of interferons during protozoan parasitic infections, many additional studies have shown that interferons can influence both in vitro and in vivo resistance to parasitic infections. Interferons, and interferon-γ in particular, have profound influences on immune responses, particularily activation of macrophages and enhancement of macrophage killing (2,3). Therefore, the tendency in research has been to search for the effects of interferons that augment or regulate immune defenses to the parasitic infections. It has now become clear, however, that interferons can have other, possibly direct, effects on parasite infections.

Toxoplasma gondii infections are an example of parasitic infections that can be influenced by both the immunoregulatory effects of interferons as well as other effects. Administration of interferon-γ to mice resulted in increased antibody production and survival when the mice were infected with T. gondii (26). This would suggest an augmentation by interferon-γ of immune defenses against the parasitic infection. However, fascinating in vitro studies by Pfefferkorn and associates (27,28) indicated that interferon-γ could also protect fibroblasts from infection with T. gondii. This was an effect of the interferon on the fibroblasts, and appeared to be related to induction of tryptophan degredation in the fibroblasts by the interferon-γ, which inhibited parasite growth (28).

Other studies have shown that malarial infections can be modulated by treatment of hosts with interferon inducers and interferons. Treatment of mice with inducers of interferon-α/β or with exogenous interferon-α/β resulted in some protection of the mice from malarial infections (29-33). The best protection occured when treatment was given at 20 hr post-infection. Pretreament (a few hr prior to infection) with interferon-γ of mice, rats or chimpanzees resulted in decreased Plasmodium berghei parasitemia (34). The effect of the interferon-γ appeared to be directed toward inhibition of extra-erythrocytic forms, but not erythrocytic forms of the parasite (34). The mechanism(s) of this effect is/are not known.

Additonal experiments have shown that Trypanosoma cruzi infections can be influenced by interferon administration. Mice that were infected with T. cruzi were treated with daily doses of interferon-α/β beginning on the day of infection (17). These mice showed a transient drop in levels of parasitemia compared to controls (17). Similar results have been seen when mice were treated with inducers of interferon-α/β (35). In addition, pretreatment of mice with interferon-γ and antibody directed against T. cruzi resulted in prophylactic protection of the mice from infection with the parasite (18, 36). Treatment of mice with antibody directed against interferon-α/β did not alter the course of infection of the mice with T. cruzi; however, the treatment protocols used were very limited (37).

In vitro mechanistic studies have been carried out suggesting that both immunological and non-immunological components are involved in protective effects of interferons against T. cruzi infections. Cultures of murine macrophages or rat heart myoblasts were protected from infection with T. cruzi when interferon-α/β was added at the time of infection (38). This protective effect appeared to result from an interaction of the interferon directly with the parasite, and not on the cells. On the other hand, murine macrophage cultures treated with interferon-γ showed an expected immune-mediated effect (39). The macrophage cultures were activated to have enhanced uptake and killing of T. cruzi after interferon-γ treatment (39). Here, the possibility of a two-step series of protection arises. Early interferon-α/β production induced by T. cruzi could protect cells from infection. As time and the infection progress and this defense is overwhelmed, immune-specific interferon-γ production could result in enhanced destruction of the T. cruzi.

Interferon-mediated resistance with regard to Trypanosoma brucei may also involve a two-step series, since interferon-α/β production appears to precede interferon-γ production in resistant strains of mice infected with Trypanosoma brucei rhodesiense (22). However, attempts to protect mice against T.b. rhodesiense using exogenous interferon-γ, or interferon inducers have not been successful (40). In addition, attempts to aggravate T.b. rhodesiense infections by treatment of mice with anti-interferon-γ have also not been successful (40). Limited protocols tried in these experiments (40), as well as the complex nature of the T.b. rhodesiense infection may make the development of appropriate protocols for treatment of the infection with interferon a difficult task to accomplish.

Additional studies with T.b. rhodesiense suggested that interferon could play some role in resistance to secondary infections that contribute to the morbidity and mortality of a primary trypanosome infection. Mice that were infected with *T.b. rhodesiense* were resistant to encephalomyocarditis-D virus-induced diabetes (41). Since the host was immunosuppressed by the *T.b. rhodesiense* infection and made no detectable antibody directed against the virus, it is likely that the interferons induced by *T.b. rhodesiense* made some contribution to resistance to the secondary virus infection (41).

In vitro studies have shown that interferons, either induced or exogenously added, enhanced the killing of intracellular *Leishmania donovani* (42,43). Other lymphokines could have interacted with interferon-γ to have mediated this enhanced killing (44).

In general, it is clear that interferons can mediate *in vitro* infections of cells with protozoan parasites. *In vivo* effects also occur, but have been more difficult to demonstrate.

THE POTENTIAL FOR CLINICAL TREATMENT OF PARASITIC DISEASES WITH INTERFERONS

Although it is apparent that interferons are induced by protozoan parasites and that interferons can enhance killing of those parasites as a host defense, the development of interferons as a therapeutic agent for parasitic diseases may be a difficult task. Therapeutic use of interferons in *in vivo* animal models has been difficult to obtain.

Interferons probably play a role in natural host resistance to parasitic infections. The immunomodulatory role of interferons, as well as the immune-mediated production of interferon-γ, would suggest that immune responses to parasites can be mediated by interferons. This has been borne out by the experiments described above. In addition, as also described above, interferons may have effects on host cells that aid in resistance that are not related to specific immune responses. Also, there may be direct effects of the interferons on the parasites. This evidence does support a role in natural resistance to parasites.

Utilizing exogenous interferons or interferon inducer treatment to control parasitic diseases may be more difficult. It is clear that interferons have immunoregulatory effects that are time, dosage and route dependent (2,3). For example, depending on the time, route and dose of interferon administered, one can either enhance or suppress antibody production (2,3). In addition, treatment of natural killer cells with interfeorn enhances their activity, while treatment of target cells with interferon makes them resistant to the action of natural killer cells (45). Therefore, developing a protocol that makes use of the appropriate effect of interferon at the appropriate time during a parasitic infection may be very difficult.

In addition, many of the protozoan parasitic infections are very immunosuppressive for the host. Therefore, components of host immune responses that may be required for interaction with exogenous interferon for effective therapy may be non-functional in the infected host. This problem would not have been observed in the *in vitro* experiments that were described above.

Despite these potential drawbacks, interferon therapy may be effective in many cases, such as control of secondary, often fatal, infections in host infected with parasites. Also, Ferreira and associates (34) suggested that interferon could be used in conjunction with other drugs for the treatment of malaria. Drugs available for the elimination of extraerythrocytic forms of malaria in the liver are often toxic (34). The use of interferon-γ together with low doses of these drugs could perhaps serve as an effective mode of treatment (34). Certainly, the use of interferons in conjunction with other biological response modifiers and chemotherapeutic agents for treatment of parasitic diseases needs to be explored. It should also be remembered that the possibility of toxic effects of combined drug therapy and interferon therapy exists, since interferons can drastically affect the metabolism of drugs (46).

In any case, the use of interferon in therapy for parasitic infections in an intruiging possibility that requires further exploration. Additional extensive testing in animal models should be carried out prior to widespread human testing. After these types of models are used, it may be that interferons and other biological response modifiers may provide an interesting and successful avenue for treatment of those parasitic diseases for which an effective chemotherapy has not been developed. The future holds the promise of fascinating possibilities for research in this area.

LATE NOTE

In recent months, Kubelka and associates (47) have performed studies to indicate that interferon-γ, through an immune mediated mechanism can affect the *in vitro* course of infection of *Schistosoma mansoni*, a non-protozoan parasite. Interferon-γ had the capacity to induce schistosmulicidal activity of murine macrophages in tissue culture (47).

ACKNOWLEDGEMENTS

Studies performed in the author's laboratory described herein were funded in part by US NASA award NCC2-214. Addional support was provided by funds administered by the University of Louisville School of Medicine Research Committee from NIH BRSG grant RR-05375.

REFERENCES

1. Isaacs, A. and Lindenmann, J. Proc. Royal Soc. B 147: 258-267, 1957.
2. Gresser, I. Cell. Immunol. 34: 406-414, 1977.
3. Vilcek, J. and DeMaeyer, E. Interferons and the Immune System. Elsevier Science Publishers B.V., Amsterdam, 1984.
4. Salvin, S.B., Nishio, J. and Shonnard, J.T. Infect. Immun. 9: 631-635, 1974.
5. Degre, M. and Rollag, H. Texas Rep. Biol. Med. 41: 388-394, 1982.
6. Sonnenfeld, G., Wirth, J., Kierszenbaum, F., DeGee, A.L.W. and Mansfield, J.M. In: The Interferon System (Eds. F. Dianzani and G. Rossi),Serono Symposium Publication Volume 24, Raven Press, New York, 1985, pp. 193-199.
7. Sonnenfeld, G., Kierszenbaum, F., Gould, C.L., DeGee, A.L.W. and Mansfield, J.M. In: The Interferon System: A Current Review, (Eds. S. Baron, F. Dianzani, G.J. Stanton and W.R. Fleischmann) University of Texas Press, Austin, 1987, In Press.
8. Mansfield, J.M. Parasitic Diseases, Vol. 1 The Immunology, Marcel Dekker, New York, 1981.
9. Freshman, M.M., Merigan, T.C., Remington, J.S., and Brownlee, I.E. Proc. Soc. Exp. Biol. Med. 123: 862-866, 1966.
10. Huang, K.-Y., Schultz, W.W. and Gordon, F.B. Science 162: 123-124, 1968.
11. Ojo-Amaize, E.A., Salimonu, L.S., Williams, A.I.O., Akinwolere, A.O., Shabo, R., Alm, G.V. and Wigzell, H. J. Immunol. 127: 2296-2300, 1981. Ojo-Amaize, A.E., Vilcek, J., Cochrane, A.H. and Nussenzweig, R.S. J. Immunol. 133: 1005-1009, 1984.
12. Rhodes-Feuillete, A., Druilhe, P., Canivet, M., Gentilini, M. and Peries, J. C.R. Acad. Sci. Paris (III) 293: 635-637, 1981.
13. Rhodes-Feuillete, A., Jaureguiberry, G., Ballet, J.J., Andriue, B., Druilhe, P., Le Bras, J., Galibert, F., and Peries, J. J. Interferon Res. 5: 159-168, 1985.
14. Rhodes-Feuillete, A., Bellosguardo, M., Druilhe, P., Ballet, J.J.,

Chousterman, S., Canivet, M. and Pieries, J. J. Interferon Res. 5: 169-178, 1985.
15. Rytel, M.W. and Marsden, P.D. Amer. J. Trop. Med. Hyg. 19: 928-931, 1970.
16. Sonnenfeld, G. and Kierszenbaum, F. Amer. J. Trop. Med. Hyg. 30: 1189-1191, 1981.
17. Kierszenbaum, F. and Sonnenfeld, G. J. Parasitol. 68: 194-198, 1982.
18. Wietzerbin, J., Garcia-Pons, F., and Plata, F. J. Interferon Res. 6 (Supplement 1): 142, 1986.
19. Mansfield, J.M. In: Parasitic Diseases, Vol. 1 The Immunology (Ed. J.M. Mansfield), Marcel Dekker, New York, 1981, pp. 167-226.
20. Talas, M. and Glaz, E.T. J. Infect. Dis. 139: 595-598, 1979.
21. Bancroft, G.J., Sutton, C.J., Morris, A.G. and Askonas, B.A. Clin. Exp. Immunol. 52: 135-143, 1983.
22. De Gee, A.L.W., Sonnenfeld, G. and Mansfield, J.M. J. Immunol. 134: 2723-2726, 1985.
23. De Gee, A.L.W., Sonnenfeld, G. and Mansfield, J.M. Submitted for Publication, 1986.
24. Carvahlo, E.M., Badaro, R., Reed, S.G., Jones, T.C. and Johnson, W.D. J. Clin. Invest. 76:2066-2069, 1985.
25. Barral-Netto, M., Reed, S.G., Sadigursky, M. and Sonnenfeld, G. Clin. Exp. Immunol. 67: In Press, 1987.
26. McCabe, R.E., Luft, B.J. and J.S. Remington. J. Infect. Dis. 150: 961-962, 1984.
27. Pfefferkorn, E.R. and Guyre, P.M. Infect. Immun. 44: 211-216, 1984.
28. Pfefferkorn, E.R. Proc. Natl. Acad. Sci., USA 81: 908-912, 1984.
29. Jahiel, R.I., Vilcek, J., Nussenzweig, R. and J. Vanderberg. Science 161: 802-804, 1968.
30. Jahiel, R.I., Nussenzweig, R.S., Vanderberg, J. and Vilcek, J. Nature 220: 710-711, 1968
31. Jahiel, R.I., Nussenzweig, R.S., Vilcek, J. and Vanderberg, J. Amer. J. Trop. Med. Hyg. 18: 823-835, 1969.
32. Jahiel, R.I., Vilcek, J. and Nussenzweig, R. Nature 227:1350-1351, 1970.
33. Sauveger, F., and Fauconnier, B. Biomedicine 29: 184-187, 1978.
34. Ferreira, A., Schofield, L., Enea, V., Schellekens, H., Van Der Meide, P., Collins, W.E., Nussenzweig, R.S. and Nussenzweig, V. Science 232: 881-884, 1986.
35. James, S.L., Kipnis, T.L., Sher, A., and Hoff, R. Infect. Immun. 35: 588-593, 1982.
36. Plata, F., Wietzerbin, J., Garcia-Pons, F., Falcoff, E. and Eisen, H. Europ. J. Immunol. 14: 930-935, 1984.
37. Quan, P.C., Rager-Zinsman, B., Wittner, M. and Tanowitz, H. J. Parasitol. 69: 164-1166, 1983.
38. Kierszenbaum, F., and Sonnenfeld, G. J. Immunol 132: 905-908, 1984.
39. Wirth, J.J., Kierszenbaum, F., Sonnenfeld, G. and Zlotnik, A. Infect. Immun. 49: 61-66, 1986.
40. DeGee, A.L.W., Mansfield, J.M., and Sonnenfeld, G. J. Parasitol. 72: In Press, 1986.
41. Gould, C.L., DeGee, A.L.W., Mansfield, J.M. and Sonnenfeld, G. J. Interferon Res. 6: In Press, 1986.

42. Herman, R. and Baron, S. Nature 226: 168-170, 1970.
43. Murray, H.W., Rubin, B.Y. and Rothermal, C.D. J. Clin. Invest. 72: 1506-1510, 1983.
44. Hoover, D.L., Finebloom, D.S., Crawford, R.M., Nacy, C.A., Gilbreath, M., and Meltzer, M.S. J. Immunol. 136: 1329-1333, 1986.
45. Trinchieri, G. and Santoli, D. J. Exp. Med. 147: 1314-1333, 1978.
46. Sonnenfeld, G., Harned, C.L. and Nerland, D.E. Texas Rep. Biol. Med. 41: 363-369, 1982.
47. Kubelka, C.F., Ruppel, A. Krammer, P.H. and Gemsa, D. Parasitol. 92: 325-336, 1986.

II

Clinical uses in cancer

The potential applications of interferons in the treatment of cancer have attracted more attention than their use in viral diseases. This does not mean that the antiviral actions should be neglected, but emphasizes the needs for new treatments of cancer. As indicated in the Introduction, there is no simple explanation on why interferons should have beneficial effects on cancer. Anti-oncogene, antigrowth, immunostimulating effects could all be involved. Tumors respond very differently to IFN therapy, but some leukemias appear to respond at a very high rate. In the one human cancer in which IFN has a dramatic and consistent effect, Hairy Cell Leukemia, the beneficial effect of type I IFNs seems to result from an effect on differentiation of hematopoietic cells (1). This mechanism may act in addition to an a antigrowth effect on the tumor cells (2). However, in solid tumors, it is unlikely that IFN alone will prove effective enough and much remains to be done to achieve responses comparable to what is seen in some leukemias.

References:

1. Michalevicz, R. and Revel, M. Proc. Natl. Acad. Sci. USA 84 :2307-2311, 1987.
2. Paganelli, K.A., Evans, S.S., Han, T. and Ozer, H. Blood 67:937-942, 1986.

8

INTERFERON TREATMENT OF NON-HODGKIN'S LYMPHOMA AND MYELOMA: A MODEL FOR BIOTHERAPY

Robert K. Oldham, M.D.

Director, Biological Therapy Institute, Hospital Drive, Franklin, TN 37064

INTRODUCTION

The concept of biological therapy is not new. However, prior to 1980, biological substances were impure mixtures derived from cell cultures or extraction. Although many different biological activities were described, including the elusive interferon, no serious studies with biotherapy were possible because of the poor quality of the biological reagents. All of this changed in the late 1970s with the use of recombinant genetics to produce highly purified biological preparations. Alpha-recombinant interferon (rIFN-alpha) became the first biological substance produced by recombinant methods for use as an anticancer medicinal in man (1). In the seven short years since the first recombinant alpha interferon molecule was prepared, a large number of these molecules have become available and are now being tested in clinical trials. Thus, rIFN-alpha is a model for the investigation of new biological substances within the fourth modality of cancer treatment, biotherapy (2,3).

Historical aspects in the development of immunotherapy have recently been reviewed (4,5). Prior to 1980, the term immunotherapy was considered to be synonymous with biological therapy. However, it is now clear that there are many biological systems and mediators which may affect cancer growth and metastases which are distinct from the immune system. Thus, biotherapy in the broader sense now refers to therapeutic use of agents derived from biological sources and/or the use of agents which affect biological responses. The term biologicals refers to agents or approaches produced from biological sources. With current technology, this involves use of recombinant genetics to isolate the gene, transfection into an appropriate producer cell, and the isolation and purification of the protein product. Agents and approaches which alter biological responses for the

M. Revel (ed.), CLINICAL ASPECTS OF INTERFERONS.

benefit of the patient have been called biological response modifiers (BRM). The use of biologicals and BRM in cancer treatment can now be termed biotherapy (2,3). In the broadest sense, biotherapy could include blood products, transplanted organs, antibiotics, growth/maturation factors and a variety of other biological agents and approaches historically included under immunotherapy.

Immunotherapy has had a checkered past. Kari Cantell said it well when describing some of the early interferon research, "Much second class research was carried out with third class preparations slightly contaminated with interferon" (6). In addition, there have been questionable approaches used by certain practitioners and even the frank quackery of some who purport to deliver "biological therapy". As a result of the huge expenditures by the National Institutes of Health in the area of molecular biology and viral oncology, tools were developed over the last 20 years which have led to the current technology of "genetic engineering." In contrast to previous attempts to develop biotherapy, we now have techniques available which can produce absolutely pure proteins identical to those found in the body. It is with these techniques that hundreds and later thousands of biological compounds and their synthetic analogs will be developed. A major value of the research done thus far with interferon is that it may be viewed as a model for the development of other biological substances as medicinals. No longer should the paradigm used for the development of chemotherapy drugs be continued. A new paradigm that will accommodate biologicals and their use in medicine is now being developed (7). In this article, I shall attempt to summarize the interferon model and shall examine the studies using interferon in the treatment of non-Hodgkin's lymphoma and myeloma.

A. NEW PARADIGM

The drug development paradigm has been described in some detail (7,8). Over the past 25 years, over 500,000 drugs have been tested as anticancer agents, but only about 40 are available commercially. Perhaps 10 of these anticancer drugs can be classed as moderately to highly effective, and the rest are only marginally effective. The current process for development of a new drug involves a very long and costly set of procedures from initial concept, extraction or synthesis and formulation, documentation of biological activity and purity, early studies in the laboratory and in experimental animals to determine the mechanism of action and toxicity, and compilation of all the preclinical data into an investigational new drug application (INDA).

Subsequent to the INDA, further preclinical work is done to expand the information based on the mechanism of action and preclinical toxicology of a new drug. In addition, with the INDA, early-phase studies are begun in the clinic to determine the biological activity in humans. Testing generally begins in phase I (toxicity tests). Phase II studies are then conducted to determine the therapeutic activity. Therapeutic doses are selected that represent the investigator's "best guess" as to the therapeutically active dose range. In addition, there are the factors of schedule and route of administration. Phase III trials to compare a new agent with existing modalities of treatment often follow. The extent of phase III trials will depend on the treatment alternatives available. Such trials may need to be extensive, randomized, and controlled in diseases where other effective therapies are available. The final step in the commercial development of a new agent is the approval of a new drug application (NDA).

This process generally takes from 5 to 10 years and may involve the expenditure of $100 million to bring a drug from concept to the commercial market. This long and expensive process is said to protect the public from toxicity and assure efficacy of new drugs. However, marked differences exist between countries as to he number of drugs available to patients and that country's view of what is in the public interest. Therefore, one can debate which rules are the "best" for the development of new drugs. Thus, such a paradigm of drug development is highly restrictive and has only worked reasonably well because of the public acceptance of the regulatory structure and because of the small number of drugs that actually break through to the clinical arena. The advent of biotherapy will put great pressure on this paradigm, and the effective development of biologicals will require changes in the regulations and the policies implicit in the development of new pharmaceuticals.

A process similar to that described for drugs is now being applied to biotherapy. Biotechnology is making available a range of biologicals (lymphokines/cytokines, monoclonal antibodies, antigens, growth and maturation factors) that are pure and well defined (9). Biotherapy involves a group of substances which are less toxic, more selective and interact with existing receptors and physiologic mechanisms and are, in effect, very different substances than drugs. These differences and the number of biologicals available (hundreds of lymphokines/cytokines and their analogs, thousands of monoclonal antibodies, etc.) that new methods of translation from the laboratory to the clinic be developed. These new methods will require innovative clinical studies, including the combining of phase I and phase II into a single study design, escalating doses within patients, and a recognition that a broader and more diverse testing system may bring more of these biologicals through to clinical practice in a shorter period of time with less expense without compromise of the public safety (10-12).

INTERFERONS: THE MODEL

Although the concept of interferon (IFN) was discovered in 1957, extensive clinical trials with these materials were not carried out until the late 1970's. The initial clinical trials were attempted with "natural" extracted interferon purified from the supernatant fractions of stimulated white blood cells. Low purity, lot-to-lot variation, complex extraction and purification methods, and expense all limited the use of these materials in clinical trials. The advent of increased interferon availability through recombinant genetics has quickly broadened testing for several recombinant alpha IFN as well as for beta and gamma recombinant interferons (13).

There are several considerations with respect to the design of phase I clinical trials for biologicals which differ markedly from those with drugs. The dose-response curve for these agents may be very broad and/or multiphasic, with peak biological effects being seen at different doses for each system affected by the interferon. For biologicals, there is a need to measure biological responses in the context of clinical trials (14-16). The pharmacokinetics after intravenous and intramuscular administration differ for IFN-alpha and poor absorption has been seen after intramuscular administration of beta and gamma interferon. Thus, the proper conduct of phase I clinical trials must take into account appropriate measurements of bioavailability, pharmaco-kinetics, biological response modification, and toxicity, all in the context of escalating doses to determine the dose-response curves for each of these effects.

NON-HODGKIN'S LYMPHOMA

In the early 1980s, the BRM program of the National Cancer Institute had access to recombinant alpha interferon through Genentech and Hoffman-LaRoche. After a few patients had been tested in a brief phase I toxicity study, my research team was allowed to proceed with the first large scale phase I trial using rIFN-alpha (17). Eighty-one patients with a variety of refractory cancers were treated in this large multiple, fixed dose, phase I trial using this new form of biotherapy. The classic toxicities of alpha interferon were seen and described in detail. The rapid reversibility of these toxicities and the lack of cumulative toxicity were important observations. Objective evidence of anti-tumor activity was seen in selected solid tumors but most particularly in non-Hodgkin's lymphoma (NHL), chronic lymphocytic leukemia (CLL), and Hodgkin's disease (HD). Because of these responses and some of the results noted from other phase I trials, we immediately embarked upon an aggressive study to document the efficacy of rIFN-alpha in NHL, CLL, hairy cell leukemia (HCL), and chronic myelogenous leukemia (CML). Our studies with HCL (18) and those of others (19) documented overall response rates in excess of 90% and ultimately led to the further studies leading to FDA approval of the NDA for rIFN-alpha for use in this indication. Studies by Talpaz et al (20) also documented at the striking therapeutic effects of this interferon in the treatment of CML.

Our own studies in NHL demonstrated, for the first time, that this diverse group of lymphomatous disorders was responsive to interferon, even when the patients had been heavily pretreated with chemotherapy and/or radiotherapy. Heavily pretreated patients received rIFN-alpha (50 million units/m) by intramuscular injection three times weekly for periods ranging from 3 months to as long as one year. Forty-five patients began the study and thirty-seven were evaluable for response. Thirteen of twenty-four (54%) of evaluable patients with low histologic grade, non-Hodgkin's lymphoma had objective responses (9 partial respones and 4 histologically documented complete responses). Two of six (33%) of patients with intermediate grade lymphoma and only one of seven (14%) with high grade lymphoma had partial responses.

For those patients that responded, the median duration was eight months. Maintenance therapy, while in complete remission appeared to prolong the duration as an anecdotal observation, but formal studies are now underway to confirm document that observation.

In a follow-up study, O'Connell et al. (21) used rIFN-alpha in moderate doses (12 million units/m 3 times per week) for 8 weeks in 16 patients with NHL and observed a 44% response rate. Of interest is the lower dose utilized and the fact that these patients, unlike our studies cited above, had not been treated with chemotherapeutic agents. Response duration was 18 weeks and probably reflects both the low dose used in the study and the lack of maintenance therapy.

These studies and other similar studies with several alpha interferons have demonstrated clearly that a majority of patients with NHL, particularly the low grade variety, do respond to interferon both for untreated patients and for those that have been heavily pretreated. The response duration appears to be in the six month to one year range with most of the responses being partial. Less than 20% have achieved a durable complete response and in neither group of patients has the role of maintenance therapy or dose been completely defined.

The mechanism of action for interferon in the lymphomas, as in other neoplasms, is still unclear. However, the bulk of the evidence indicates that alpha interferon is acting as an antiproliferative agent. None of the immunological studies have demonstrated a solid correlation between changes in immunological function and response rates. In addition, low doses of interferon are most effective as an immuno-modulator and higher doses often suppress those same immune responses. Finally, the dose response curve in non-Hodgkin's lymphoma would suggest that higher doses induce more responses and some patients who have failed to response or who relapse with lower doses can be reinduced with higher doess of alpha interferon. Some of the studies will be discussed later for cutaneous T-cell lymphoma.

While these studies are of interest, the real payoff for alpha interferon in non-Hodgkin's lymphoma may come in combination with chemotherapy. Studies are underway at Stanford to induce a deep response with combination chemotherapy and then utilize maintenance alpha interferon in an attempt to maintain that level of response or deepen it further. These studies, as well as studies using other forms of biotherapy, such as interleukin-2/activated cells, should be monitored closely since there is the real possibility that biotherapy will lead to effective new treatments for patients with non-Hodgkin's lymphoma.

CUTANEOUS T-CELL LYMPHOMA

CTCL or mycosis fungoides and the associated Sezary's syndrome are T-cell lymphomas which predominantly involve the skin which also may involve peripheral blood, lymph nodes, spleen, and visceral organs. Like NHL, these disorders tend to be indolent disorders with a chronic progressive course characterized by the infiltration of skin and lymph nodes with malignant T-cells. Median survivals of 8-10 years have been reported.

A large number of therapeutic approaches are available or have been tried for these disorders. Topical therapy including nitrogen mustard, psoralen (with ultraviolet A irradiation) and total body electron beam irradiation have all been tried. However, these approaches are palliative since the systemic spread of the malignant T-cell ultimately occurs in nearly all patients. Systemic chemotherapy, including very aggressive combination chemotherapy, has been tried but has again proven palliative. Ultimately, these patients have recurrence and progression of their disease. As the disease generalizes from the skin to involve lymph nodes, bone marrow, peripheral blood, and visceral organs, control is more difficult and most patients with CTCL ultimately die of the disease or from an infectious complication of CTCL and/or its treatment.

We initiated a trial of rIFN-alpha in 1982 based on the responses seen during the phase I trial in patients with NHL and CTCL. The results were dramatic. High dose rIFN-alpha (50 million units/m) was given intramuscularly three times a week to patients with advanced CTCL refractory to chemotherapy and radiotherapy. Of the 20 patients treated all were evaluable and 9 had objective responses. These partial remissions lasted from 3 months to more than 25 months with a median of 5 months duration. Although no patient had a complete response, several of the patients had very striking partial responses with resolution of cutaneous and extracutaneous lesions. In addition, five patients had minor or mixed responses, indicating some therapeutic effect but without achieving the standard of a partial response. This objective response rate of 45% was higher than some members of our research team had seen with previous experimental therapies including VP-16 and high dose Methotrexate, serotherapy (anti-thymosin globulin and monoclonal antibody), and the treatment appears to be as active or more active than experimental therapies reported by others including 2d-doxycoframycin, retinoic acid and monoclonal antibody.

A major finding of these lymphoma studies was the ability to reinduce response in patients who were showing evidence of progression on lower doses by simply giving higher doses of rIFN-alpha. In a subset of patients where relapse was occurring or treatment did not appear to be effective, increasing the initial dose or a tolerable maintenance dose could again induce shrinkage of cutaneous lesions. These data, along with the complete lack of any immunological function correlating with response, would indicate that rIFN-alpha was acting predominantly by antiproliferative mechanism of action.

Follow-up studies are underway to intergrate interferon therapy into the palliative treatment regimens for CTCL. In addition, there are now several experimental therapies (cited above) that have some efficacy in this disorder. Thus, it would appear the future is bright for multi-modality chemotherapy, radiotherapy, and biotherapy in the systemic treatment and perhaps eradication of malignant T-cells in this disorder (22).

MULTIPLE MYELOMA

Multiple myeloma was among the earliest disorders treated with the impure preparations of Cantell interferon. As early as 1979 these preparations were reported as active in myeloma with objective response in the bone marrow, plasma cell population as well as decrements and circulating myeloma proteins. Follow-up studies seem to confirm the activity of Cantell interferon, but as these studies progressed, response rates tended to fall (23,24). In spite of these responses, there was little evidence of effect on survival as a result of interferon treatment.

As a result of these early studies and as a result of the enthusiasm derived from the *in vitro* activity of interferon on cultured myeloma cells (25), multiple trials of rIFN-alpha, both from Hoffman-LaRoche and Schering, were rapidly initiated. These studies are summarized in Table I and confirm the activity of alpha interferon in patients with myeloma. Response rates have varied from 15-50% but, on average, a response rate of 20-30% may be expected in this disorder. Studies are now underway combining interferon with Melphalan and Prednisone in the treatment of patients not previously exposed to chemotherapy (25). rIFN-alpha appears to be an active biotherapeutic agent in multiple myeloma but only an undefined subset of patients. The initial studies seem to indicate that patients with IgG myeloma were less responsive and studies are underway to determine whether tumor load, protein level, isotype, and/or previous treatment are important predictors of response in patients with melanoma treated with interferon.

SUMMARY AND NEW DIRECTIONS

From the studies sited above and from others still being conducted, it is clear that leukocyte and rIFN alpha have activity in NHL (26-29), CTCL (30), and multiple myeloma (25). Formal phase III studies are underway with rIFN alpha using it with chemotherapy and radiotherapy and as both induction and maintenance to determine what its role might be in the multimodality treatment of these diseases. Clearly, as a single agent, it is as effective or more effective than existing drugs in similar clinical settings. Of major interest is the less severe toxicity and the lack of cumulative toxicity seen in these studies. The response rate for new drugs in phase I studies is about 2% (31). Thus, the alpha interferon phase I response rates (all cancers) of approximately 10% and the more encouraging studies in phase II in these three disorders makes it clear that interferon has and will have a continuing role in the treatment of NHL, CTCL and multiple myeloma. The major question that remains is how to intergrate this into multimodality therapy and how far "up front" should interferon be in these treatment regimens.

Few data are available for beta interferon (32) and gamma interferon (33). In the phase I studies that have been reported, very few patients with these three diseases have been treated and thus, the level of activity in these disorders is unknown. However, it is notable that beta interferon causes less hematological toxicity than alpha interferon, and this may predict for less activity in hematological and lymphatous disorders (32). A great deal of research needs to be done with beta interferon to determine whether it has activity in these diseases and the potential for synergism when used together with alpha or gamma interferon is very real, based on _in vitro_ studies (34).

Studies with gamma interferon are somewhat further along. Sporadic responses to gamma interferon have been seen in patients with these disorders. However, no formal phase II trials of gamma interferon in NHL, CTCL or multiple myeloma have been reported. Gamma interferon appears to have much broader biological activity than alpha or beta interferon and indeed its maturational effects on hematologic cells may prove to be a major feature of its clinical activity. The physiology, biochemistry and early chemical trials have all been reviewed in detail in a recent manuscript by Bonnem and Oldham (33). The summary data from the clinical trials to date show responses in 36/392 (9%) treated gamma interferon patients in phase I trials. Given the 2% or less response rate for drugs in similar trials (31), this is an encouraging figure.

Synergistic antiproliferative activity has been reported for recombinant alpha plus recombinant gamma interferon (34,35). These _in vitro_ studies have shown considerable synergy and warrant a translation to the clinic. Already, combinations of alpha and gamma interferon, beta and alpha interferon, and beta and gamma interferon as well as gamma interferon and tumor necrosis factor are beginning. Unfortunately, all of these studies are being done using a fixed ratio combinations, a situation which does not fit the _in vitro_ antiproliferative activity of these agents which can vary considerably from cell line to cell line using interferons as test biologicals. In fact, in analyzing the data for _in vitro_ antiproliferative activities, the dose response curves indicate that each cell line has a different level of sensitivity to the antiproliferative effect of the individual interferons as well as individualistic response to the combinations. Thus, a more rational trial design would be to take the interferons to the clinic in rational combinations based on *in vitro* assays (36).

NHL, CTCL and MM represent excellent signal tumors for further testing of alpha, beta, and gamma interferon. Based on the studies to date as well as studies in animal models and *in vitro* assays, it is apparent that combinations of these agents and combinations of these agents with drugs and radiation are the wave of the future. The rational integration of these combinations in clinical practice will be difficult and will come much more slowly if they are studied agent by agent and in combination by fixed ratio after fixed ratio to try to make that integration. By contrast, improved *in vitro* assays to assess the individual and combined effects of biologicals, of interferons, other biologicals, chemotherapy and radiotherapy may lead to the design of individualistic and rational clinical trials where each individual patient may have his experimental opportunity maximized during the limited time he may have available for such experimental trials. The current design of escalating doses between patients guarantees that large numbers of patients on experimental protocols are getting subtherapeutic doses. Only by escalating doses within individual patients using dosage regimens based on some rational other than empiricism, can we design clinical trials to maximally benefit our patients with these disorders (36).

REFERENCES

1. Goeddel, D., Yelverton, E., Ulrich, A., et al. Nature (London) 287:411-416, 1980.
2. Oldham, R.K. Cancer Treat. Rep. 68:221-232, 1984.
3. Oldham, R.K. J. Cell. Physiol. Supplement. 4:91-99, 1986.
4. Oldham, R.K. and Smalley, R.V. J. Biol. Response Modif. 2:1-37, 1983.
5. Oldham, R.K. J. Natl. Cancer Inst. 70:789-796, 1983.
6. Cantell, K. *In*: Interferon I. (Eds., D. Burke, K. Cantell, E. De Maeyer, M. Landy, M. Revel, and J. Vilcek) New York Academic Press, 1981.
7. Oldham, R.K. J. Biol. Response Modif. 2:199-206, 1983.
8. Oldham, R.K. *In*: Interferon VI. (Eds. D. Burke, K. Cantell, I. Gresser, E. De Maeyer, M. Landy, M. Revel and J. Vilcek) Academic Press. pp. 127-143, 1985.
9. Oldham, R.K. and Smalley, R.V. *In*: Cancer Principles and Practice of Oncology. (Eds. DeVita, Hellman, and Rosenberg) J. B. Lippincott. pp. 2223-2245, 1985.
10. Oldham, R.K. Cancer Investigation. 3:53-70, 1985.
11. Oldham, R.K. J. Biol. Response Modif. 4:211-212,

1985.
12. Oldham, R.K. In: Interferons IV. (Eds. N.B. Finter and R.K. Oldham). Elsevier Science Publishers B.V. pp. 235-249, 1985.
13. Oldham, R.K. Hospital Practice. 20:72-91, 1985.
14. Maluish, A.E., Leavitt, R., Sherwin, S.A., Oldham R.K., Herberman R.B. J. Biol. Response Modif. 2:470-481, 1983.
15. Smalley, R.V. and Oldham, R.K. J. Biol. Response Modif. 2:401-408, 1983.
16. Smalley, R.V. and Oldham, R.K. Drugs Exptl. Clin. Res. 12:31-39, 1986.
17. Sherwin, S.A., Knost, J. A., Fein, S., Abrams, P.G., Foon, K.A., Ochs, J.J., Schoenberger, C., Maluish, A.W., and Oldham, R.K. J. Am. Med. Assoc. 248:2461-2466, 1982.
18. Foon, K.A., Maluish, A.E., Abrams, P.G., Wrightington, S. Stevenson, H.C., Alarif, A., Fer, M.F., Overton, W.R., Poole, M., Schnipper, E.F., Jaffe, E.S. and Herberman, R.B. Am. J. Med. 80:351-356, 1986.
19. Quesada, J.R., Gutterman, J.U. and Hersh, E.M. Cancer 57:1678-1680, 1986.
20. Talpaz, M., Kantarjian, H.M., McCredie, K., Trujillo, J.M., Keating, M.J. and Gutterman, J.U. 314(17):1065-1069, 1986.
21. O'Connell, M.J., Colgan, J.P., Oken, M.M., et al. J. Clin. Oncol. 4:128-136, 1985.
22. Mellstedt, H., Bjorkholm, M., Johansson, B. Lancet 1:245-248, 1979.
23. Gutterman, J.U., Blumenschein, G.R., Alexanian, R. Ann. Intern. Med. 93:399-406, 1980.
24. Mellstedt, H., Aahre, A., Bjorkholm, M. New York, Elsevier-North Holland, Inc. pp. 387, 391, 1982.
25. Cooper, R.M. Seminars in Oncol. vol. 13, No. 3, Suppl.2, pp. 13-20, 1986.
26. Merigan, T.C., Sikora, K., Bredden, J.H., et al. New Engl. J. Med. 299:1449-1453, 1978.
27. Louie, A.C., Gallagher, J.G., Sikora, K., et al. Blood. 58:712-718, 1981.
28. Horning, S.J., Merigan,T.C., Krown, S.E., et al. Cancer 56:1305-1310, 1985.
29. Leavitt, R.D., Kaplan, R., Ozer, H., et al. Blood, 64:182a, 1984.
30. Bunn, P.A., Foon, K.A., Ihde, D.C., et al. Ann. Med. Med. 101:484-487, 1984.
31. Estey, E., Hoth, D., Leyland,-Jones, B., et al. Proc. Am. Soc. Clin. Res. 3:35, 1984.
32. Hu, E. and Horning, S.J. J. Biol. Response Modif. 6(2), 1987, In press.
33. Bonnem, E.M and Oldham, R.K. J. Biol. Response Modif. 6(3), 1987, In press.
34. Czarniecki, L.W., Fennie, C.W., Powers, D.B., et al. J. Virol. 49:490-496, 1984.

35. Hubbell, H.R., Craft, J.A., Leibowitz, P.J. and Gillespie, D.H. J. Biol. Response Modif. 1987, In press.
36. Oldham, R.K.: Biotherapy: Principles and Practice. Raven Press, 1987. (In press)

9

THE ROLE OF INTERFERON ALPHA IN THE TREATMENT OF HAIRY CELL LEUKEMIA (HCL)

J.R. Quesada

Department of Clinical Immunology and Biological Therapy, M.D. Anderson Hospital and Tumor Institute, 6723 Bertner Avenue, Houston, Texas 77030

ABSTRACT

The results of treatment with alpha interferons in more than 90 patients with HCL, treated over the last four years are summarized. Interferon alpha has become an established effective therapy for HCL and has resulted in decreased morbidity and improved prognosis, for patients afflicted with this disorder.

INTRODUCTION

Until recently, the therapeutic options for patients with hairy cell leukemia (HCL) were very limited. The traditional primary therapy consisted of splenectomy because of the splenomegaly which frequently accompanies this disease. Long-term control of the disease after splenectomy was rare and improvement of the bone marrow leukemic infiltration seldomly occurred [1,2]. For patients who failed to respond to this procedure, and for those who relapsed, therapeutic option were not optimistic. The poor regenerative capacity of the bone marrow in patients with HCL precluded the use of chemotherapeutic agents. Although individual patients obtained very satisfactory responses, the infectious complications related to myelosuppression actually reduced survival [1].

The natural history of hairy cell leukemia (HCL) is variable, with most typical cases running a slow chronic course over a span of years. Occasionally, patients with HCL may harbor the disease indolently for several years, sustaining a well-balanced host-disease relationship that suggests that HCL may, in those instances, still be subject to normal cell growth regulatory mechanisms. Further, description of a spontaneous remission following infection indicates that certain stimulatory or inhibitory factors may govern the behavior of the malignant cells.

Hairy cell leukemia is a malignant disease of mononuclear cells which infiltrate predominantly the liver, spleen and the bone marrow, resulting in pancytopenia and immunologic deficiencies. The latter two abnormalities result in a high incidence of infections, which has been the main cause of morbidity and mortality in these patients [1,2].

The etiology of the disease or the precise mechanism(s) responsible for the pancytopenia have not been elucidated. A deficient

M. Revel (ed.), CLINICAL ASPECTS OF INTERFERONS.

bone marrow hematopoiesis secondary to infiltration by leukemic cells is partly responsible. However, because hematopoiesis is regulated by cells of the immune system, the immunologic deficits described in HCL may play a role in the defective bone marrow function [3]. Hematopoiesis is regulated by complex cellular interactions with mononuclear phagocytes, T lymphocytes, and stromal cells of the hematopoietic organs. T cells and macrophages have been described to produce colony-stimulating activity that sustained in vitro growth of granulocyte and monocyte colonies and is likely an important regulatory factor of hematopoiesis in vivo. It is unresolved whether the immunologic defects described in HCL play a significant role in the defective hematopoiesis of patients afflicted with this disorder, but they give rational grounds to attempt to restore or modulate some of these anomalies with a therapeutic intent [3]. Some experimental approaches have supported a role for biologic therapy of HCL. Leukapheresis and transfusion of mononuclear-enriched fractions from siblings have resulted in improvement of cytopenias. The removal of a large number of circulating malignant cells or putative cellular or humoral inhibitory influences could explain some of the beneficial effects obtained with leukapheresis. The transfusion of allogeneic donor mononuclear cell fraction has also resulted in hematologic improvement. Of interest has been the prolonged duration of response obtained with this method of treatment. To explain these effects, we have suggested favorable cellular interactions between the normal mononuclear cells and the viable hematopoietic cells resulting in production of colony stimulating activity or transfer of other molecular information [3].

ALPHA INTERFERONS IN HCL

The antitumor activity of alpha interferon (IFNα) demonstrated in malignancies of B cell origin (lymphoma, multiple myeloma) [4] lead to the initiation of a pilot study in patients with HCL. In this regard, the demonstration of immunoglobulin gene rearrangement on hairy cells had confirmed the B cell ineage of HCL [5]. Plasma cell-related antigens in the membrane of hairy cells place HCL close to the end of the maturation pathway of B lymphocytes [6]. However, hairy cells can express both monocyte-like characteristics and the receptor for interleukin-2 (IL-2) which suggest a quite unique stage of differentiation [5]. Interferon was also of interest because of its potentiating activities in several functions of the immune system [3]. We began in July 1982 treatment of selected HCL patients with a partially purified preparation of interferon alpha (IFNα). Of 7 patients treated with IFNα (Finnish Red Cross Blood Center) at a dose of 3 million units daily, 3 achieved complete remissions and 4 partial remissions [7]. In contrast to other modalities of treatment, interferon α resulted in a reduction of HCL infiltration in the bone marrow with attendant improvement or restoration of hematologic parameters in the peripheral blood. The response criteria was not limited to the peripheral blood parameters, but included the disappearance (complete remission) or more than 50% reduction (partial remission) of bone marrow HCL infiltrate.

The results of that preliminary study have now been extensively confirmed using purified recombinant alpha interferons (rIFNα) [8-11]. The results obtained with rIFNα excluded the involvement of lymphokines other than IFNα which are contained in the partially purified preparation. rIFNα in contrast to IFNα contains a single purified species of IFNα. The results in the first 30 patients treated at M.D. Anderson Hospital with rIFNα have been recently published [8]. Nine CR and 17 PR were obtained. The other four patients had hematologic improvement of clinical significance, but did not achieve remission status as defined. In this study, we included 7 patients with splenomegaly (and therefore candidates for splenectomy). The incidence of CR was significantly higher ($p = < 0.01$) in previously untreated patients (5 of 7) than in those in whom splenectomy had been performed (4 of 23).

None of 23 patients treated with IFNα relapsed while on treatment. Median duration of remission is exceeding 20 mo. It has become evident that treatment for less than 6 months may be insufficient because there are patients who continue to improve beyond this point and remissions have been obtained as late as 10 or 12 months after initiation of treatment. A small number of patients have relapsed while receiving rIFNα treatment, which was coincidental with the development of high titers of neutralizing antibodies to this agent (unpublished observations).

Discontinuation of treatment results in steady increase in the bone marrow leukemic infiltrate in the majority of patients. Approximately 20% of 42 patients sustain their remission status for periods ranging from 4+ to 24+ months (median 10+ mo) after discontinuation of treatment. Conversely, another 30% of the patients have required reinitiation of treatment due to progressive decline of the granulocyte count, which has proven to be the most sensitive indicator of disease relapse. Reinstitution of treatment has resulted in reinduction of remission in all patients completing 3 mo of treatment.

The optimal length of treatment is still unknown. Our data suggest that prolonged therapy beyond one year may be superior, and that further reduction of the leukemic infiltrate can be accomplished in some patients leading to larger number of patients achieving complete remission. For some individuals, periodic administration (every 6 to 12 months) may prove satisfactory in the long-term control of the disease. Nevertheless, there are patients who will sustain long-term remissions without further treatment.

Laboratory or clinical parameters to guide the length of therapy or its reinstitution must be sought and defined. Similarly, the optimal dose of alpha interferon requires better definition. There is no indication to use doses higher than 3 MU/day in the treatment of HCL. Actually, recent studies have demonstrated therapeutic activity in doses as low as 0.1 to 1 MU/day but follow up of these patients is still short to determine the quality of their responses [12]. It seems reasonable to suggest that maintenance therapy for two or more years at small, non-toxic doses may prove ideal.

The toxicity of low doses of alpha IFNs is minor except for unusually sensitive individuals in whom moderate to severe fatigue or fever may require dose or schedule modifications. In patients with severe neutropenia, IFN may further decrease the WBC and neutrophile count in the first 2-4 weeks of treatment. Approximately 20% of our patients have complicated their induction treatment with infections or with fever of unclear origin that merited prescription of antibiotics. Platelets rarely decreased in individuals with pretherapy counts above $50,000/mm^3$. Two patients who received IFNα concomitantly with a triple-drug antibiotic treatment (Isoniazid, Ethambutol, Ritampicin) for mycobacterial infections developed a distal neuropathy of considerable degree. We suspected an adverse drug interaction with the antibiotics to explain this rare complication.

We have speculated on the mechanisms responsible for the antitumor activity of interferon α and the ensuing benefit in the hematologic parameters in these patients [7]. A direct antitumor effect on the malignant cell is possible because interferon has been shown to have potent antiproliferative effect against lymphoma cell lines in vitro and antitumor effects in patients with malignant lymphomas. Induction of differentiation of the malignant hairy cells remains as an intriguing possibility because interferon α is known to induce differentiation of erythroleukemia and myeloid-leukemia cell lines in vitro.

Participation of the immune system cannot be dismissed. Restoration of cell mediated immunity and natural killer cells activity occurs after treatment with interferon has been reported. Whether interferon can overcome some of the immunologic deficits in patients with HCL and, in turn, favor the sustenance of remission has not yet been adequately studied.

Endogenous Alpha Interferons in HCL

The effects of IFNα in patients with HCL are somehow reminiscent of the results obtained when a deficient hormone or vitamin is replaced with exogenous products. Recently, we have studied 19 patients in different stages of disease activity and remission for production of IFNα by peripheral blood mononuclear cells (MNC) [14]. Among the different groups of patients, a severe deficiency in IFNα production was identified in eight of eight patients with active disease and in six of six patients who had achieved PR after treatment with IFNα or rIFNα, and had treatment discontinued for a minimum of 6 months. These patients were deficient despite normal immunophenotype of circulating MNC. In contrast, 5 patients who achieved CR after IFNα treatment, splenectomy or infection produced amounts of IFNα similar to those of the control population. These data suggest that a deficiency of IFNα production may be involved in the physiopathogenesis of HCL; that endogenous production of IFNα may bear relevance to the induction and sustenance of remission after splenectomy, or "spontaneous" remissions and, that relapse after induction of PR may be associated to an incomplete restoration of endogenous production of IFNα. The nature of the defect requires further study.

CONCLUSIONS

Interferon has become in a relative short period of time an established treatment for patients with HCL. The U.S. Food and Drug Administration conferred approval in April 1986 simultaneously with several other countries. More recently, deoxycoformycin in low doses has proven to be also very effective treatment for this rare type of leukemia [15]. Moreover, splenectomy may also result in good long-term control of the disease without additional systemic therapy. The particular indications for each of these therapeutic choices require further experience and identification of prognostic factors or biologic indicators to assist in the clinical decision. Table I presents a preliminary set of indications and relative contraindications to each of treatment strategies, including observation without therapy. I continue to follow approximately 12 patients in whom no therapy has been required for months to years after either splenectomy or spontaneous remission of the disease. The indications and relative contraindications for splenectomy in newly diagnosed patients are derived, in part from the data of Golomb and colleagues [2] who have best studied the outcome of this procedure.

The results with IFNα have already decreased significantly the morbidity and projections indicate also a considerable impact on survival. Conceivably, a judicious combination of these therapeutic options may provide optimal therapy and a potential cure. Furthermore, HCL offers a suitable human model of disease from which further understanding of the mode of action of IFNα in the regulation of cell growth, cell differentiation and modulation of hematopoiesis can be learned.

REFERENCES

1. Flandrin, G., Sigaux, F., Sebahoun, G., and Bouffette, P. Sem. Oncol. 11 (Suppl. 2):458-471, 1984.
2. Golomb, H.M., Ratain, M.J., and Vardiman, V.W. in DeVita, V.T., Hellman, S., and Rosenberg, S.A. eds: Advances in Oncology 1986. J.B. Lippincott Philadelphia pp. 311-322.
3. Quesada, J.R., Hersh, E.M., and Gutterman, J.U. Sem. Oncol. 11 (Suppl. 2):507-510, 1984.
4. Gutterman, J.U., Blumenschein, G.R., Alexanian, R., et al. Ann. Intern. Med. 93:399, 1980.
5. Korsmeyer, S.J., Green, W.C., Cossman, J., et al. Proc. Natl. Acad. Sci. 80:4522-4526, 1983.
6. Anderson, K.C., Boyd, A.W., Fisher, D.C., et al. Blood 65:620, 1985.
7. Quesada, J.R., Reuben, J.M., Manning, J.T., et al. N. Engl. J. Med. 310:15-18, 1984.
8. Quesada, J.R., Hersh, E.M., Manning, J., et al. Blood 68:493-497, 1986.
9. Ratain, M.V., Golomb, H.M., Vardiman, J.W., et al. Blood 65:644-648, 1985.
10. Jacobs, A.D., Chaplin, R.E., Golde, D.W. Blood 65:1017-1020, 1985.
11. Flandrin, G., Siguax, F., Castaigne, S., et al. Blood 67:817, 1986.
12. Gastl, G., Denz, H., Abbrederis, C., et al. Onkologie 8:143-144, 1985.
13. Foon, K.A., Maluish, A.E., Abrams, P.G., et al. Am. J. Med. 80:351, 1980.
14. Lepe-Zuniga, J.L., Quesada, J.R., et al. Blood 68 (Suppl. 1):202a, 1986.
15. Spiers, A.S.D., Parekh, S.J., and Bishop, M.B. J. Clin. Oncol. 2(12):1336, 1984.

TABLE I

TREATMENT	RELATIVE INDICATIONS	RELATIVE CONTRAINDICATIONS
OBSERVATION	BM cellularity < 50%. 1-2 cytopenias.	Repeated infections. Blood products dependency.
SPLENECTOMY	Massive splenomegaly. Hypersplenism. Spleen rupture. Spleen infarct.	Normal size or minimally enlarged spleen. Splenomegaly with BM cellularity > 85%.
ALPHA INTERFERON	Splenectomy relapse. Splenomegaly with BM cellularity > 85%. Active infections. Normal size or minimally enlarged spleen.	Antibodies to IFNα. Cost ($10.00/MU).
DEOXY-COFORMYCIN	IFNα failures. Antibodies to IFNα.	Controlled drug. Requires participation in cancer centers. Renal dysfunction. Active infections.
RADIOTHERAPY	Bone lesions.	

BM = bone marrow

10

CLINICAL STUDIES OF ALPHA AND GAMMA INTERFERONS IN CHRONIC MYELOGENOUS LEUKEMIA

Moshe Talpaz,[*+] Razelle Kurzrock,[*+] Hagop Kantarjian,[+] Jordan U. Gutterman[*]

THE DEPARTMENTS OF CLINICAL IMMUNOLOGY[*] AND HEMATOLOGY[+], THE UNIVERSITY OF TEXAS M.D. ANDERSON HOSPITAL AND TUMOR INSTITUTE, 1515 Holcombe Blvd., Houston, Texas 77030, USA

ABSTRACT

Since 1981, we have conducted a series of clinical trials with interferons for the treatment of Philadelphia chromosome-positive chronic myelogenous leukemia (CML Ph'). The interferons examined are partially pure alpha interferon, followed by recombinant alpha A interferon (Roche), recombinant gamma interferon, and most recently, combinations of alpha and gamma interferons. Here we summarize what we consider to be the major concepts emerging from those trials.

INTRODUCTION

In patients with chronic myelogenous leukemia that is Philadelphia chromosome-positive (CML Ph'), the disease is characterized by a multistep evolutionary course starting with a benign phase lasting on average of 3.5 years and terminating in a blastic crisis. Chemotherapy directed to altering the relentless course of this disease has been unsuccessful, although supralethal chemotherapy or total body irradiation or both followed by allogeneic bone marrow transplantation turned out to be curative in a small number of patients (1,2).

Recent evidence indicates that most patients with CML in the early phase have a reservoir of normal stem cells (3,4). The presence of this putative reservoir was further demonstrated in CML patients by restoration of normal stem cell growth with aggressive

M. Revel (ed.), CLINICAL ASPECTS OF INTERFERONS. Copyright 1988.

chemotherapy, an effect that was, however, invariably transient (5). Therapy objectives may therefore diverge from merely controlling the excessive myeloid proliferation in CML for trying to restore the suppressed normal stem cells.

The interferons, by virtue of their ability to restore stable, normal stem cell growth in the bone marrow of a small number of CML Ph' patients, may provide the basis for rational combined therapy (chemotherapy and biologic therapy) for this disease (6). In this chapter, we review evolution of the therapeutic trials with interferons in CML patients and discuss potential further directions of biologic therapies.

RESULTS:

ALPHA INTERFERONS IN EARLY BENIGN-PHASE CML

Partially pure alpha interferon (IFN-α) (Finnish Red Cross, Helsinki, Finland) was first demonstrated to have considerable cytoreductive activity in a pilot study of 7 patients with CML Ph' in benign phase (7). The study was then expanded to include 51 patients who had not been treated or who had undergone single-agent chemotherapy for fewer than six months (8). The patients received daily intramuscular injections of IFN-α at doses ranging from 3 to 9 x10^6 units (u) during the induction phase, the quality of the patient's response and the tolerance to the drug then dictating the maintenance dosage ranging from 3 x 10^6 u to 9 x 10^6 u daily. Of the 51 patients, 36 (71%) achieved complete hematologic remissions, (CR) (criteria for response shown in Table 1)

Table 1. Response Criteria for Interferon Therapy in Chronic Myelogenous Leukemia*

Hematologic Remission Status	WBC Decline	Splenomegaly
Complete	$<9 \times 10^3$ /μl**	None
Partial	$\geq$50% to $\leq 20 \times 10^3$ /μl and/or —>	Persistence

* Treatment failure was defined as > partial Hematologic remission; **With normal differential

five patients (10%) had partial hematologic remissions,and 10 patients (20%) failed to respond (Table 2).

Table 2. Results of Alpha Interferon Therapy in 5l patients with Benign-Phase Chronic Myelogenous Leukemia

Number of Patients in Risk Category* (Percentage)

Category	Total		Low		Intermediate		High	
Complete hematologic remission	36 (71)		19 (83)		9 (60)		7 (64)	
Philadelphia chromosome Status								
100%	16 (31)	40%	5 (22)	61%	4 (27)	33%	6 (55)	9%
35% - 95%	14 (27)		10 (43)		3 (20)		1 (9)	
5% - 34%	4 (8)		2 (9)		2 (13)		-	
0%	2 (4)		2. (9)		-		-	
Partial hematologic remission	5 (I0)		2 (9)		3 (20)		-	
Resistant disease	10 (20)		2 (9)		3 (20)		4 (66)	
	51 (100)		23 (100)		15 (100)		11 (100)	

* Pretreatment characteristics to determine risk category were not available for two patients:One had resistant disease and died in accelerated phase, and one achieved complete hematologc remission but no suppression of the Philadelphia chromosome. (Reprinted with permission from ref. 12).

The application of a prognostic model for CML Ph' developed at our institute (9) was useful for selecting patient groups with a high response rate to IFN-α. Not surprisingly, patients with a predicted favorable outcome of chemotherapy the low-risk group fared better with IFN-α than did the intermediate and high-risk groups.

Varying degrees of Ph' clone suppression and partial restoration of normal karyotype cell growth were observed in 20 (39%) of the 51 patients, or 55% of the patients in CR. This phenomenon, too, was more commonly observed in the low-risk group (Table 2). In contrast to the frequent, though transient, cytogenetic responses to aggressive chemotherapy (10), the cytogenetic response in IFN-α was durable in two thirds of the responding patients and lasted for a median of 10+ months (range 3+ to 30+ months). Complete suppression of the ph' cells on more than one examination was seen in two patients.

Patients receiving IFN-α experienced a median of more than three years of remission. Disease relapses were observed, however, beginning with the second year of therapy. Most patients relapsed during a benign phase in which they were specifically resistant to IFN-α, but they were still exquisitely sensitive to hydroxyurea. In two such patients, we studied the levels of IFN-α receptors in the CML cells and induction of the enzymes 2'5' oligoisoadenylate synthetase (2'5'A) during the course of therapy (Fig. 1). A decline in the induction of 2'5'A was seen in these two patients during the resistant phase without a change in IFN-α binding to cellular receptor. We postulated, therefore, that cellular changes that occur beyond receptor binding probably confer secondary resistance to IFN-α in the CML cells (11).

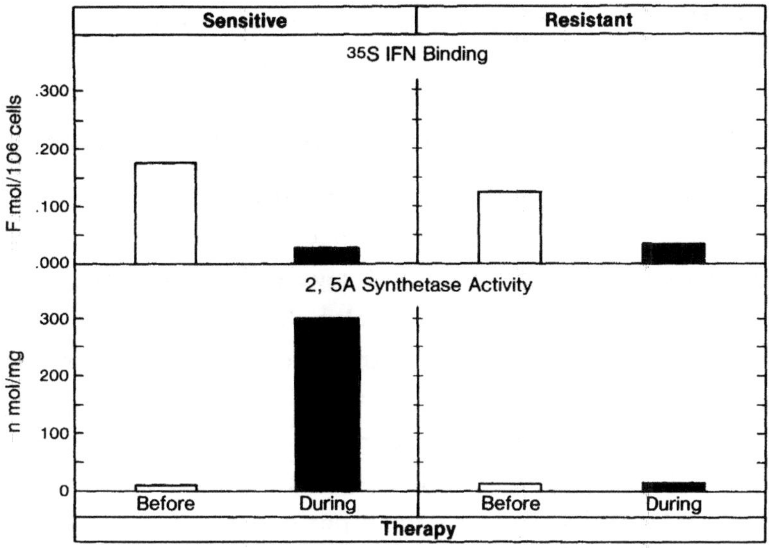

Figure 1. Binding of ^{35}S-labeled rIFN-αA and 2',5'-oligoisoadenylade synthelase (2'5'A) activity in peripheral blood cells isolated from a patient with CML in both IFN-sensitive and - resistant stages of the disease. ☐ levels of ^{35}S-labeled IFN binding or 2'5'A activity before therapy; ■ levels during IFN therapy (Reprinted with permission from ref. 11).

Thirteen patients developed blastic crises during the median follow-up period of 3.2 years. Unlike the typically more frequent occurrence of myeloid blast crisis than lymphoid blastic crisis, we observed the reverse phenomenon among patients treated with IFN-α, two third of these patients experiencing lymphoid blast crisis. Blastic crisis, uncommon among patients who continued to respond to IFN-α, occurred only in three such patients and was lymphoid in all three.

Seventy-six percent of the patients are projected to be alive three years after the initiation of therapy. Analysis of survival within the various risk groups demonstrated a 95% projected three year survival in the low-risk group, compared with 80% and 26% for the intermediate and high-risk groups respectively ($p<0.001$).

EFFECT OF IFN-α IN PATIENTS WITH LONG-STANDING CML AND THROMBOCYTOSIS

To examine the effect of IFN-α in patients with CML Ph' complicated by thrombocytosis, we studied nine patients with either long-standing or accelerated CML and severe thrombocytosis ($>1\times10^6$ platelets/μl). Most of these patients also demonstrated clinical sequelae to the thrombocytosis with thromboembolic or hemorrhagic complications (12).

Daily intramuscular administration of 9×10^6 units of IFN-α caused a rapid decrease in the nine patients' high platelet count (Table 3) but failed to suppress their peripheral white blood cells. We concluded that patients with long-standing or accelerated CML have a different clinical response to INF-α than those with early-stage CML, and that platelet control by IFN-α is obtained by a different mechanism than the control of excessive granulocytes.

Table 3. Clinical Course of Patients with Chronic Myelogenous Leukemia and Thrombocytosis Treated with Alpha Interferon

Patient	Changes in Platelet Counts		Changes in Leukocyte Counts		Duration of Therapy
	Pretherapy	End of Ther.	Pretherapy	End of Ther.	
	$\times 10^6$ /μl		$\times 10^3$ /μl		Days
1	1.66	0.994	11.2	42.98	71
2	1.28	0.628	48.2	61.9	19
3	1.18	0.257	18.4	256.0	15
4	1.39	0.245	55.4	164.0	15
5	2.79	0.573	6.1	26.2	85
6	1.90	0.465	8.4	11.6	>300
7	1.345	0.295	28.4	17.0	>143
8	2.295	0.7	48.7	70.0	35
9	1.548	0.5	34.3	56.1	23
Mean±SE	1.71±0.53	0.52±0.24	28.8±18.9	78.7±0.9	...

CLINICAL TRIAL OF RECOMBINANT HUMAN INTERFERON ALPHA

As highly purified interferons produced by genetic techniques became available, we conducted a study with recombinant alpha A interferon (rIFN-α A) (Roferon® Roche, Nutley, NJ). Preliminary results of this study (6) demonstrated that of 17 patients with minimally treated, early-stage, benign-phase CML Ph',12 achieved complete and two achieved partial hematologic remissions. These findings essentially confirmed studies with partially pure IFN-α, further demonstrating that cytoreductive activity is an inherent characteristic of the IFN-α molecule, and that a single species of IFN-α is, at similar doses, as active in CML as the multispecies IFN-α. The study was then expanded to include patients with long-standing disease of two years or more and again demonstrated that, in this subset of patients, the effect of IFN-α is only marginal.

Perhaps the most significant finding of this study was the induction of durable, complete cytogenetic remissions, which were seen in four of the first 17 patients for periods ranging from more than six to more than 21 months. Thus, IFN-α administered alone can completely suppress the Ph' clone and restore the growth of presumably "normal" cells. The achievement of "true" durable, complete remission in CML was a novel finding.

To further confirm the complete cytogenetic remissions, cells from two of these patients were also studied with a molecular probe specific for the chromosome 22 breakage cluster region (bcr). Since more than 90% of the patients who have CML display the Ph' chromosome as the result of a 9:22 translocation, this constitutes a hallmark of the disease. The c-abl oncogene, located on chromosome 9, and the bcr gene, located on chromosome 22, are juxtaposed in CML Ph' (13,14), producing a bcr/abl chimeric mRNA transcript (15) and a chimeric fusion protein (16). With the introduction of molecular probes capable of identifying this alteration, molecular analysis, along with cytogenetic analysis, may be used to monitor IFN therapies, intensive chemotherapy, and other treatment modalities throughout the disease course in patients with CML. We used southern blot analyses, with a 3' bcr probe (Oncogene Science, Mineola, Long Island, NY) to confirm the complete suppression of the Ph' chromosome at the DNA level. Complete disappearance of rearranged restriction fragments of the bcr gene, which were a characteristic of the disease before rIFN α-A therapy, was accompanied by the restoration of normal bone marrow and achievement of durable, complete cytogenetic remissions for 21 and 15 months, respectively, in two patients with CML Ph' CML (Fig. 2) (17).

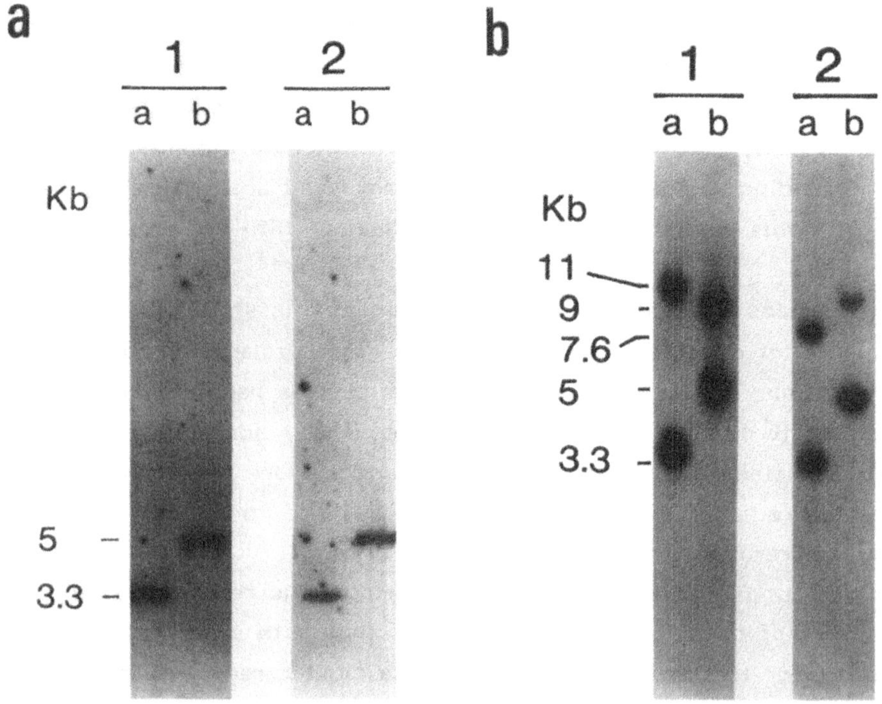

Fig. 2. a) Southern blot analysis demonstrating the normal (3.3-kb) BamHI (lane a) and the 5-kb BglII (lane b) hybridization fragments as well as the rearranged bands in two patients with Philadelphia chromosome-positive chronic myelogenous leukemia before therapy. b) In same patients after more than six months of recombinant alpha A interferon therapy, only the normal hybridization fragments can be detected, a finding that coincided with cytogenetic remission. (Reprinted with permission from ref. 17).

EFFECTS OF ALPHA INTERFERONS ON CYTOGENETIC CLONAL EVOLUTION

One of the characteristics of CML is the evolution of new cell clones that bear multiple chromosomal changes in addition to the Ph' translocation. These changes imply an impending blastic transformation. We were able to demonstrate durable suppression of these secondary clones in several patients, having initially observed and described it in two patients (18), and having since then seen the mechanism in five other patients with the use of either IFN-α or rIFN-α A. An example for such a suppression is shown in Table 4 (19).

Table 4. Philadelphia Chromosome-Positive Chronic Myelogenous Leukemia Patients with Cytogenetic Clonal Evolution: Hematologic and Cytogenetic Changes during the Course of Alpha Interferon

Peripheral Blood			Bone Marrow Cytogenetics		
Date	IFN Dose*	WBC**	Immature Cells	% of Primary Ph' Clone§	% of Secondary Ph' Clone¶
12/82	9	116.0	14	100	0
3/83	3-6	16.3	42	100	0
5/83	6-9	10.2	30	100	10
11/83	9	17.1	39	100	10
2/84	6	13.8	21	100	45
6/84	9	6.8	22	no mitotic figures	...
11/84	9	6.9	9	100	0
4/85	..	8.0	7	100	0
9/85	..	8.9	7	100	5
1/86	..	6.6	9	insufficient metaphases	...
6/86	9	3.6	3	100	0

* x 10^6 u/day; ** x 10^3 /µl;
§ 46,XY,t[q+;22q-]+22q-; ¶ 48,XY,+8,t[9q+;22q-]+22q-
*** In 16b metaphases.

TOXICITY OF ALPHA INTERFERON

The patients' toxic reactions were similar with the use of both the partially pure and the recombinant alpha interferons.

Neurologic toxic reactions were the most significant, and although most of the patients maintained near-normal performance status, occasionally a patient had to be taken off the study because of neurologic problems. One severe, acute, treatment-limiting toxicity was the development of severe musculoskeletal pain in 12% of the patients receiving IFN-α. This toxic effect seems to be unique to CML patients treated with alpha interferons and coincided with lack of response. Thus, the reaction did not lead to treatment interruption in patients responding to therapy.

STUDIES WITH GAMMA INTERFERON

Although gamma interferon (IFN-γ) is a molecule structurally distinct from IFN-α, this agent suppresses hemopoietic progenitor cells in vitro in a fashion similar to that of IFN-α (20,21). Therefore we began a study of rIFN-γ (Genentech, South San Francisco, CA) at doses of 0.25-0.5 $mg/m^2/day$ administered intramuscularly ($5-10x10^6$ μ/m^2 respectively) in patients with benign-phase CML Ph'. Twenty-six of 30 patients who entered the study were evaluable. Six patients have achieved a complete and four a partial hematologic response (Table 5).

Table 5. Changes in Hematologic Parameters among Philadelphia Chromosome-Positive Chronic Myelogenous Leukemia Patients Responding to Recombinant Gamma Interferon

	Type of Response	Pretherapy Median (range)	Posttherapy* Median (range)
White Blood Cells ($x10^3$ / μl)	PR	80 (25-99)	14.5 (10-19)
	CR	54 (14-98)	3.5 (2.0-7.3)
Platelets ($x10^3$ / μl)	PR	660 (191-848)	300 (231-580)
	CR	495 (322-857)	290 (75-250)

Abbreviations: PR partial remission; CR complete remission;
* at time of maximal response.

Currently, the median follow-up period of patients who are in complete remission is 7.5 months (range 5 to 12 months). No relapses have occurred among the complete responders and, so far, five patients have experienced cytogenetic improvement with emergence of 5% to 45% diploid cells in the bone marrow. Fever and flu-like symptoms were the most common side effects, but patients often developed partial tolerance after about a week. The majority of patients tolerated the therapy with little change in their performance status.

DISCUSSION

In a series of studies began in 1981, we demonstrated that the interferons (alpha and gamma) have considerable antitumor activity in CML Ph'. Highlights of these studies were the observations that IFN-α induces hematologic remission in about 70% of patients with early-benign phase CML, and that about 50% of the responders also achieve varying degrees of cytogenetic response, which was stable in two-thirds of the cytogenetically responding patients.

The effect of IFN-α on the outcome of CML cannot be assessed because of the brief follow-up period, but emerging patterns include a higher activity of IFN-α in CML patients with favorable prognoses, those in a biologically early disease stage; a clinical course that compares favorably with that of the commonly used chemotherapeutic agents; and the emergence of a subset of patients who might experience a sustained, complete cytogenetic and clinical response.

Alpha interferon seems to be active in suppressing the malignant evolution in the myeloid compartment, this is manifested by the low incidence of myeloid blastic crisis in CML. Moreover, therapy with IFN-α seems to fail to control the pathways leading to lymphoid blastic crisis. The latter occurs despite clinical response to IFN-α, perhaps as the result of a more primitive stem cell involvement with the disease process.

Complete cytogenetic response should become the major objective of therapy with interferons and several concepts should be contemplated in the future in an effort to maximize the cytogenetic response:

1. Establishing a system to study *in vitro* the existence of normal stem cells in CML patients and to evaluate *in vitro* various combinations that will provide a differential growth advantage of the "normal" karyotype stem cells over the Ph' clone.
2. Directing treatment to patients who have adequate "reservoirs" of normal stem cells.
3. Focusing clinical work on examinining the role of high-dose therapy ($\geq 10 \times 10^6$ units/m^2 daily) with IFN-α in an effort to maximize the cytogenetic response.
4. Maximizing the activity of IFN-γ in inducing a cytogenetic response by administering it after a cytoreductive course of IFN-α ,or after a cytoreductive course of chemotherapy. The two interferons might also be administered concomitantly.
5. Expanding understanding of biochemical mechanisms that confer cellular sensitivity or resistance to IFN to help design therapy that circumvents this resistance and augments IFN activity.

REFERENCES

1. Speck B., Bortin M.M., Champlin R., Goldman J.M., Herzig R.H., McGlave P.B., Messner H.A., Weiner R.S. and Rimm, A.A. Lancet 1: 665-668, 1984.
2. McGlave, P.B., Arthur D.C., Weisdorf D., Kim T., Goldman A., Hurd D.D., Ramsay N.K.C. and Kersey J.H. Blood 63: 219-222, 1984.
3. McGlave P., Mamus S., Vilen B. and Dewald G. Exp. Haematology (in press).
4. Coulombel L., Kalousek D.K., Eaves C.J., Gupta C.M. and Eaves A.C. N. Engl. J. Med. 308: 1493-8, 1983.
5. Singer J.W., Arlin Z.A., Najfield V., Adamson J.W., Kempin S.J., Clarkson B.D. and Fialkow P.J. Blood 56: 356-360, 1980.

6. Talpaz M., Kantarjian H.M., McCredie K., Trujillo J.M., Keating M.J. and Gutterman J.U. N. Engl. J. Med. 314: 1065-1069, 1986.
7. Talpaz M., McCredie K.B., Mavligit G.M. and Gutterman J.U. Blood 62: 689-692, 1983.
8. Talpaz M., Kantarjian H.M., McCredie K.B., Keating M.J., Trujillo J. and Gutterman J.U. Blood 69: 1280-1288, 1987.
9. Kantarjian H.M., Smith T.L., McCredie K.B., Keating M.J., Walters R.S., Talpaz M., Hester J.P., Bligham G., Gehan E. and Freireich E.J. Blood 66: 1326-1335, 1985.
10. Kantarjian H.M., Vellekoop L., McCredie K.B., Keating M.J., Hester J., Smith T., Barlogie B., Trujillo J. and Freireich E.J. J Clin Oncol 3: 192-200, 1985.
11. Rosenblum M., Maxwell B., Talpaz M., Kelleher P., McCredie K. and Gutterman J. Cancer Res 46: 4848-4852, 1986.
12. Talpaz M., Mavligit G., Keating M., Walters R.S. and Gutterman J.U. Ann Intern Med 99: 789-792, 1983.
13. Heisterkamp N., Stephenson J.R., Groffen J., Hansen P.F., de Klein A., Bartram C.R. and Grosveld G. Nature 306: 239, 1983.
14. Groffen J., Stephenson J.R., Heisterkamp N., de Klein A., Bartram C.R. and Grosveld G. Cell 36: 93-99, 1984.
15. Sam K., Heisterkamp N., Grosveld G., de Klein A., Verma R., Coleman M., Dosik H. and Groffen J. N Engl J Med 313: 1429-1432, 1985.
16. Kloetzer W., Kurzrock R., Smith L., Talpaz M., Spiller M., Gutterman J. and Arlinghaus R. Virology 140: 230-238, 1985.
17. Yoffe G., Blick M., Kantarjian H., Gutterman J., Spitzer G. and Talpaz M. Blood 69:961-963, 1987.
18. Talpaz M., Trujillo J.M., Hittelman A.W.N., Keating M.J. and Gutterman J.U. Br J Hematol 60: 619-624, 1985.
19. Talpaz M., Kantarjian H., Trujillo J. and Gutterman J.U. Cancer Bulletin 38: 267-268, 1986.
20. Klimpel G.R., Fleischmann Jr. W.R. and Klimpel K.D. J Immunol 129: 76-79, 1982.
21. Raefsky E.L., Platanias L.C., Zoumbos N.C. and Young N.S. J Immunol 135: 2507-2512, 1982.
22. Kurzrock R., Talpaz M., Kantarjian H., Saks S. and Gutterman J.U. Blood 68 (suppl. 1): 225a, abstract, 1986.

11

INTERFERON THERAPY OF CHRONIC LYMPHOCYTIC LEUKEMIA AND VIRAL INFECTIONS IN LEUKEMIA PATIENTS

KENNETH A. FOON, M.D. AND ROBERT PETER GALE, M.D., Ph.D.

Department of Medicine, Division of Hematology and Oncology, University of Michigan, Ann Arbor, Michigan and the Department of Medicine, Division of Hematology and Oncology, and the Transplantation Biology Unit, University of California, Los Angeles

ABSTRACT

Chronic lymphocytic leukemia is a hematologic malignancy characterized by proliferation and accumulation of relatively mature appearing lymphocytes. Most cases involve a clonal proliferation of B-lymphocytes. The disease is usually stable over months to years and transformation to a more aggressive disease state only rarely occur. More typically the disease slowly evolves to a point where it suppresses the production of normal blood cells leading to complications of anemia and infection. In the later stages of the disease it is very resistant to therapy and new therapies are needed. Crude and purified alpha interferon preparations have not been active in patients with advanced forms of the disease. However, preliminary results do suggest that low-doses of interferon in early stages of the disease may be efficacious. Interferon alpha has not yet been shown to have a role in the treatment of viral infections associated with leukemia.

INTRODUCTION

Interferon was the term originally applied to a soluble factor that was recognized by it's ability to induce interference against viral infection of chick chorio-allantoic membrane by influenza A virus (1). It has subsequently been shown to be a family of closely related proteins and glycoproteins which in addition to antiviral activity, are potent regulators of cellular function and structure and possess direct anti-proliferative activities. These latter properties underly the current interest in interferon as an anti-cancer agent.

Three major species of human interferon are recognized and are designated interferon-α, interferon-β, and interferon-γ (2). Interferon-α is produced by leukocytes (B cells, T cells and macrophages) upon exposure to B cell mitogens, viruses, foreign cells or tumor cells. Interferon-β is produced by fibroblasts upon exposure to viruses or foreign nucleic acids. Interferon-γ is produced by T-lymphocytes upon stimulation with T-cell mitogens, specific antigens, or interleukin-2 (3). By use of recombinant DNA techniques, complete nucleotide sequences for interferon α, β and γ have been defined and amino acids sequences derived.

The genes recognized to code for interferon-α have been assigned to chromosome 9 (4). Sixteen distinct sequences for interferon-α have been described. Each is approximately 166 amino acids in length with an additional 20 amino acid secretory peptides present on the amino-terminal end. The human genes differ by approximately 10% in nucleotide sequence and 15-20% in amino acid sequence (5). Two recombinant human interferons, αA and αD comprise over 60% of interferons present after buffy coats stimulation and have been extensively studied (6). While they possess different antiviral and antiproliferative activity in vitro, similar in vivo effects on immune effector cells have been observed (6). The interferon-α used in the first human clinical trials was obtained from Sendai virus stimulated buffy coat leukocytes and represented 1% purity (10^6 units/mg protein) (7). Refinement in purification methods by use of high performance liquid chromatography, 2 dimensional polyacrylamide gel electrophoresis and immuno-affinity chromatography has allowed purification to homogeneity (10^8 units/mg protein) (8-10). The use of recombinant DNA techniques with splicing of the interferon-α gene into E. coli has further allowed for pure single species interferon-α in larger quantities.

Industrial scale production of interferon-β and γ has only recently been accomplished and clinical trials are limited in number. On the other hand, interferon-α has been extensively studied for the past decade in both basic science and clinical research and it is among the most potent biologic agents ever administered to man. While antitumor activity has been seen both in vitro and in vivo and in some solid malignancies (breast cancer, renal cell cancer, Kaposi's sarcoma, bladder cancer, ovarian cancer and melanoma) (11,12), the most impressive responses have occurred in the hematologic malignancies (13). A summary of the results for patients with chronic lymphocytic leukemia are presented.

CHRONIC LYMPHOCYTIC LEUKEMIA

Chronic lymphocytic leukemia is a hematologic malignancy characterized by proliferation and accumulation of relatively mature-appearing lymphocytes (14). Most cases involve a clonal proliferation of B-lymphocytes. Chronic lymphocytic leukemia typically occurs in persons over 50 years (median age, 60 years) and effects males more than females at a ratio of 2:1 (14). The disease is usually stable over months to years but transformation to a more aggressive disease state does occur. Alkylating agents, radiation therapy, and corticosteroids are commonly used to treat patients, although few data show that survival is substantially improved. Crude and purified interferon-α preparations were reported to be moderately active in patients with advanced chronic lymphocytic leukemia (15-17), but this was controversial (18-20). In our phase II trial at the National Cancer Institute 19 patients were treated with both high (50×10^6 units per square meter of body surface area intramuscularly) and low dose (5×10^6 units per square meter of body surface area intramuscularly) recombinant interferon-α (Roferon®) 3 times weekly (20). Eighteen patients were evaluable for response and only 2 brief partial responses were seen (Table 1). Five patients appeared to have an acceleration of disease while receiving recombinant interferon-α. This finding is in marked contrast to responses in patients with chemotherapy refractory low grade non-Hodgkin's lymphoma (21), cutaneous T cell lymphoma (22), and hairy cell leukemia (23). In one preliminary report, 8 patients with untreated early stage disease were treated with recombinant interferon-α, 2 X 10^6 units per square meter three times weekly (24). An objective response was described in all patients suggesting that interferon-α might be useful in previously untreated patients with low tumor burden.

Table 1. Results of Interferon-α Trials of Chronic Lymphocytic Leukemia

	Number Patients	Complete Response	Partial Response	Percent Response	References
NCI Study	18	0	2	11	20
Published literature	39	0	6	15	15-19

TOXICITY OF INTERFERON

The major dose-limiting toxicity of interferon was a flu-like syndrome consisting of malaise, fatigue, anorexia, and in rare cases, mental confusion. These were clearly dose-dependent toxic effects, since they were relieved by a dose reduction. In our National Cancer Institute studies the average duration of therapy was 4 weeks at the 100% dose of 50 x 10^6 units per square meter of body surface area and 9 weeks at 50% dose. Fatigue was the most common reason for dose reduction and all except one patient required at least 50% reduction. Febrile responses were noted in all patients but were almost always most severe after the first dose and were rarely a problem after the first week of therapy. The peak temperatures ranged from 38-42°C. Hematologic toxicity was not dose limiting.

Patients with chronic lymphocytic leukemia had an unexpected toxicity with 7 of the 19 patients demonstrating reactivation of perioral herpes simplex lesions during treatment; two of sufficient severity to require discontinuation of interferon therapy and treatment with acyclovir (14).

VIRAL INFECTIONS

Minimal data exists regarding the effects of interferon in leukemia patients with either herpes infections or cytomegalovirus. In our experience with CLL patients who were treated with recombinant interferon-α (20), one third of the patients developed severe oral herpes simplex virus infections. In a number of cases, we were forced to stop the interferon and treat these patients with acyclovir. In another study where interferon-α was given to patients prior to surgical decompression of the trigeminal nerve, it appeared to precipitate and accentuate herpes simplex virus infections (25). In a recent report, treatment with 3×10^6 units three times weekly of recombinant interferon-α resulted in moderate suppression and decreased duration of infection in patients with recurrent genital herpes simplex virus infections (26). Interferon-α in combination with acyclovir ointment was effective in healing herpetic dendritic keratitis (27). These patients, of course, did not have leukemia and were not immunosuppressed. In a case report, an acute myeloblastic leukemia patient treated with interferon-α for a severe cutaneous herpes simplex viral infection had a dramatic recovery (28).

There has not been extensive experience with interferon-α treatment of established cytomegalovirus in leukemia patients, however, there has been experience with interferon-α in bone marrow and renal transplantation. These patients are profoundly immunosuppressed and have a high incidence of cytomegalovirus infections. Prophylactic interferon-α therapy (3×10^6 units three times weekly for six weeks) prior to renal transplantation delayed cytomegalovirus excretion (29) and was an effective prophylaxis against serious cytomegalovirus infections (30). In contrast, interferon-α plus high-dose acyclovir was ineffective against cytomegaloviral pneumonia after bone marrow transplantation (31). In a phase I trial, treatment of marrow transplant patients with recombinant interferon-α did not prove to be efficacious with all four patients dying of cytomegalovirus pneumonia (32).

DISCUSSION

The issue as to whether interferon has any role in the therapy of chronic lymphocytic leukemia remains open. In our studies at the National Cancer Institute there was a minimal response in two patients and perhaps even a progression of disease in a number of patients who were treated with either 50 X 10^6 or 5 X 10^6 units per square meter body surface area of recombinant interferon-α. In another preliminary study, responses were reported in untreated patients with early stages of disease. Future studies for patients with chronic lymphocytic leukemia should include patients with earlier stages of disease and doses of interferon that have been proven to have biological response modifying effects. The preliminary results from Montserrat and co-workers (24) support this.

The toxicities with interferon at high doses have been troublesome. The flu-like syndrome and high fever can be alleviated by initiating therapy with low dose interferon and then escalating the dose. This "tachyphylaxis" appears to be a consistent finding in most patients. However, the severe fatigue following high doses of interferon is not alleviated by these methods. It seems clear from these studies that acetaminophen should be given routinely to all patients at the institution of interferon therapy to alleviate the fever. Patients with severe cardiac disease and those where high temperature is life-threatening should not be placed on interferon at high doses. Renal and hepatic functions should be monitored routinely since occassionally patients will have a dose limiting toxicity to one of these organs. Lowering of blood counts did not appear to be due to direct bone marrow suppression but is important when considering combined chemotherapy and interferon trials because the chemotherapy doses are modified according to the peripheral blood counts. It has been suggested that because chemotherapy is immunosuppressive and interferons may be immunostimulatory, chemotherapy should be given intermittently and the interferon continuously.

REFERENCES

1. Isaacs, A., Lindenman, J. Proc R Soc Lond Biol. 147:258-267, 1957.
2. Stewart, II, W.E., Blalock, J.E., Burke, D.C., et. al. J Immunol (letters to the Editor), 125:2353, 1980.
3. Fleischmann, W.R., Klimpel, G.R., Tyring, S.K., Voss, W.R., Baron, S. In: Novel Approaches to Cancer Chemotherapy, (Ed. P.S. Sunkara), Academic Press Inc., Orlando, FL, 1984; pp. 1-22.
4. Sehgal, P.B. Biochem Biophys Acta. 695:17-33, 1982.
5. Borden, E.C. Cancer. 54:2770-2776, 1984.
6. Hawkins, M.J., Borden, E.C., Merritt, J.A., et. al. J Clin Oncol. 2:221-226, 1984.
7. Cantell, K., Hirvonen, S. Texas Rep Biol Med. 35:138-144, 1977.
8. Pestka, S. Scientific America. 249:37-43, 1983.
9. Lin, L.S., Stewart, II, W.E. Methods Enzymol. 78:481, 1981.
10. Berg, K., Heron, I. Methods Enzymol. 78:487, 1981.
11. Kirkwood, J.M., Ernstoff, M.S. J Clin Oncol. 2:336-352, 1984.
12. Bonnem, E.M., Spiegel, R.J. J Biol Resp Modif. 3:580-598, 1984.
13. Roth, M.S., Foon, K.A. Am J Med. 81:871-882, 1986.
14. Gale, R.P., Foon, K.A. Ann Intern Med. 103:101-120, 1985.
15. Gutterman, J.U., Blumenschein, G.R., Alexanian, R. Ann Intern Med. 93:399-406, 1980.
16. O'Connell, M.J., Colgan, J.P., Oken, M.M., Ritts, R.E., Kay, N.E., Itri, L.M. J Clin Oncol. 4:128-136, 1986.
17. Misset, J.L., Mathé, G., Gastiabury, J., et. al. Biomedicine. 36:112-116, 1982.
18. Horning, S.J., Merigan, T.C., Korwn, S.E., et. al. Cancer. 56:1305-1310, 1985.
19. Schuloff, R.S., Lloyd, M.J., Stallings, J.J., et. al. J Biol Resp Mod. 4:310-323, 1985.
20. Foon, K.A., Bottino, G., Abrams, P.G. Am J Med. 78:216-220, 1985.
21. Foon, K.A., Sherwin, S.A., Abrams, P.G. N Engl J Med. 311:1148-1152, 1984.
22. Bunn, P.A., Foon, K.A., Ihde, D.C., et. al. Ann Intern Med. 101:484-487, 1984.
23. Foon, K.A., Maluish, A.E., Abrams, P.G., et. al. Am J Med. 80:351-356, 1986.
24. Montserrat, E., Vinolas, N., Urbano-Ispizua, A., Ribera, J.M., Gallart, T., Rozman, C. Blood. 68(suppl 1):227, 1986.
25. Ho, M., Pazin, G.J., Armstrong, J.A., Haverkos, H.S., Dummer, J.S., Jannetta, P.J. J Infec Dis. 150:867-872, 1984.
26. Kuhls, T.L., Sacher, J., Pineda, E., et. al. J Infec Dis. 154:437-442, 1986.
27. Meyers, J.D., Day, L.M., Lum, L.G., Sullivan, K.M. J Infec Dis. 148:551-556, 1983.
28. de Koning, E.W.J., van Bijsterveld, O.P., Cantell, K. Arch Ophthal. 101:1866-1868, 1983.
29. Cheeseman, S.H., Rubin, R.H., Stewart, J.A., et. al. N Engl J Med. 300:1345-1349, 1979.
30. Hirsch, M.S., Schooley, R.T., Cosimi, A.B., et. al. N Engl J

Med. 308:1489-1493, 1983.
31. Wade, J.C., McGuffin, R.W., Springmeyer, S.C., Newton, B., Singer, J.W., Meyers, J.D. J Infec Dis. 140:557-562, 1983.
32. Shalev, Y., Berrebi, A., Green, L., et. al. Arch Dermatol. 120:922-926, 1984.

12

INTERFERONS IN THE TREATMENT OF METASTATIC RENAL CELL CARCINOMA (RCC)

Jorge R. Quesada

Department of Clinical Immunology and Biological Therapy, M.D. Anderson Hospital and Tumor Institute, 1515 Holcombe Avenue, Houston, Texas 77030

ABSTRACT

The status of systemic therapy for metastatic RCC is reviewed with emphasis on the results of treatment with interferons and its combinations. The use of interferon in RCC represents a modest step forward in the treatment of this highly malignant disease.

INTRODUCTION

There is no effective systemic treatment for patients with metastatic renal cell carcinoma (RCC). Survival expectancy for the approximately 20,000 new patients to be diagnosed in the United States in one year remains dismal. Close to 80% of the newly diagnosed patients will present either regional or metastatic disease at the time of diagnosis and approximately 50% of them will die within the first 12 months. Hormonotherapy, chemotherapy and immunotherapy have yielded in general, very poor results. Recent reviews [1,2] revealed a low response rate for these modalities of treatment in clinical studies using present conventional methods of assessment of tumor response. Despite some encouraging results prior to 1970, there is at present little evidence in support of the benefit of hormonal manipulation in RCC. A review of several reports totaling more than 400 patients treated with either female or male hormones since 1971 revealed an objective response rate of 2% [2]. Therefore, the hormonal dependency of human RCC has not been confirmed by these studies, despite well-documented animal models in which hormonal therapy can significantly modify the course of the disease.

Similarly, no clear effective chemotherapeutic agent or regimen has been demonstrated in the treatment of metastatic RCC [1,2]. Response rates to traditional chemotherapeutic agents used individually are approximately 10%. Similar response rates have been obtained with newer chemotherapeutic agents including: Epidoxorubicin, dianhydrogalactitol, methyl GAG, PALA, AMSA and others [3]. Vinblastine has often been regarded as the most active single agent in the treatment of RCC. Recent randomized studies in which Vinblastine has been included have failed to demonstrate antitumor activity superior to other chemotherapeutic agents. Combination chemotherapy has not improved the poor efficacy of single agents [1,2].

M. Revel (ed.), CLINICAL ASPECTS OF INTERFERONS. Copyright 1988.

The ethiopathogenesis of RCC remains unknown. The malignant transformation seems to occur in the mature proximal convoluted tubular epithelium and a variety of carcinogenic factors (ionizing radiation, tobacco, thorotroast, lead and cadmium compounds, asbestos) have been implicated. More recently, oncogenic viruses or cellular oncogenes have been proposed in the events leading to the malignant transformation. Support for this proposal derives from the reported cases of family pedigrees with multiple instances of RCC, the increased incidence of RCC in certain genetically-linked disorders (Von-Hippel Lindau Syndrome, polycystic kidney disease), identification of chromosomal translocations, experimental animal models and expression of cellular oncogenes in fresh tumor samples [3].

A role for the immune system in the control of RCC has been long suspected. Diverse in vitro models have demonstrated tumor-specific immune cell-mediated cytotoxicity towards malignant cells from RCC [4]. The immunocompetence of patients with advanced disease is progressively impaired and their prognosis seemingly correlated with the functional status of the immune system as measured by a panel of immunologic tests [5]. These observations lead to the use of diverse methods aimed to enhance the host's immune defenses. Objective tumor responses have been reported using non-specific immunotherapy (BCG, Corinebacterium parvum) and active specific immunotherapy (tumor vaccines). More recently, the description of a set of cell surface antigens of human RCC offers the promise of passive immunotherapy with tumor-specific monoclonal antibodies. Attempts to transfer tumor-specific cell-mediated immunity lead to the use of impurified cellular products including immune RNA and Transfer Factor, which resulted in occasional transient tumor responses [4].

The lack of effective therapy, the immunologic background suggesting that RCC was amenable to biologic control and an anecdotal response in a patient who received an interferon inducer (Poly I:C) [S. Krown, Memorial Sloan Kettering, NY; personal communication] prompted us in 1981 to explore the antitumor activity of partially purified alpha interferon (IFNα) in metastatic RCC.

ALPHA INTERFERONS

Our original study reported on the activity of partially purified IFNα (State Serum Institute, Finnish Red Cross Center in 19 selected patients using a dose of 3 million units (MU) per day by i.m. injection [6]. Five patients (26%) showed partial responses in the lungs or mediastinum. In addition, two other patients obtained a minor response and in three patients mixed effects were observed (decrease in the size of some lesions with simultaneous increase of existing lesions). These findings were subsequently confirmed in a larger series of patients [7] and supported by work from other investigators using other types or sources of IFNα and diverse treatment designs (Table I).

Subsequently, we confirmed that the antitumor activity resided in the IFNα molecule itself by using a highly purified recombinant DNA-derived IFNα (rIFNα) [8]. Twenty-nine percent of 41 patients treated with rIFNα at doses ranging from 10 to 20 MU/m^2 achieved remission. However, none of the 15 patients who received 2 MU/m^2

responded to this treatment. The reasons for the requirement for larger doses of the rIFNα as compared to the partially purified preparation are unclear. Perhaps the IFNα contained additional subspecies of IFNα or substances other than interferon that contribute to its antitumor effects. Other studies using purified alpha interferons have confirmed antitumor activity against RCC (Table II).

Data from these studies suggest that the response rate to IFNα may vary with the type of interferon, dose, route and schedule of administration. Low doses (1 MU) or less frequent administration may be suboptimal. In our experience, the length of treatment is important. Some patients may require several months of treatment before a response is obtained. Studies designed to be terminated in less than two months may be insufficient to document objective tumor responses. Selection of patients also seem to influence the response rate. In our series, most of the responsive patients were male patients in good performance status, have undergone resection of the primary tumor, metastases were confined to the chest and had overall small tumor load. The response rate of patients with metastases confined to the lung or mediastinum can be as high as 40% but, with rate exceptions IFNα has no activity against unresected primary RCC, recurrent tumors, retroperitoneal, brain or liver metastases.

Therapy with alpha interferons in metastatic RCC has not shown yet an impact on survival in these studies. We have reported that survival for responsive patients compared favorably to that of the non-responsive ones [7]. However, we emphasized that such results may be due to variables other than the IFNα treatment. Still, considering the lack of activity of available agents in the treatment of RCC, the use of IFNα is at least promisory. The consistent, though modest antitumor activity of alpha interferons reported in these studies suggest a role for these biologic agents in future treatment approaches for RCC. These results have encouraged newer clinical studies using other types of interferons as well as preliminary pilot studies using combinations of alpha interferons with either other interferons, other biologic response modifiers, or chemotherapeutic agents.

BETA AND GAMMA INTERFERONS

Considerably less experience exists using beta (IFNβ) or gamma (IFNγ) interferons in RCC. Preliminary data using a rIFNβ intravenously, twice weekly in doses ranging from 0.1 to 600 MU/m^2 showed 3 partial remissions of 11 patients treated (27%). Treatment was in general, well tolerated and enhancement of monocyte cytotoxicity and natural killer cell activity was demonstrated [9].

Recently, our group has published the results of two sequential phase II studies of rIFNγ in 33 patients with metastatic RCC [10]. Fifteen patients received rIFNγ by daily i.m. injection in doses ranging from 0.25 to 1.0 mg/m^2. One of 14 evaluable patients (7%) achieving a partial remission. Similarly, one of 16 patients (6%) who received rIFNγ intravenously by continuous infusion (0.01 to 0.05 mg/m^2 obtained partial remission. In a few of the patients, tumor progression occurred rapidly within the first 4 weeks of treatment raising the possibility that rIFNγ could accelerate the growth of established lesions [10].

The incidence of clinical toxicity was similar for both rIFNγ studies. Maximum tolerated doses were considerably lower for the continuous i.v. infusion (0.05 mg/m^2) than for the i.m. schedule (1.0 mg/m^2). The toxicities differed from that observed with alpha interferons in the more frequent occurrence of headaches, protracted fever, frequency and severity of abnormal liver function tests and hypertryglyceridemia [10].

The differences observed between IFNα and IFNγ may be related to the biologic and molecular characteristics of these proteins, but clinical variables in the populations studied cannot be ruled out.

INTERFERON COMBINATIONS

We initiated several pilot projects combining IFNα with other biologic response modifiers in an effort to increase the response rate, or the incidence of complete remissions or the duration of responses. All of these studies were based on laboratory observations where synergism between the agents used had been demonstrated [11-13].

Alpha and gamma interferons:

The combination of rIFNα or rIFNβ with rIFNγ has been shown synergistic in laboratory models [11]. We treated 35 patients with RCC with simultaneous i.m. administration of rIFNα and rIFNγ [3]. The first 10 patients were given a unit: unit dose ratio (2 MU/m^2 of each interferon) which resulted in treatment discontinuation in 50% of the patients due to severe toxicity and no objective responses. Subsequently, 25 patients received a mg: mg dose ratio (0.01 mg/m^2 of each interferon) equivalent to 2 MU of rIFNα and 0.2 MU of rIFNγ. This treatment regime was well tolerated and only 2 patients (8%) had treatment discontinued because of toxicity. Three patients obtained PR (12%). Interestingly, 10 other patients had minor responses (40%) but did not achieve PR.

IFNα and other biologic response modifiers:

Eleven patients received a combination of IFNα (3 MU/day) and Diflouromethylornithine (DFMO) at doses ranging from 6 to 12 gm/m^2/day i.v. based on synergism demonstrated in vitro [12]. Two partial remissions were observed (18%). IFNα toxicity was similar to previous experience. DFMO-related toxicities were predominantly decrease audition and anemia [3].

In a different study, IFNα (3 MU/day) and double stranded RNA (dsRNA) in doses ranging from 10 to 80 mg/m^2/day i.v. were used. Two partial remissions were obtained in 12 patients treated (16%) with no additive toxicity related to dsRNA [3].

IFNα AND CHEMOTHERAPY

Various chemotherapeutic agents have synergistic activity with IFNα in vitro and in animal models [14,15]. These results prompted the use of these types of combinations in the clinic, despite the poor activity of the chemotherapy drugs and the modest activity of IFNα. Table 3 shows the results of four such studies. No clinical evidence of therapeutic synergisms have been observed, whereas toxicity has been enhanced [16-18]. Tolerance to the myelosuppressive effects of chemotherapy was also impaired in the combination studies.

More recently, we observed four PR and 3 minor responses in 13 patients treated with FAMP (5-fluorouracil, hydroxyrubicin, mitomycin C and cis platinum). Historical data with FAMP in RCC showed marginal activity. Furthermore. 3 additional patients had symptomatic responses and 3 had a quiescent state. None of these patients showed progression of the diseases for at least 3 months of follow-up observation (Dexeus F, unpublished). All patients had in common a history of prior exposure to diverse interferon programs. Although anectdotal in nature, these results have prompted a randomized clinical study between FAMP versus IFNα followed by FAMP. The contention is that the apparent increased sensitivity to the chemotherapy is in virtue of the prior exposure to IFNα which might have modified somehow the biologic behavior of otherwise resistant malignant cells.

CONCLUSIONS AND PROSPECTS

The use of interferons represent a modest advance in the treatment of metastatic RCC. A small selected proportion of patients can obtain temporary benefit from optimal doses and schedules of administration. More importantly, the application of IFNs in RCC has further confirmed the sensitivity of this tumor to biologic response modifiers, which had already been suspected from preliminary work with immunotherapeutic agents. Conceivably, the responses observed in those early trials might have been associated with the induction of cytokines, such as IFNα. Newer biologic agents, including Tumor Necrosis Factor (TNF) and Interleukin-2 (IL-2) are promisory. A recent study administering high doses of IL-2 and activated autologous killer cells has reported activity in RCC [19]. Overall, these data indicate a role of biologic agents in the therapeutic options against RCC and new studies should be enthusiastically pursued.

REFERENCES

1. Hrushesky, W.J., and Murphy, G.P. J. Surg. Oncol. 9:277-278, 1986.
2. Harris, D.T. Sem. Oncol. 10:422-430, 1983.
3. Quesada, J.R. Sem. Oncol. (in press).
4. Peehl, D.M., and Stamey, T.A. J. Urol. 135:897-904, 1986.
5. Hersh, E.M., Wallace, S., Johnson, D.E., and Bracken, R.B. In Johnson, D.E. and Samuels, M.L., eds: Cancer of the Genitourinary Tract 1979, pp. 47-56.
6. Quesada, J.R., Swanson, D.A., Trindade, A., Gutterman, J.U. Cancer Res. 43:940-947, 1983.
7. Quesada, J.R., Swanson, D.A., and Gutterman, J.U. J. Clin. Oncol. 3(8):1086-1092, 1985.
8. Quesada, J.R., Rios, A., Swanson, D.A., Trown, P., and Gutterman, J.U. J. Clin. Oncol. 3(11):1522-1528, 1985.
9. Rinehart, J., Neidhart, J., Meyer, M., Hiner, C., and Young, D. Proc. Am. Soc. Clin. Oncol. 4:224, 1985.
10. Quesada, J.R., Kurzrock, R., Sherwin, S.A., and Gutterman J.U. J. Biol. Resp. Modif. (in press).
11. Czarniecki, C.W., Fennie, C.W., Powers, D.B. J. Virol. 49:490-496, 1983.
12. Rosenblum, M.G., and Gutterman, J.U. Cancer Res. 44:2339-2340, 1984.
13. Hubbell, H.R., Liu, R.S., and Maxwell, B.L. Cancer Res. 44:3252-3257, 1984.

14. Balkwill, F.R., and Moodie, E.M. Cancer Res. 44:904-908, 1984.
15. Aapro, M.S., Alberts, D.S., and Salmon, S.E. Cancer Chemother. Pharmacol. 10:161-166, 1983.
16. Figlin, R.A., deKernion, J.B., Maldazys, J., and Sarna, J. Cancer Treat. Rpt. 69(3):263-267, 1985.
17. Muss, H.B., Welander, C., Caponera, M., Reavis, K., Cruz, J.M., Cooper, R., Jackson, D.V. Jr., Richards, F., II., Stuart, J.J., Spurr, C.L., White, D.R., Zekan, P.J., and Capizzi, R.L. Cancer Treat. Rpt. 69(6):721-722, 1985.
18. Creagan, E.T., Kovach, J.S., Long, H.J., Richardson, R.L. J. Clin. Oncol. 4(3):408-413, 1986.
19. Rosenberg, S.A., Lotze, M.T., Muul, L.M., N. Engl. J. Med. 313: 1485-1492, 1985.

TABLE I

NATURAL ALPHA INTERFERONS IN METASTATIC RENAL CELL CARCINOMA

TYPE	TREATMENT SCHEDULE (X 10^6 units)	NO. EVAL. PATIENTS	CR/PR	%	INVESTIGATOR
IFNα	3/daily	50	13	26	Quesada '83, '85
IFNα	3 x 5d/wk	43	7	16.5	DeKernion '83
IFNα	1/d x 4 wk	14	0	0	Kirkwood '84
	10/d x 4 wk	16	2	12.5	
IFNα	3/daily	11	3	27	Edsmyr '85

TABLE II

PURIFIED ALPHA INTERFERONS IN METASTATIC RENAL CELL CARCINOMA

TYPE	TREATMENT SCHEDULE ($X 10^6$ units)	NO. EVAL. PATIENTS	CR/PR	%	INVESTIGATOR
IFNαLy	$5/m^2$ tiw	33	5	15	Neidhart '84
IFNαLy	$3/m^2$ tiw x 6 wk	21	1	5	Vugrin '85
IFNαLy	3-6/daily	56	11	19	Kimura '84
IFNαLy	5/daily	73	17	23	Umeda '86
rIFNαA	$2/m^2$/daily $20/m^2$/daily	15 41	0 12	0 29	Quesada '85
rIFNαA	3-36/daily	108	15	14	Umeda '86
rIFNα_2	6-10 x 3-5/wk x 2 mo	45	8	18	Umeda '86
rIFNα_2	$2/m^2$ tiw (sq) $30/m^2$/d x 5d q 2-3 wk (i.v.)	10 10	0 1	0 10	Kempf '85

TABLE III

INTERFERON ALPHA COMBINATIONS WITH CHEMOTHERAPY IN

METASTATIC RENAL CELL CARCINOMA

REGIME	NO. PTS.	CR/PR	%	INVESTIGATOR
IFNα + Vinblastine	23	2	13	Figlin
IFNα + Vinblastine	11	1	9	Quesada
IFNα + Hydroxyrubicin	15	0	0	Muss
rIFNα + BCNU	9	2	22	Creagan

13

INTERFERONS IN BREAST CANCER.

T. DORVAL*, P. BEUZEBOC*, E. FALCOFF**, E. GARCIA-GIRALT*, M. JOUVE*, T. PALANGIE*, P. POUILLART*.

Institut Curie, * Service de Médecine Oncologique, ** Section de Biologie. 26, rue d'Ulm - 75231 PARIS. FRANCE.

ABSTRACT

Because of its antiviral, immunomodulating and antiproliferative properties, Interferon was tested as an antitumor agent.

As in vitro and in vivo experiments showed a possible role for Interferons in treating malignant tumors, phase II trials were activated.

A review of the clinical trials applied to the treatment of metastatic breast cancer does not demonstrate any efficiency of Interferon and furthers studies are required.

1. INTRODUCTION

The use of conventional drugs applied in various combination chemotherapy programs combined or not with hormonotherapy has improved the response rate, response duration and survival in patients with metastatic breast cancer (1). However, once those patients have become refractory or have relapsed on that therapy, salvage regimens are very disappointing and second responses to treatment, when observed, are usually of very short duration.

As in vitro and more recently in vivo studies showed a possible role for interferons in treating malignant diseases, phase II trials were activated in various tumors, including metastatic breast cancer.

Interferon has demonstrated an antitumor activity against breast cancer ; in vitro, experiments showed that IFN inhibited human breast cancer cell lines (2) and induced an extension of all phases of the cell cycle. The same inhibiting action was observed against

M. Revel (ed.), CLINICAL ASPECTS OF INTERFERONS. Copyright 1988.

mouse mammary tumors, human breast cancer xenografts in nude mice and breast cancer in patients, to some degree(3, 4, 5, 6).
Based upon these observations, phase II trials were activated in an attempt to define the efficiency of the different forms of interferon in breast cancer (results summurized in table 1).

2. NATURAL LEUKOCYTE INTERFERON

Jordan U. Gutterman and al., treated 17 patients with recurrent metastatic breast cancer who failed conventional therapy. Interferon was given at a dose of 3 to 9 x 10^6 I.U. intramuscularly daily (7). 5 out of the 11 patients who received 3 x 10^6 IU daily achieved a significant response (4 partial responses, 1 response less than 50%. Two out of the 6 patients who received Interferon at a dose of 9 x 10^6 IU daily achieved a partial remission.
Time to response varied from 6 to 12 weeks and response duration from 8 to 60 weeks (median 27 weeks).
The authors noticed a positive correlation between prior response to chemo or hormonotherapy and response to Interferon and between site of metastatic tumor (soft tissue) and response to Interferon.
These encouraging results were confirmed by Z. Maricic and al. (8) who observed 3 major responses out of 4 patients treated with 1 x 10^6 IU daily intramuscularly ; Borden and al. in an extensive phase II trial (9) reported about 23 patients receiving 3 x 10^6IU Interferon intramus cularly daily ; only 5 patients were responders,none of them was complete responder. Mean time to response was 6 weeks and response duration ranged from 14 to 176 days with a median of 59 days.
Dose escalation did not provide any obvious benefit in terms of response rate or response duration ; after achieving partial response, patients were randomized to discontinue Interferon or to continue therapy ; patients who stopped therapy had evidence of shorter response duration.
Older age correlated with response to Interferon.

3. LYMPHOBLASTOID INTERFERON

Because of very limited series, few informations appear available for lymphoblastoid Interferon in breast cancer.

Gregory P. Sarna entered 18 patients in a phase II trial (10). Patients received 30 x 10^6 I.U./m² intramuscularly weekly. None out of the 15 evaluable patients achieved a complete or partial response.

Silver and al. (11) reported a trial of low dose (2 x 10^6I.U./m² daily for 10 days) versus high dose (5 x 10^6 I.U./m² escalating by 5 x 10^6I.U. per day daily for 10 days) in 27 patients. One patient only achieved partial response.

4. FIBROBLAST INTERFERON

Two contradictory trials do not allow a precise idea of the role of IFN β in the treatment of metastatic breast cancer.

Quesada and al. (12) reported an objective antitumor effect of Interferon in 3 out of 6 patients who received 3 to 6 x 10^6 I.U. B IFN daily. Responses (1 major and 2 minor responses) occured in soft tissue and lymph node metastasis.

On the other hand, U. Bruntsch and al. (13) did not observe any response in nine patients treated with 6 x 10^6 IU of β IFN daily.

5. RECOMBINANT INTERFERON

5.1. Recombinant leukocyte A Interferon.

Four trials are in accordance to recognize a limited role for r INF αA in treating metastatic breast cancer.

Stephen A. Sherwin and al. (14) report about 29 patients who received 50 x 10^6 I.U./m² of r IFN αA intramuscularly three times a week. Of 17 evaluable patients, one had stable disease, 16 had evidence of tumor progression.

A.Nethersell and al. (15) observed 2 responders out of 12 evaluable patients entered in a phase II trial of r IFN α A given at a dose of 20 x 10^6 I.U. three times per week. Response to Interferon did not correlate with prior response to chemo or hormonotherapy.

Quesada and al. in a collaborative phase I - II trial of IFN α A (16) observed one partial remission among 23 patients treated in a

dose escalating schedule from 3 to 50 x 10^6 IU of r IFN α A.
K. Sikora and al. (17) report 3 partial responses out of 15 patients who were administered r IFN α A at a dose of 20 x 10^6 IU daily or 50 x 10^6 IU three times a week.

5.2. Recombinant leukocyte 2 Interferon.

Lenzhofer and al. (18) entered 10 patients in a phase II trial of r IFN α 2 given at a dose of 10 x 10^6 I.U. daily. Out of 7 evaluable patients, no response was reported.

N. Padmanabhan and al. (19) observes no response in 14 patients who received either 2 x 10^6 I.U./m2/day 3 times a week or 50 x 10^6 I.U./m2 5 consecutive days every 3 weeks.

A more extensive phase II trial with a dose of 30 to 50 x 10^6 I.U./m2 given 5 consecutive days every 2 or 3 weeks or 2 x 10^6 IU/m2 given 3 times a week was reported by H.B. Muss and al (20). All 33 patients had failed conventional therapy and no response was observed.

6. IMMUNE INTERFERON

Series are to now too limited to get any suitable information about Interferon gamma and breast cancer ; nevertheless, phase I trials do not demonstrate any major action of Interferon gamma in metastatic breast cancer.

7. INTERFERON AND HORMONE RECEPTORS

Because of its potential properties of induction of differenciation, we have examined the evolution of hormone receptors in 6 patients who received Interferon beta at a dose of 6 x 10^6 I.U. every 5 days (21). Levels of receptors were measured before and after therapy was applied.

The level of estrogen receptors increased from 2 to 10 fold in the 2 tested patients ; the level of progesteron receptors increased in 5 of 6 tested patients.

Further observations are required to conclude about a possible modulation of hormonal receptors in breast cancer by Interferons and a possible correlation between hormone receptors modulation and response to Interferons.

8. CONCLUSION

Exept those reported by Gutterman and al., the results of the different clinical trials testing the activity of the different forms of Interferon in breast cancer are very poor with a very low response rate and response duration.

Because of too limited series, questions remain still unanswered about the characteristics of responders (size and site of metastasis ; hormonal status of patient and tumor).

As conventional phase II trials of Interferons have been largely negative, further studies would be activated to determine a possible role for Interferons in combination with cytotoxic drugs or as an adjuvant treatment after radical surgery.

Table 1. Interferon and breast cancer.

INTERFERON	DOSE-SCHEDULE	Nb of Pts	CR	PR	AUTHOR
NATURAL	3 to $9x10^6$ IU IM/daily	17	0	7	Gutterman
NATURAL	1 x 10^6 IU IM daily	4	0	3	Maricic
NATURAL	3 X 10^6 IU IM daily	23	0	5	Borden
LYMPHOBLASTOID	30 x 10^6 IU IM weekly	15	0	0	P. Sarna
LYMPHOBLASTOID	2x10 IU^6IM daily 10days	27	0	1	Silver
	vs 5x10 IU^6IM daily 10days				
FIBROBLAST	3 x 10^6 IU daily	6	0	1	Quesada
	or 6 x 10^6 IU daily				
FIBROBLAST	$6x10^6$ to $60x10^6$ daily	9	0	0	Bruntsch
r IFNαA	$50x10^6$ IU/m² Tiw	17	0	0	Sterwin
r IFNαA	$20x10^6$ IU IM daily	12	0	2	Nethusell
	or $50x10^6$ IU IM Tiw				
r IFNαA	3 to $50x10^6$IU IM daily	23	0	1	Quesada
r IFNαA	$20x10^6$ IU IM daily	15	0	3	Sikora
	or $50x10^6$ IU Tiw				
r IFNα2b	$10x10^6$ IU IM daily	0	0	0	Lonzhoffer
r IFNα2b	$2x10^6$IU/m² IM Tiw	14	0	0	Padmanabhan
	or $50x10^6$IU/m² 5 days				
r IFNα2b	30 to $50x10^6$IU/m² 5days	33	0	0	B. Muss
	or 2 x 10^6 IU/m² Tiw				

REFERENCES

1. HENDERSON, I.C., CANELLOS, G.P. New Engl. J. Med. 302: 78-90, 1980.

2. BALKWILL, F., WATLING, D., TAYLOR-PAPADIMITRION, J. Int. J. Cancer. 22: 258-265, 1978.

3. BALKWILL,F., TAYLOR-PAPADIMITRION, J., FANTES, K.H. et al. Eur. J. Cancer. 16: 569-573, 1980.

4. BORDEN, E.C., BALKWILL, F.R. Cancer 53: 785-789, 1984.

5. KIRKWOOD,J.M., ERNSTOFF, M.S. J. Clin. Oncol. 2: 336-352, 1984.

6. ERIK de CLERCQ, ZHEN-XI-ZANG, KRIS HUYGEN and LEYTEN, R. J.N.C.I. 69: 653-657, 1982.

7. GUTTERMAN, J.U., BLUMENSCHEIN, G.R., ALEXANIAN, R., and al. Ann. of Int. Med. 93: 399-406, 1980.

8. MARICIC, Z., NOLA, P., IKIC, D., SMUDJ, K. and al. J. Cancer Res. Clin. Oncol. 101: 317-323, 1981.

9. BORDEN, E.C., HOLLAND, J.F., DAO, T.L. and al. Annals of inter. Medecine 97: 1-6, 1982.

10. SARMA, G.P. and FIGLIN, R.A. Cancer Treat. Reports. 69: 547-549, 1985.

11. SILVER, HKB, CONNORS, J., SALINAS, F.A. Antiviral Res. 1:90, 1983

12. QUESADA, J.R., GUTTERMAN, J.U. and HERSH, E.M. J. of Interferon Research. 4: 593-599, 1982.

13. BRUNTSCH, U., GROOS, G., TIGGES, F.J. and al. Cancer Chemother. Pharmacol. 13: 39-42, 1984.

14. SHERWIN, S.A., MAYER, D.M., OCHS, J.J. and al. Annals of Inter. Med. 98: 598-602, 1983.

15. NETHERSELL, A., SMEDLEY, H., KATRAK, M. AND al. Br. J. Cancer. 49: 615-620, 1984

16. QUESADA, J.R., HAWKINS, M., HORNING, S. and al. The Am. J. of Med. 77: 427-432, 1984.

17. SIKORA, K., SMEDLEY, H.M., KATRAK, M. 13th International Congress of chemo therapy. Vienna, Austria. 2: 266/36.

18. LENZHOFER, R., MICKSCHE, M., DITTRICH, C. and al. Drugs exptl. Clin. Res. 7: 463-470, 1984.

19. PADMANABHAN, N., BALKWILL, F.R., BODMER, J.G. and al. Br. J. Cancer. 51: 55-60, 1985.

20. MUSS, H.B., KEMPF, R.A., MARTINO, S. and al. Journal of Clin. Oncol. 9: 1012-1016, 1984.

21. POUILLART, P., PALANGIE, T., JOUVE, M.,GARCIA-GIRALT, E. and al. Eur. J. Cancer Clin. Oncol. 10: 929-935, 1982.

14

OSTEOSARCOMA MANAGEMENT AND INTERFERON

H. STRANDER[1], H.C.F. BAUER[2], O. BROSJÖ[2], L-Å. BROSTRÖM[2], S. EINHORN[1], U. NILSONNE[2] and O.S. NILSSON[2].

[1]Department of Oncology (Radiumhemmet) and [2]Department of Orthopedic Surgery, Karolinska Hospital, S-104 01 Stockholm, Sweden.

ABSTRACT

Human osteosarcoma is a malignant disease which requires multimodality treatment including surgery and adjuvant therapy. In this article we have summarized various aspects pertinent to IFN therapy of human osteosarcoma. Different experimental approaches are described but we emphasize that IFN treatment should not be used as the primary single treatment in osteosarcoma. Osteosarcoma exerts biological effects which makes it possible to study biological parameters extensively. Such a biological approach might lead to an extensive combination treatment schedule and a diversified programme for the treatment of human osteosarcoma patients.

INTRODUCTION

Osteosarcomas are highly malignant tumors of rare occurrence, still, they constitute the most common type of primary bone sarcoma (1). They affect all ages but occur most frequently in young people. After having established the clinical use of high dose exogenous natural leukocyte interferon (IFN) therapy in the early 1970s, we initiated an adjuvant IFN osteosarcoma trial on a wide number of patients at the Karolinska Hospital (2-11). Over the years our results have been compared with those obtained by other groups employing adjuvant chemotherapy, and have also been related to concurrent and historical non-adjuvant controls. A group from the

M. Revel (ed.), CLINICAL ASPECTS OF INTERFERONS.

National Institutes of Health, USA, visited us during the trial and a consensus report concerning the ethical justification of the trial was agreed upon. At the same time as the IFN adjuvant trial was initiated, other groups started to employ adjuvant chemotherapy in the treatment of osteosarcoma (12-15). Gradually, the chemotherapeutic approach has been intensified, the drugs have been given prior to surgery and combination chemotherapy has been employed for prolonged periods of time (16-20).

It still remains to be definitely established whether adjuvant chemotherapy or IFN treatment have improved the long-time survival rate of osteosarcoma patients. To complicate matters the natural course of osteosarcoma appears to have changed inasmuch the survival rate has increased over the last 10 years irrespective of the therapy given. Therefore, it has been difficult to draw final conclusions concerning the efficacy of adjuvant chemotherapy; randomized studies employing various protocols have not been performed until recently and their follow-up periods are, as yet, too short to allow definite conclusions concerning long-term survival.

The antitumor effects of IFN on various malignancies have been clearly established (10-11,21-28), and they have therefore by now been introduced as antitumor drugs in several countries. Whether the osteosarcoma disease is also IFN sensitive remains to be proven, but the summary below will provide some indications. The present article attempts to summarize the experience collected over the last 15 years concerning IFN effects on osteosarcomas in vitro, in experimental models, and in patients. Hopefully, such information will allow us to optimize IFN treatment in this particular disease.

IFN TREATMENT OF OSTEOSARCOMA CELLS IN VITRO

Murine osteosarcoma cells have been found to be sensitive to IFNs in tissue culture (29-31), but they have been less sensitive to the growth-inhibitory effect of IFNs than other malignant cell lines. It has also been demonstrated that human osteosarcoma cells can be inhibited in their growth in tissue culture (32-33). The antiproliferative effect of IFNβ might be stronger than that of IFNα (34). IFNγ has been shown in some systems to potentiate both IFNα and IFNβ and it has also been shown to have an antiproliferative effect of its own on human osteosarcoma cells (35-36; Strander unpublished data).

Both IFNα and IFNβ have been combined with chemotherapeutic agents in vitro in the treatment of osteosarcoma cells and in both cases the efficacy has been increased leading to additive or synergistic combination effects (37-38). Interestingly, it has been reported that virus-infected osteosarcoma cells in the murine system can lose some of their sensitivity to IFN (39).

MURINE MODELS

In 1978 Glasgow and co-workers reported on the antitumor effects of IFN on murine osteosarcoma in vivo (40-41). Daily treatment with 3-6x10^4 U of crude MuIFN type I preparations completely inhibited or delayed the appearance of tumors in mice inoculated sc with murine osteosarcoma. This effect was successfully inhibited with anti-IFN antibody, suggesting that IFN itself was the biologically active component of this highly impure IFN preparation. Tumor development was also strikingly inhibited in animals treated with 6x10^2 U of type II IFN and was achieved with 100-fold less IFN type II than type I IFN. Although the IFN preparations in these early experiments were extremely impure, the results suggested that IFNγ had a stronger antitumor effect, per antiviral unit, than IFNα/β.

Onehundredfifty μg poly ICL, a potent IFN inducer in primates, has also been reported to have a significant antitumor effect on murine osteosarcoma (42). This effect, however, has only been seen after intratumoral injections, starting immediately after tumor transplantation. If poly-ICL injections are initiated after development of palpable tumors, or given at a distance from the site of tumor inoculation, no antitumor effect is seen. In a different experimental model, Glasgow and Kern (41) induced metastatic osteosarcoma nodules in the lungs of C57BL/6 mice by intravenous injection of a suspension of murine osteosarcoma cells. Daily administration of 5x10^4 to 10^5 U of MuIFN type I reduced the tumor mass in the lungs and the number of tumor nodules present in histopathologic sections. However, if the IFN therapy was not initiated until 7 days after cell inoculation, no antitumor effect was seen. In these experiments IFN was given ip, suggesting that osteosarcomas may be inhibited by systemic treatment.

To summarize, osteosarcoma models in mice and rats have been established in order to test IFN efficacy under experimental conditions. In the rat model the results have been negative (43) while a murine osteosarcoma model in mice has successfully established that IFN treatment can be made effective if initiated early in the development of the disease (44). It has also been established in a murine osteosarcoma model that optimal results are achieved on repeated intratumor treatment prior to extensive tumor growth (41). In yet another murine system it has been established that animals bearing plutonium-induced osteosarcoma require extremely high doses of IFN in order to show tumor inhibition (45).

THE NUDE MOUSE MODEL

Balkwill and co-workers reported in 1980 that the development of human breast cancer cell lines in nude mice can be inhibited by HuIFNα(Namalwa). In subsequent experiments, a dose of $2x10^5$ IU/day was needed to arrest the growth of such established tumors. IFN is relatively species specific, and observed antitumor effects by IFN in nude mice can therefore be regarded as direct effects of the HuIFN on the human tumor cells, and not as an indirect mechanism mediated by the host cells (46-48).

A report on the effect of IFN on a human osteosarcoma cell line transplanted to nude mice was presented in 1983 by Masuda and co-workers (49). The growth of the xenograft was strikingly inhibited by ip injections of $5x10^4$ IU of human leukocyte IFN (HuIFNα) given 3 times weekly. Since 1981 we have established human osteosarcomas in nude mice from fresh osteosarcoma surgical specimens, and different osteosarcomas have been transplanted in serial passage. Examination of histologic appearance and nuclear DNA content by flow cytometry has shown that the tumors retain their basic tumor characteristics even after 3 years of passage in mice (50).

In a series of experiments we have examined the susceptibility of these osteosarcoma xenografts to growth inhibition induced by natural HuIFNα. HuIFNα treatment was given as daily sc injections at a distance from the growing tumor and it was started after measurable growth had occurred. All tumors could be growth-inhibited by HuIFNα. The lowest inhibitory dose ranged from 10^5-10^6 IU/day (51,52). IFN had to be given daily to attain growth arrest, and growth resumed upon reduction of the IFN dose. As opposed to results obtained in experiments with murine osteosarcomas, the growth inhibition appeared to be independent of tumor size.

Although the growth inhibition was dramatic in these experiments and could be maintained for several months, there was surprisingly little change in the histologic appearance of the growth-inhibited tumors. As previously noted by Balkwill in her studies on breast cancer xenografts (46), the tumors from IFN treated mice looked like miniatures of tumors from control mice. However, in 4 osteosarcomas the tumor tissue became mineralized and was partly or completely replaced by normal bone and bone marrow. This picture of histologic differentiation was seen only in the 4 xenografts that were sensitive to the lowest doses of IFN. Interestingly, the less proliferative osteosarcoma xenografts appeared to be more sensitive to the growth-inhibiting effect of IFN. Hence, untreated control tumors of these xenografts had lower proportions of S-phase cells (Table 1) and lower growth rates than xenografts that were less sensitive to IFN (52).

Table 1. Sensitivity to HuIFN and S-phase proportions of 12 human osteosarcoma xenografts.

Inhibitory dose (IU/day)	No of tumors	S-phase cells (%)
$\leq 2 \times 10^5$	5/12	14 ± 5.4
1×10^6	6/12	22 ± 5.4
$> 1 \times 10^6$	1/12	28

This tumor model has been employed to compare the antitumor effect of recombinant and natural HuIFNα. Although the natural HuIFNα appeared to be somewhat more effective per antiviral unit, the sensitivity of the individual tumors to IFN appeared to be the most important factor determining the antiproliferative effect (Bauer et al., unpublished data). The antitumor effects of HuIFNα and HuIFNγ combinations, both natural and recombinant, were also studied (53) and it was found that both natural and recombinant HuIFNγ preparations inhibit osteosarcoma xenografts at approximately the same dose range as IFNα. A clear additive growth-inhibiting effect was seen when IFNα and IFNγ were given simultaneously. In vitro studies have suggested that these IFNs might have a synergistic antitumor effect, but we have not been able to demonstrate synergism in our nude mouse experiments.

In conclusion, these experiments demonstrate that human osteosarcoma xenografts can be growth-inhibited with human IFN. The required growth-inhibitory dose varies considerably for different tumors, reflecting the heterogeneity of this tumor type. The doses needed to achieve growth inhibition appear to be substantially higher than the doses given to patients. On the basis of these experiments it may be argued that if IFN can be delivered to the tumor site in sufficiently high doses a significant antitumor effect can be expected.

IMMUNOLOGICAL ASPECTS

IFN receptors on human osteosarcoma cells have been demonstrated to vary between different cell lines. Monoclonal antibodies towards osteosarcoma antigens have been established (54-56) and MHC class 2 antigens have been induced on human osteosarcoma cells by IFNγ (57). IFN can render lymphocytes cytotoxic to allogeneic tumor cells, but it has not been possible to affect these systems in patients with osteosarcoma (58). IFNs have been shown to exert some effects on the immune system of osteosarcoma patients (59). Of special interest is the finding that the NK activity in the peripheral blood is increased by IFN therapy (see 59). Several other immune parameters have also been studied (60-61).

ADJUVANT IFN TREATMENT OF HUMAN OSTEOSARCOMA

A clinical adjuvant IFN osteosarcoma trial in central Sweden was initiated in 1971 (see 10,62,63) and the results have been updated several times since (for most recent updates see 7,8). Patients have been divided into 4 different groups: a) An IFN treated group, consisting of patients with osteosarcoma without metastases on admission, treated at the Karolinska Hospital as a consecutive series over the last 16 years. b) A non-adjuvant group, consisting of patients treated between 1971 and 1976 elsewhere in Sweden, which has been employed as a concurrent control (adjuvant treatment was not given before 1977 for patients in Sweden). c) A chemotherapy group comprising patients treated in Sweden outside the Karolinska Hospital between 1977 and 1980. No patients treated after 1980 have yet been analysed, but it should be emphasized that the members of the Scandinavian sarcoma team have now treated such patients with an extensive chemotherapy schedule according to Rosen (20) and that this group is therefore going to be analyzed separately. In the near future the extended group will be compared to the IFN treated group. d) The historical control group, is made up of 35 patients, seen on a consecutive basis at the Karolinska Hospital before 1971.

The IFN preparations used in this trial consist of natural IFNα prepared according to two methods (64-66). The IFN has been administered by the im route. During the first month after the diagnostic open biopsy the IFN has been given in a dose of $3x10^6$ U im daily and therafter 3 times per week for a further 17 months. Based on the results obtained in the nude mice we have intensified the schedule, and since January 1985 all patients are being treated with daily IFN therapy continuously. If patients do not complain of serious side effects we continue this treatment beyond 18 months after surgery. The side effects in connection with this treatment have been reported in extenso previously and are typical for those brought about by low-dose natural IFNα therapy (67-70). However, no patients have had to stop IFN therapy due to extensive side effects.

Different prognostic variables have been analyzed in the various groups, the most important being treatment delay, size of the tumor on admission, pathological findings including histologic subtype and grade, location of tumors and type of surgery. On the basis of these studies we have been able to conclude that the IFN-treated group cannot be directly compared to the historical control group since the latter comprises patients with less favourable prognostic variables

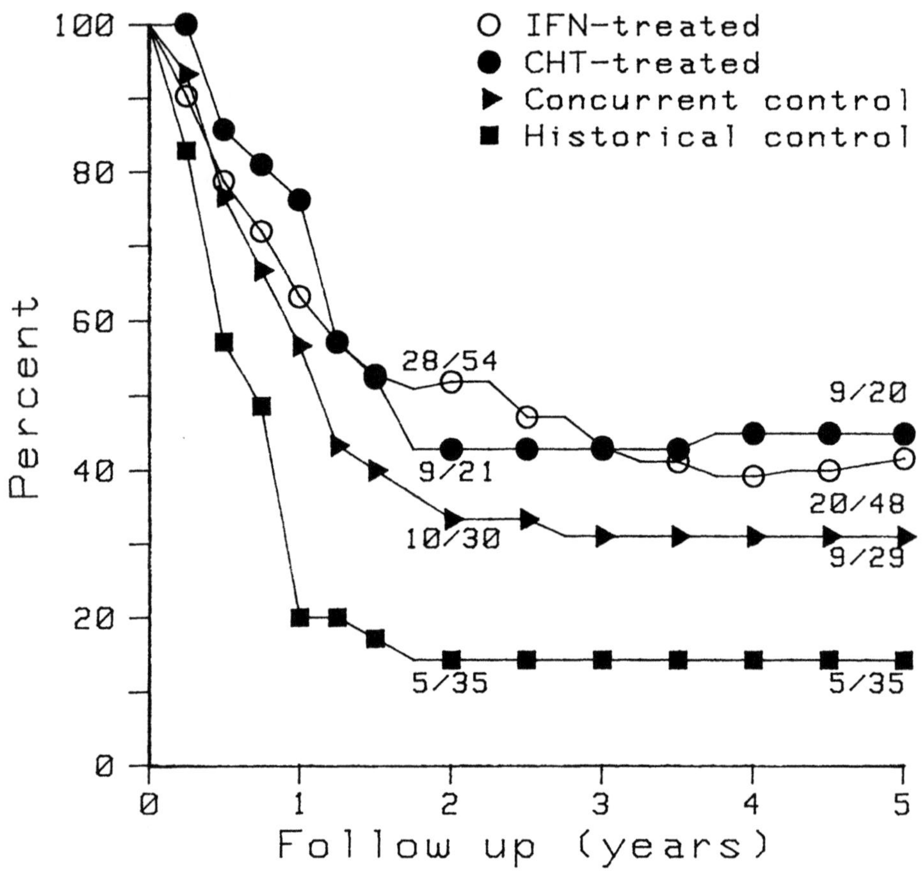

Figure 1. Mestastasis free survival rate in the four groups.

(2,3). On the other hand the 2 concurrent control groups can be employed according to the prognostic analysis.

The development of metastases and the survival rate in these 4 groups of patients can be seen from Figs. 1 and 2. However, too few patients have yet been analyzed in order to draw reliable conclusions and the differences are not statistically significant. Moreover, it will be interesting to study the long-term effects of IFN therapy since there are indications that chemotherapy prolongs survival of the patient but might be inefficient for long-term cure (71-72).

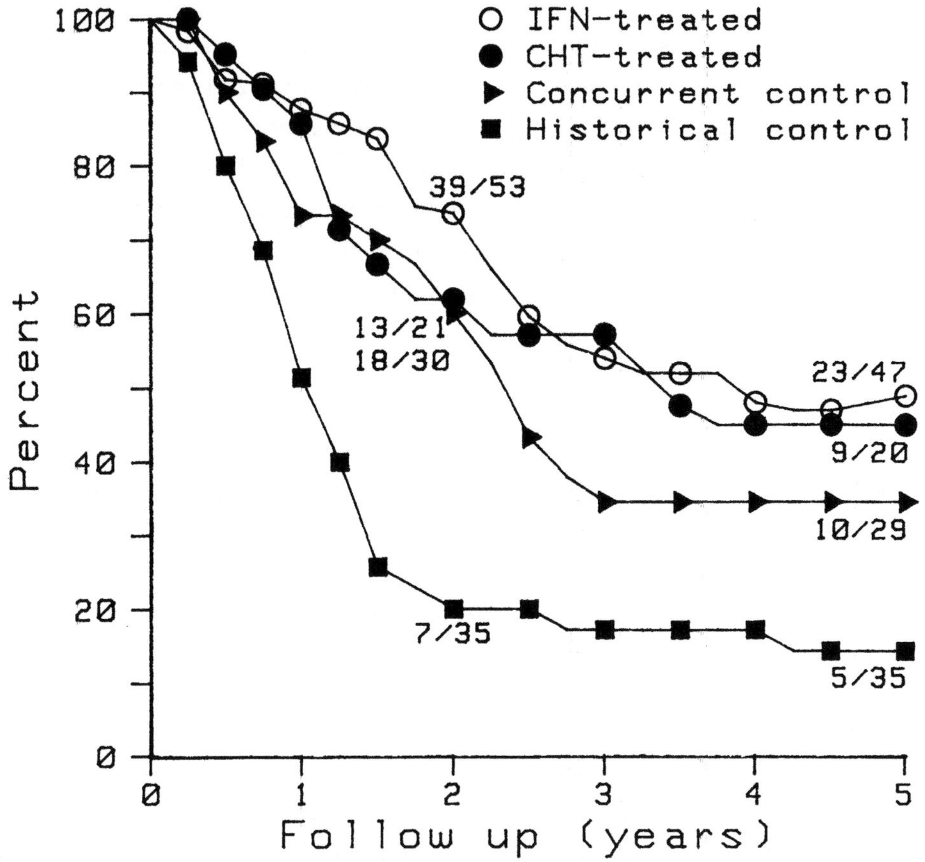

Figure 2. Survival rate in the four groups.

There are other findings in this trial that indicate an IFN effect. Extremely late development of metastases in several patients and especially the occurrence of multiple cerebral metastases as the first sign of tumor spread in one patient are such examples (7,8). Tumor site does not seem as important as tumor size for prognosis of IFN-treated patients (73). For further considerations it is of interest that IFNs can affect the amount of hormonal receptors on cells (74). It will be important to extend the follow-up period of the treated patients, to add new control patients to this trial and to analyze all recent osteosarcoma cases in Sweden (from 1980 onwards).

Of special interest in this trial is the finding that patients receiving IFN have fewer symptoms typical of acute viral infections than their family members. This has been demonstrated both through interviews and viral serology. Serologically diagnozed infections were registered during IFN therapy among the patients where metastases developed at some time either during or after IFN therapy (70). In metastatic patients, clinical viral infections (herpes simplex, herpes zoster, rubella and measles) occured during ongoing IFN therapy, while patients without signs of recurrent disease rarely showed clinical or serologic signs of viral infection. This may indicate resistance to IFN in patients that developed metastases during continuous IFN treatment.

COMBINATION THERAPY OF IFNβ AND CHEMOTHERAPY IN OSTEOSARCOMA PATIENTS

In West Germany a multicenter trial has been performed on 100 patients with osteosarcoma who were treated for 12 months according to a chemotherapeutic protocol (75-76). At 40 months of observation the expected continuous disease-free survival rate of 71 evaluable patients was 55%. When patients with fatal chemotherapy toxicity and local recurrence were removed from the analysis the corresponding rate was 56%. In 1980, another study was initiated where the same multicenter group doubled the methotrexate dose compared to the previous trial. Also, there was randomization between different chemotherapeutic regimens. Furthermore, some randomly selected patients received IFNβ for a short period of time in addition to chemotherapy. No differences were found between the chemotherapy groups or between the groups receiving and not receiving IFN, but the follow-up time for this trial is still too short to allow firm conclusions about long-term results. IFNβ was given during an extremely short period in this particular trial and based on present data it would be advisable to give larger doses during more prolonged periods of time.

The experimental data cited above argue perhaps for the use of IFNβ rather than IFNα in the treatment of human osteosarcoma. It has also been shown in experimental systems that it is possible to combine IFN with chemotherapy (see above) and the combinations of IFNα/β with IFNγ should be explored (77).

TREATMENT OF METASTATIC DISEASE

In Japan, Ito et al. (78-79) have treated 3 patients with pulmonary metastases of osteosarcoma with iv or im administration of natural human leukocyte IFN. In 2 cases the tumor mass decreased temporarily 6 and 8 months after IFN treatment and the serum alkaline phosphatase level returned to normal. In one case there was no effect at all on the growing tumors. Caparros et al. (80) have treated 11 pati- ents with metastatic osteosarcoma with human natural leukocyte IFNα daily for a minimum of 30 days. The doses were escalated from $3x10^6$ to 10^7 IU per day after 1-3 weeks of treatment or the patients were started directly on 10^7 IU per day. These patients were late in their disease and all except one had previously received chemotherapy. Metastases were localized to the lungs or to the skeleton. No objective reponses were noted.

We have employed a combination of natural IFNα and irradiation on metastatic disease (6). One previously untreated patient with biopsy-confirmed metastases received a combination of irradiation, 20 Gy over both lungs, and daily IFN treatment ($3x10^6$ IU im). This patient has survived for 6 years and is still being treated with IFN. Clearly, such a combination is of interest, not least because in vitro results have demonstrated additive or synergistic effects when IFN is combined with irradiation (81).

DISCUSSION

As yet, there is no firm evidence that IFN therapy can successfully be applied as a standard single therapy for osteosarcoma patients. Furthermore, it is still debated how efficient the use of adjuvant chemotherapy and lung irradiation are in the treatment of this disease (71-72). Of particular interest is the finding that the prognosis for non-adjuvant treated concurrent patients is more favourable than for the patients included in the historical groups (2,3,73,82). Since almost all trials employing adjuvant treatment have been non-randomized it is difficult, at present, to conclude how efficient these particular treatments are. It is clear from the results presented that we need more investigations on the biological effects of IFN treatment of osteosarcoma before we can optimize treatment. If the IFN system per se is not to be used in the future for osteosarcoma, biological investigations employing this particular system might give us information as to how to proceed with the use of other biological response modifiers in malignancies.

Most clinicians agree nowadays that osteosarcoma patients should be treated with differentiated surgery (63,83-84), i.e. resections should be made whenever possible. On the other hand, it is important to emphasize that the surgical modality should not be influenced by the adjuvant treatment in respect of the effort to achieve radical removal of the tumor. All possibilities of radical treatment must be used to attain optimal effects.

Recently, the genetics of the osteosarcoma disease has become of particular interest due to the changes found in the chromosomes of osteosarcomas occurring in patients with retinoblastoma (85-88). Other recent findings of interest are that the ground substance surrounding osteosarcoma cells differs from what is normally found (89) and that various growth factors are produced by bone sarcomas (90-95). It is possible that autocrine secretion of such factors might overcome normal growth limitations (96-98). The importance of oncogen effects also has to be emphasized, and here IFNs could be employed in various ways since they have been shown to influence such systems (99). The genetics of the IFN system should be explored (100) and the effects on differentiation more clearly established (101).

Perhaps these latter systems will allow the testing of tumor cells from various patients and in this way enable us to select the best treatment for a particular malignant disease. As for now, schedules for IFN treatment of various malignancies are based on purely empirical data (102). The relationship of chemotherapy to the IFN system is largely unexplored (103). Hopefully, however, all the various biological parameters in human osteosarcoma can be used for the development of a more efficient antitumor therapy, whether it be irradiation, chemotherapy or biological response modifier therapy.

Interdisciplinary discussion is helpful (104). Finally, it should be emphasized, that the various treatment modalities are not mutually exclusive.

ACKNOWLEDGEMENTS

This study was supported by grants from The Jonas Björnberg Memorial Foundation, the Cancer Society of Stockholm, the Swedish Cancer Association, and the Karolinska Institute Funds.

REFERENCES

1. Dahlin, D.C. and Coventry, M.B. J. Bone Joint Surg. (Am). 49:101-110, 1967.
2. Broström, L-Å., Aparisi, T., Ingimarsson, S., Lagergren, C., Nilsonne, U., Strander, H. and Söderberg, G. Intern. J. Rad. Oncol. Biol. Phys. 6:1717-1721, 1980.
3. Broström, L-Å., Aparisi, T., Ingimarsson, S., Lagergren, C., Nilsonne, U., Strander, H. and Söderberg, G. Intern. J. Rad. Oncol. Biol. Phys. 6:1711-1715, 1980.
4. Broström, L-Å., Ingimarsson, S., Strander, H. and Söderberg, G. Acta Orothop. Scand. 51:589-594, 1980.
5. Ingimarsson, S., Cantell, K., Carlström, G., Dalton, B., Paucker, K. and Strander, H. Acta Med. Scand. 209:17-19, 1981.
6. Strander, H., Einhorn, S., Ingimarsson, S., Aparisi, T., Broström, L-Å., Nilsonne, U., Söderberg, G. and Midander, J. Third Ann. Intern. Cong. Interferon Res., Nov. 1-3, Miami, 1982.
7. Strander, H., Aparisi, T., Broström, L-Å., Einhorn, S., Ingimarsson, S., Nilsonne, U. and Söderberg, G. In: Current Concepts of Diagnosis and Treatment of Bone and Soft Tissue Tumors. Springer Verlag, 1984, pp. 119-130.
8. Strander, H., Aparisi, T., Broström, L-A., Einhorn, S., Ingimarsson, S., Nilsonne, U. and Söderberg, G. In: Interferon - Research, Clinical Application and Regulatory Considerations (Eds. K.Z. Zoon, P.D. Noguchi and T-Y. Lin), Elsevier, New York, 1984, pp. 247-253.
9. Strander, H. In: Interferon IV (Eds. N. Finter and R.K. Oldham), Elesevier Science Publ., Amsterdam 1985, pp. 135-143.
10. Strander, H. Adv. Cancer Res. 46:1-265, 1986.
11. Strander, H. Texas Rep. Biol. Med., in press, 1987.
12. Jaffe, N., Frei, E. III, Traggis, D. and Bishop, Y. N. Eng. J. Med. 291:994-997, 1974.
13. Cortes, E.P., Holland, J.F., Wang, J.J., Sinks, L.F., Blom, J., Senn, H., Bank, A. and Glidewell, O. N. Eng. J. Med. 291:998-1000, 1974.

14. Sutow, W.W., Sullivan, M.P. and Fernbach, D.J. Proc. Am. Assoc. Cancer Res. and ASCO 15:172, 1974.
15. Malawer, M.M., Abelson, H.T. and Suit, H.D. In: Cancer - Principles and Practice of Oncology (Eds. V.T. DeVita, Jr., S. Hellman and S.A. Rosenberg), 2nd ed., Lippincott, Philadelphia, 1985, pp. 1293-1342.
16. Rosen, G. and Nirenberg, A. Cancer Treat. Rep. 66:1687-1697, 1982.
17. Rosen, G. and Nirenberg, A. Progr. Clin. Biol. Res. 201:39-51, 1985.
18. Jaffe, N. Cancer Treat. Res. 24:223-233, 1985.
19. Goorin, A.M., Delorey, M., Gelber, R.O., Price, K., Vawler, G., Jaffe, N., Watts, H., Link, M., Frei, E. III and Abelson, H.T. Cancer Treat. Symp. 3:155-159, 1985.
20. Rosen, G. Recent Results Cancer Res. 103:148-157, 1986.
21. Strander, H. and Einhorn, S. Phil. Trans. R. Soc. Lond. B299:113-117, 1982.
22. Strander, H. and Einhorn, S. Am. J. Clin. Oncol. 5:297-301, 1982.
23. Strander, H. In: Interferon - From Molecular Biology to Clinical Application (Eds. D. Burke and A. Morris), Cambridge University Press, Cambridge, 1983, pp.7-33.
24. Borden, E. Interferon 5:43-83, 1983.
25. Kirkwood, J.M. and Ernstoff, M.S. J. Clin. Oncol. 2:336-352, 1984.
26. Strander, H. In: The Biology of the Interferon System 1984. (Eds. H. Kirchner and H. Schellekens), Elsevier Biomedical Div., Amsterdam, 1984, pp. 489-493.
27. Stringfellow, D.A. (Ed.) Clinical Application of Interferons and Their Inducers, 2nd ed, Marcel Dekker, Inc., 1986.
28. Cantell, K. Immunobiol. 172:231-242, 1986.
29. Delbruck, H.G., Allouche, M. and Jasmin, C. Biomedicine Expr. 33:239-241, 1980.
30. Degré, M. Acta Pathol. Microbiol. Scand. B. 88:219-223, 1980.
31. Ito, M. and Buffett, R.F. J. Nat. Ca. Inst. 66:819-825, 1981.
32. Strander, H. and Einhorn, S. Int. J. Cancer 19:468-473, 1977.
33. Einhorn, S. and Strander, H. In: Human Interferon (Eds. W.R. Stinebring and P.J. Chapple), Plenum Publ. Corp., 1978, pp. 159-174.
34. Einhorn, S. and Strander, H. J. Gen. Virol. 35:573-577, 1977.
35. Rubin, B.J. and Gupta, S.L. Proc. Nat. Acad. Sci. USA 77:5928-5932, 1980.
36. Vastola, K.A., Diviak, L.F. and O'Malley, J.A. Proc. Am. Assoc. Cancer Res. 24:650, 1983.
37. Broström, L-Å. Cancer Lett. 10:83-90, 1980.
38. Dannecker, G., Treuner, J., Buck, J., Dilger, M. and Niethammer, D. J. Interf. Res. 5:541-550, 1985.
39. Degrén, M. and Glasgow, L.A. Acta Pathol. Microbiol. Immunol. Scand. B91:389-394, 1983.
40. Glasgow, L.A., Crane, J.L., Kern, E.R. and Youngner, J.S. Cancer Treat. Rep. 62:1881-1888, 1978.
41. Glasgow, L.A. and Kern. E.R. J. Nat. Cancer Inst. 67:207-212, 1981.
42. Davies, M.E. and Field, A.K. J. Interf. Res. 3:89-95, 1983.

43. Jasmin, C., Allouche, M., Judde, J.G. and Delbruck, H. Proc. Am. Assoc. Cancer Res. 22:261, 1981.
44. Satomi, H. Nippon Seikegeka Gakkai Zasshi, 57: 199-213, 1983.
45. Taylor, G.N., Kern. E.R., Braaten, B., Glasgow, L.A., Wrenn, M.E., Atherton, D.A., Mays, C.W., Charrier, K. and Ayoroa, G. J. Nat. Ca. Inst. 72:1137-1140, 1984.
46. Balkwill, F.R., Moodie, E.M., Freedman, C. and Fantes, K.H. Int. J. Cancer 30:231-235, 1982.
47. Gresser, I. and Tovey, M. Biochem. Biophys. Acta 516:231-247, 1978.
48. Riviere, Y. and Hovanessian, A.G. Cancer Res. 43:4596-4599, 1983.
49. Masuda, S., Fukuama, H. and Beppu, Y. Eur. J. Cancer Clin. Oncol. 19:1521-1528, 1983.
50. Bauer, H.C.F., Brosjö, O., Broström, L-Å., Nilsson, O.S., Reinholt, P.P. and Tribukait, B. Europ. J. Cancer Clin. Oncol. 2:821-830, 1986.
51. Brosjö, O., Bauer, H.C.F., Broström, L-Å., Nilsonne, U., Nilsson, O.S., Reinholt, F.P., Strander, H. and Tribukait, B. Cancer Res. 45:5598-5602, 1985.
52. Brosjö, O., Bauer, H.C.F., Broström, L-Å., Nilsson, O.S., Reinholt, F.P. and Tribukait, B. Cancer Res. 46:258-262, 1987.
53. Brosjö, O., Bauer, H.C.F., Nilsson, O.S. and Strander, H. In: The Biology of the Interferon System (Eds. K. Cantell and H. Schellekens), 1987, in press.
54. Embleton, M.J. Immunol. Ser. 23:181-207, 1984.
55. Embleton. M.J., Pimm, M.U. and Baldwin, R.W. In: Basic Mechanisms and Clinical Treatment of Tumor Metastasis (Eds. M. Torisu and T. Yoshida), Academic Press, New York, 1985, pp. 485-506.
56. Bruland, D., Fodstad, D., Funderud, S. and Pihl, A. Int. J. Cancer 38:27-31, 1986.
57. Shaw, A.R., Chan, J.K., Reid, S. and Seehafer, J. J. Nat. Ca. Inst. 74:1261-1268, 1985.
58. Vánky, F.T., Argov, S.A., Einhorn, S.A. and Klein, E. J. Exp. Med. 151:1151-1165, 1980.
59. Einhorn, S., Blomgren, H., Jarstrand, C., Strander, H. and Wasserman, J. In: The Biology of the Interferon System 1983 (Eds. E. de Maeyer and H. Schellekens), Elsevier Science Publ., Amsterdam, 1983, pp. 347-352.
60. Einhorn, S., Ling. O., Einhorn, N. and Strander, H. J. Clin. Lab. Immunol., 1987, in press.
61. Einhorn, S., Ling. P., Einhorn, N., Strander, H. and Wasserman, J. Cancer Immunol. Immunother., 1987, in press.
62. Strander, H., Cantell, K., Carlström, G. and Jakobsson, P.Å. J. Nat. Cancer Inst. 51:733-742, 1973.
63. Nilsonne, U. Sem. Hop. Paris. 58:1764-1766, 1982.
64. Strander, H. and Cantell. K. Ann. Med. Exp. Fenn. 44:265-273, 1966.
65. Cantell, K., Hirvonen, S., Kauppinen, H-L. and Myllylä, G. Methods Enzymol. 78:29-38, 1981.
66. Cantell, K., Hirvonen, S. and Koistinen, C. Methods Enzymol. 78:499-505, 1981.

67. Ingimarsson, S., Cantell, K. and Strander, H. J. Inf. Dis. 140:560-563, 1979.
68. Ingimarsson, S., Bergström, K., Broström, L-Å., Cantell, K. and Strander, H. Acta Med. Scand. 208:155-159, 1980.
69. Ingimarsson, S., Cantell, K. and Strander, H. In: Interferon - Properties and Clinical Uses (Eds. A. Khan, N.O. Hill and G.L. Dorn)., Leland Fikes Found. Press, Dallas, 1980, pp. 633-641.
70. Ingimarsson, S., Cantell, K., Carlström, G. and Strander, H. Intern. J. Cancer 26:395-399, 1980.
71. Lange, B. and Levine, A.S. Cancer Treat. Rep. 66:1699-1704, 1982.
72. Carter, S.K. Drugs 31:337-367, 1986.
73. Broström, L-Å., Strander, H. and Nilsonne, U. Clin. Orthop. 167:250-254, 1982.
74. Dimitrov, N.V., Meyer, C., Strander, H., Einhorn, S. and Cantell, K. Ann. Clin. Lab. Sci. 14:32-29, 1984.
75. Purfurst, C., Beron, G., Toragler, S., Kotz, R., Salzer-Kuntschik, M. and Winkler, K. Klin. Pädiatr. 197:233-238, 1985.
76. Winkler, K., Beron, G., Kotz, R., Salzer-Kuntschik, M., Beck, J., Beck, W., Brandeis, W., Ebell, W., Erttmann, R., Göbel, U., Havers, W., Henze, G., Hinderfeld, L., Höcker, P., Jobke, A., Jürgens, H., Kabisch, H., Landbeck, G., Preusser, P., Prindull, G., Ramach, W., Ritter, J., Sekera, J., Treuner, J. and Wüst, G. J. Clin. Oncol. 2:617-624, 1984.
77. Kurzrock, R., Rosenblum, M.G., Quesada, J.R., Sherwin, S.A., Itri, L.M. and Gutterman, J.U. J. Clin. Oncol. 4:1677-1683, 1986.
78. Ito, H., Murakami, K., Yanagawa, T., Ban, S., Sawamusa, H., Sakakida, K., Matsuo, A., Imanishi, J. and Kishida, T. Cancer 46:1562-1565, 1980.
79. Imanishi, J. and Kishida, T. Tex. Rep. Biol. Med. 41:647-652, 1982.
80. Caparros, B., Rosen, G. and Cunningham-Rundles, S. Proc. Am. Assoc. Cancer Res. 23:121, 1982.
81. Mossman, K.L., Hil, L.T. and Dritschilo, A. J. Nat. Med. Assoc. 74:1083-1087, 1982.
82. Taylor, W., Ivins, J., Dahlin, D., Edmonson, J. and Pritchard, D. Proc. Mayo Clin. 53:695-700, 1978.
83. Nilsonne, U. Clin. Orthop. 191:21-26, 1984.
84. Eckardt, J.J., Eilber, F.R., Grant, T.T., Mirra, J.M., Weisenberger, T.H. and Dorey, F.J. Cancer Treat. Symp. 3:117-130, 1985.
85. Matsunaga, E. J. Nat. Ca. Inst. 65:47-51, 1980.
86. Gilman, P.A., Wang. N., Fan, S.F., Reede, J., Khan, A. and Leventhal, B.G. Cancer Genet. Cytogenet. 17:123-132, 1985.
87. Hansen, M.F., Konfos, A., Gallie, B.L., Phillips, R.A., Fodstad, O., Brugger, A., Gedde-Dahl, T. and Cavenee, W.K. Proc. Nat. Acad. Sci. USA 82:6216-6220, 1985.
88. Dryja, T.P., Rapaport, J.M., Epstein, J., Goorin, A.M., Weichselbaum, R., Konfos, A. and Cavenee, W.K. Am. J. Hum. Genet. 38:59-66, 1986.
89. Shapiro, F.D. and Eyre, D.R. J. Nat. Ca. Inst. 69:1009-1016, 1982.

90. Heldin, C.H., Westermark, B. and Wasteson, A. Exp. Cell. Res. 136:255-261, 1981.
91. Blatt, J., White, C., Dienes, S., Friedman, H. and Foley, T.P.Jr. Bioch. Bioph. Res. Comm. 123:373-376, 1984.
92. Graves, D.T., Owen, A.J., Barth, R.K., Tempst, P., Winoto, A., Fors, L., Hood, L.E. and Antoniades, H.N., Science, 226:972-974, 1984.
93. Ross, R., Raines, E.W. and Bowen-Pope, D.F. Cell 46:155-169, 1986.
94. Weich, H.A., Sebald, W., Schairer, H.U., and Hoppe, J., FEBS Lett. 198:344-348, 1986.
95. Heldin, C.H., Johnsson, A., Wennergren, S., Wernstedt, C., Betsholtz, C. and Westermark, B. Nature 319:511-514, 1986.
96. Graves, D.T., Owen, A.J. and Antoniades, H.N. Cancer Res. 43:83-87, 1983.
97. Betsholtz, C., Westermark, B. and Heldin, C.H. Cell 39:447-457, 1984.
98. Westermark, B. and Heldin, C-H. Med Oncol. & Tumor Pharmacother. 3:177-183, 1986.
99. Samid, D., Flessate, D.M., Greene, J.J., Chang, E.H. and Friedman, R.M. Progr. Clin. Biol. Res. 202:203-209, 1985.
100. Strander, H. Ann. New York Acad. Sci. 350:168-170, 1980.
101. Rossi, G.B. Interferon 6:31-68, 1985.
102. Goldstein, D. and Laszlo, J. Cancer Res. 46:4315-4329, 1986.
103. Ernberg, J., Einhorn, S., Strander, H. and Klein, G. Biomedicine 35:190-193, 1981.
104. Taylor, W.F., Ivins, J.C., Pritchard, D.J., Dahlin, D.C., Gilchrist, G.S. and Edmonson, J.H. Mayo Clin. Proc. 60:91-104, 1985.

15

CLINICAL USE OF INTERFERONS IN THE TREATMENT OF MALIGNANT BRAIN TUMORS

M. NAGAI

Department of Neurosurgery, Dokkyo University School of Medicine, Mibu, Tochigi 321-02, Japan

ABSTRACT

A phase II study was conducted to evaluate clinical effectiveness of α, β and γ-type interferons (IFN) in the treatment of patients with malignant brain tumors. Of 46 cases of glioblastoma and malignant astrocytoma treated, satisfactory therapeutic responses were noted in 13 cases (28.3%) and also 3 of 7 cases of medulloblastoma. There was no significant difference in response rate between systemic administration and local (intratumoral or intrathecal) administration, nor between patients of fresh cases and those of recurrent tumors. The response rate was highest with β-IFN but did not significantly differ from that with α-IFN. With the γ-type IFN, efficacy in reduction of tumor size was low but immunologic parameters showed greater response.

As for side-effects, fever was fairly frequent but only rarely serious. There were only mild degrees of bone marrow suppression and liver dysfunction, and impairment of renal function was scarcely encountered. The results thus indicate that IFN therapy can be appraised as a useful adjunct to treatment of malignant brain tumors. Its application in combination with radiotherapy and chemotherapy may enhance the antitumor efficacy. The data also suggest its possible usefulness in long-term maintenance therapy for malignant brain tumors.

M. Revel (ed.), CLINICAL ASPECTS OF INTERFERONS.

INTRODUCTION

Since the late 1970s there has been a rapid increase in the number of reports on the clinical use of interferon (IFN) in cases of malignancy. In Japan, a project team was organized in 1978 for the study of clinical applications of IFN under the guidance of the Ministry of Health and Welfare. The course and outcome of investigations in this field that ensued are reviewed in detail in the reports of Yamazaki (1),(2). Joining the project team in 1979 we have taken charge of evaluating the clinical efficacy of IFN in malignant brain tumors. The clinical trial of IFN began with IFN-β of natural type (Toray Ind., Inc.) in Japan, where this type of IFN was the first to be produced on an industrial scale. Later, other types of IFN also became available, rendering comparative sutdies of IFNs feasible (3),(4),(5).

This report summarized the results of phase II studies of IFNs while reviewing the relevant reports so far published. Some comments are also made on the prospect of the clinical trial of IFNs.

SUBJECTS AND METHODS

A total of 53 cases of primary malignant brain tumor were used, including 46 cases of malignant glioma (glioblastoma and malignant astrocytoma) and 7 cases of medulloblastoma, in which there was a measurable residual tumor. This report deals mainly with the former of these groups.

In 12 of the 46 cases of malignant glioma there was a history of using more than 1 type of IFN (e.g., IFN-β initially and then IFN-γ during a subsequent episode of recurrence). Actually, 34 patients in all were included in this series. There were 16 males and 18 females who were in an age range of 5-74 years (mean: 40 years) and of whom 18 had glioblastoma and 16 had malignant astrocytoma. The tumor was fresh in 20 and recurrent in 14 of these patients.

In this study IFN-α, IFN-β and IFN-γ, each of natural and recombinant types, were used, as listed in Table 1. The dose employed was 2×10^{6}I.U./m^2 (except for L-IFN-γ, which was administerd in one-tenth the amount) which is equivalent to approximately 3×10^{6}I.U./body

Table 1. Interferon preparations used in phase II Studies.

HFIF:	human fibroblast interferon (natural β), (Toray). sp. act., 10^7
GKT-β:	recombinant interferon-β, (Kyowa-Toray). sp. act., >4×10^7
HLBI:	human lymphoblastoid interferon (natural α), (Wellcome). sp. act., 10^7
rIFN-αA:	recombinant leukocyte-A interferon, (Roche-Takeda). sp. act., 2-4×10^7
L-IFN-γ:	human leukocyte interferon-γ (natural γ), (Green Cross). sp. act., 1-2×10^7
TRP-2:	recombinant interferon-γ, (Takeda-Roche) sp. act., 5-9×10^6
GI-3:	recombinant interferon-γ, (Toray-Daiichi). sp. act., >1×10^7

(sp. act.: specific activity, I.U./mg protein)

for adults. In children or poor-risk patients 1/10-1/3 of the dose was given initially and then the dose was increased gradually with precautions against possible adverse side-reactions. As a rule, each type of IFN was administered daily (except for sundays) for 4 weeks or longer. In patients who had previously been on radio- or chemotherapy, IFN therapy was initiated at an interval of 4 weeks or longer to eliminate the after-effect of the preceding therapy.

For systemic administration, the intramuscular route was employed for IFN-α and a type of IFN-γ (GI-3). The intravenous route (drip infusion in 1 hr in 100 ml of physiological saline) was used for the administration of IFN-β and other types of IFN-γ, which fail to attain elevated blood levels following intramuscular injection. Since, for all these types of IFN, systemic administration resulted in no or few increased IFN titer in CSF, local administration was also undertaken: by intratumoral injection through the Ommaya's reservoir or through a catheter with its tip positioned within the ventricle, or by injection through lumbar puncture. The dose was the same as for systemic administration and given as a solution in 1-2 ml saline.

The efficacy of IFN was assessed at the end of a 4-week course of therapy according to the criteria of the Ministry of Health and Welfare in Japan for the evaluation of the clinical effect of cancer chemotherapy on solid tumors as follows:

1) Markedly effective (complete remission, CR): Disappearance of a

measurable or assessable lesion;

2) Effective (partial remission, PR): Over 50% tumor reduction or improvement, with no aggravation of secondary lesions;
3) Unchanged (no change, NC): Under 50% tumor reduction or under 25% exacerbation, with no aggravation of secondary lesions;
4) Progression (progressive disease, PD): Over 25% tumor enlargement with or without exacerbation of other lesions.

In cases where there were exacerbations of neurological signs and symptoms and/or deterioration in performance status, the response was not rated as remission even if a reduction in the tumor volume was evident.

RESULTS

Response rate.

The response rates obtained with different IFN products, both naturally occurring and derived by recombinant DNA technology, in 46 cases of glioblastoma or malignant astrocytoma are given in Table 2. The overall response rate for this group was 28.3% and highest response rate of 40.0% was attained with HFIF.

Table 2. Response rates in cases of glioblastoma or malignant astrocytoma.

IFNs	CR	PR	NC	PD	total	CR+PR/total (%)
HFIF	1	7	5	7	20	8/20 (40.0%)
GKT-β	0	1	2	0	3	1/3 (33.3%)
HLBI	0	1	2	0	3	1/3 (33.3%)
rIFN-αA	0	2	7	1	10	2/10 (20.0%)
L-IFN-γ	0	0	4	1	5	0/5 (0.0%)
TRP-2	0	1	2	2	5	1/5 (20.0%)
total	1	12	22	11	46	13/46 (28.3%)

Table 3 shows the response rates by mode of administration in cases of glioblastoma and malignant astrocytoma. In the group of local administration, the response rate was 31.6%, compared to 20.2% with systemic administration, the difference being not statistically significant.

Table 4 compares the response rates in fresh cases of glioblastoma or malignant astrocytoma where IFN was first administered post-

Table 3. Response rates by route of administration in cases of glioblastoma and malignant astrocytoma.

Route	CR	PR	NC	PD	total	CR+PR/total (%)	
local	1	5	8	5	19	6/19 (31.6%)	N.S.
systemic	0	3	4	8	15	3/15 (20.0%)	

Table 4. Response rates by history of cases of glioblastoma and malignant astrocytoma.

	CR	PR	NC	PD	total	CR+PR/total (%)	
fresh	0	6	7	7	20	6/20 (30.0%)	N.S.
recurrent	1	2	5	6	14	3/14 (21.4%)	

Table 5. Response rates in cases of medulloblastoma.

IFNs	CR	PR	NC	PD	total	CR+PR/total (%)
HFIF	2	1	1	0	4	
GKT-β	0	0	1	0	1	
HLBI	0	0	1	0	1	
TRP-2	0	0	1	0	1	
total	2	1	4	0	7	3/7 (42.9%)

operatively and in those cases of recurrent tumor. No significant difference was present between these 2 groups.

The response rate was as high as 42.9% in a group of 7 cases of mdulloblastoma (Table 5). Of note is the higher response rate attained even by systemic administration.

Overall, however, the response to IFN was not enough to increase survival to a significant extent.

Case report.

A representative responder is illustrated below.

Case Male, aged 30 years (Fig. 1).

The patient received radiation therapy in combination with chemotherapy after partial removal of a malignant astrocytoma in the right frontoparietal region. As the residual tumor was still of a considerable size and exhibited a tendency to enlargement two months after completion of the combined treatment, IFN therapy was started. CT scans revealed formation of a large cyst (at center

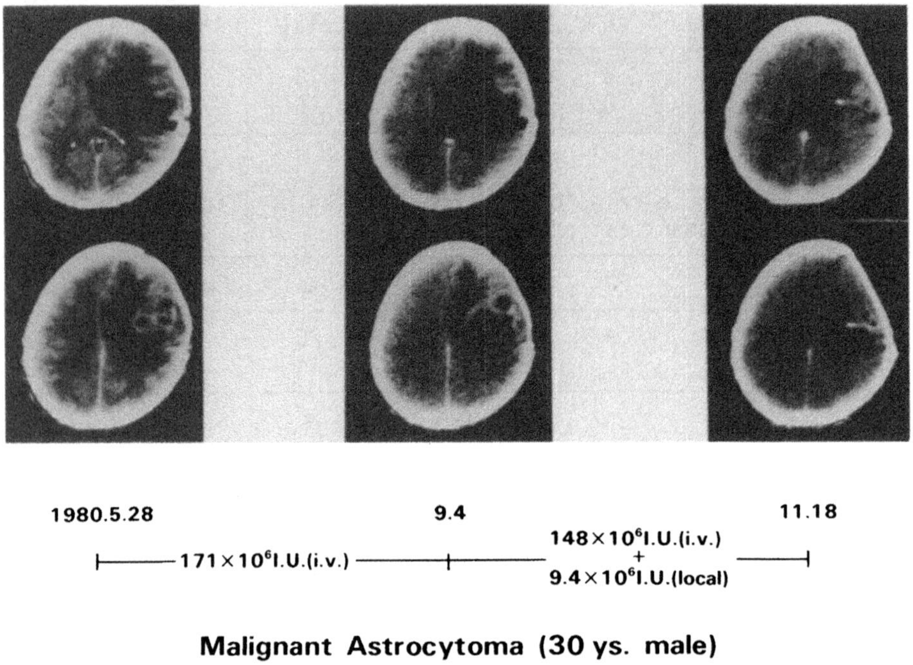

Fig. 1. CT pictures (contrast enhanced) of Case, corresponding to the interferon therapy.

in Fig. 1) after three months of systemic HFIF therapy by the intravenous route, thus suggesting some clinical responsiveness of the lesion to the treatment. In view of this, the local injection of HFIF was added via the catheter of an Ommaya reservoir passed into the cyst, and, as a result, the tumor shadow disappeared in 2.5 months with an improvement of Karnovsky performance index from 30% to 70%. The patient is in complete remission at present, six years after the treatment.

Adverse reactions.

Table 6 compares the adverse reactions to different IFN preparations used. Such reactions tended to be a little more frequent with recombinant than with natural products: the incidence of fever was high as 75-100% with the former, compared to 25-50% with the latter products, although a fever of 39°C or above was rare, if any,

Table 6. Adverse side-reactions to interferons.

	HFIF (N=30)	GKT-β (N=4)	HLBI (N=8)	rIFN-α (N=10)	L-IFN-γ (N=6)	TRP-2 (N-6)
Fever	48.3%	75.0%	25.0%	90.0%	100%	100%
Chills	13.8	25.0	12.5	20.0	0	16.7
Lassitude	13.8	50.0	25.0	20.0	33.3	33.3
Anorexia	0	25.0	0	0	16.7	33.3
Headache	6.9	25.0	0	10.0	16.7	33.3
Vomiting	3.4	0	0	30.0	0	16.7
Diarrhea	0	0	0	10.0	0	0
Myalgia	0	25.0	0	0	0	0
Leucopenia	20.7	100	37.5	80.0	33.3	50.0
Anemia	0	0	12.5	40.0	0	0
Thrombocytopenia	0	0	12.5	50.0	0	0
Hepatic dysfunction	24.1	50.0	37.5	70.0	33.3	50.0
Myocardial ischemic change	0	0	0	10.0	0	0

and could be prevented by pretreatment with acetaminophen. Chills due to fever, malaise, anorexia and headache were each observed in 10 to 50% of cases.

The incidence of leukocytopenia as a reaction to IFN therapy was 20-40% with naturally occurring products, compared to 50-100% with recombinant products. A rise in serum GOT and GPT was also observed in 25-40% and 33-70% respectively of cases. All these reactions were mild enough to reverse in 3-14 days of rest. Thrombocytopenia and anemia, both of mild severity, developed in 50% and 40% respectively of cases treated with recombinant IFN-α preparation.

In general, such adverse reactions were less frequent with local than with systemic administration of IFN. Incidentally, local infection at the site of placement of the Ommaya reservoir occurred in 2 adults and 3 children.

DISCUSSION

In 1979 we embarked on the phase I and phase II studies of IFN and have since reported its efficacy in the treatment of malignant brain tumors (3),(4),(5). At about the same time, Salford et al. (6) reported their experience with leukocyte IFN in 2 cases of recurrent malignant glioma. These authors administered IFN intratumorously as we did and achieved no reduction in tumor size but an increase in central necrosis with a tendency to tumor encapsulation and a decline in grading of disease.

In a phase I trial of leukocyte IFN given intramuscularly, Boëthius et al. (7) could observe increased survival of only 1 of 12 patients so treated. Hirakawa et al. (8) also achieved partial remission in 2 of 8 cases treated intramuscularly with leukocyte IFN. Nakagawa et al. (9) found leukocyte IFN effective in 2 of 13 cases when administered locally.

In 1985, 2 reports appeared concerning the phase I study of IFN. Obbens et al. (10) reported that they obtained slight tumor regression in 1 of 3 patients with grade III astrocytoma treated intratumorously with leukocyte IFN and that the other 2 patients had stable disease. These authors also gave leukocyte IFN by intraventricular injection to 5 patients with leptomeningeal metastases (1 with pineal region tumor, 1 with large cell lymphoma and 3 with metastatic tumor) and achieved "complete remission" or eradication of malignant cells in 4 of these patients.

In the other report, which was made by Bogdahn et al. (11), IFN-β given by intravenous infusion was described as being well tolerated clinically with no more than mild toxicity in 3 cases of low grade astrocytoma; no mention was made of the antitumor effect of the preparation used.

Then came a phase II study made by Mahaley et al. (12). In this trial lymphoblastoid IFN was administered systemically to 17 patients with recurrent malignant glioma: by intravenous infusion to 8 and by intramuscular injection to 9 patients. The results were quite satisfactory with a favorable response elicited in 7 of these patients.

Duff et al. (13) administered IFN-β (natural type) by both method of intravenous infusion and intratumorous injection to 12 patients with recurrent glioblastoma multiforme. In these patients the IFN preparation presented no problem as to its tolerance or toxicity but instead produced no overt increase in mean survival time.

Table 7. Cooperative phase II studies of treatment of glioblastoma and malignant astrocytoma by various preparations of IFN in Japan.

IFNs	No. of evaluated case	No. of CR+PR	Response rate (%)	Refer-rence
HFIF	57	8	14.0	(14)
rIFN-αA	37	4	11.0	(15)
HLBI	51	13	25.5	(16)
rIFN-α2	26	7	26.9	(17)

In Japan, IFN preparations have been subjected to full scale phase II trials in patients with malignant glioma (14),(15),(16),(17). Some of the trials with which we were concerned are listed in Table 7.

The previous studies cited above, including ours, show that IFN is of some but no satisfactory benefit in the treatment of glioblastoma or malignant astrocytoma. Contrary to the initial enthusiastic expectation, there seems to be a limitation to the clinical efficacy of IFN. Noteworthy in this connection is the high incidence of response of grade NC (no change) (Table 2), a response which involves suppression (cytostatic effect) of tumor growth for some time even if culminating in no reduction in tumor size. Indeed, Mahaley (12) defined the favorable response as a 25% or more reduction in tumor size or, alternatively, NC of 12 weeks or longer duration. The response rate in our series would be higher should it be determined by such criteria. On the other hand, the response rate would be lower in series consisting solely of patients with glioblastoma, excluding those with malignant astrocytoma, as in those reported by Duff (13). Further investigation is necessary in this respect.

In some patients in our series IFN was administered locally to determine its direct effect on tumor growth. It made no significant difference in response rate whether IFN was administerd locally or systemically, as shown in Table 3. It must be noted, on the other hand, that CR was achieved in a patient treated locally with IFN and that IFN given intrathecally was effective against dissemination of medulloblastoma (18). Obbens (10) also made mention of a marked response of leptomeningeal metastases elicited by IFN given intraventricularly.

IFN-α was administered by the intramuscular route and IFN-β by intraveous drip infusion in view of the differing pharmacokinetics (19),(20),(21),(22) of those types of IFN. No significant difference was noted in overall clinical response rate between these types of IFN. Incidentally, IFN-γ was administered systemically in most cases. Contrary to expectation, IFN-γ proved to cause no marked reduction in tumor size. However, it affected certain immunologic parameters to a greater extent than the α or β type. We have also placed emphasis on such an indirect action of IFN-γ on immunologic parameters (23).

Documented adverse side-reactions to IFN are generally mild enough to be well tolerable. Fever as such is frequent and may be associated with flu-like symptoms, as in our series. Even convulsive seizure may rarely produced (7),(12),(14),(15). Ischemic myocardial changes have occurred in 1 case (15) aside from those reported by us.

The hematological or hepatic toxicity of IFN is often mild enough to reverse upon temporary cessation of therapy in most cases. The central nervous system toxicity of IFN has also been documented (24). Mahaley (12) frequently (9/17) observed reactions of this category, such as confusion, loss of memory and lethargy. We have excluded these reactions on the ground that they are indistinguishable from the symptoms and signs of brain tumor per se. Infection associated with local IFN therapy has been observed in some isolated cases (9),(10),(13).

In summary, adverse side-reactions to IFN appear to be milder than those to the conventional chemotherapeutic agents. This will render IFN therapy relatively easy to perform.

One of the fields in which IFN can have a potential clinical value is combination therapy with radiation or anti-cancer drugs of malignant brain tumors. It has already been reported (25),(26),(27) that IFN given in conjunction with radiation is useful in inhibiting growth of glioma. A phase III trial of IFN given in combination radiation and ACNU [1-(4-amino-2-methyl-5-pyrimidinyl)methyl-3-(2-chloroethyl)-3-nitrosourea hydrochloride] in cases of malignant glioma is now under way in Japan.

Long-term maintenance therapy following remission induction will constitute an important aspect of IFN therapy.

CONCLUSION

IFN elicited a favorable response in 28.3% of 46 cases of glioblastoma and malignant astrocytoma and also in 3 of 7 cases of medulloblastoma. No serious side-effects were observed. IFN therapy appears to provide a useful adjunct in the treatment of malignant brain tumors. A properly planned combination therapy with radiation or chemotherapeutic agents will be needed to enhance the clinical efficacy of IFN.

REFERENCES

1. Yamazaki, S. Jap. J. Med. Sci. Biol. 36: 261-271, 1983.
2. Yamazaki, S. Jap. J. Med. Sci. Biol. 37: 209-223, 1984.
3. Nagai, M., Arai, T., Kohno, S. and Kohase, M. In: The Clinical Potential of Interferons (Eds. Kono, R. and Vilcek, R.J.), Univ. of Tokyo Press, 1982, pp. 257-273.
4. Nagai, M., Arai, T., Kohno, S. and Kohase, M. Texas Rep. Biol. Med. 41: 693-698, 1982.
5. Nagai, M. and Arai, T. Neurosurg. Rev. 7: 55-64, 1984.
6. Salford,L.G., Strömbald, L.-G., Nordström, C.-H., Hornmark-Stenstam, B., et al. Acta Neurochir. 56: 130-131, 1981.
7. Boëthius, J., Blimgran, H., Collins, V.P., Greitz, T. and Strander H. Acta Neurochir. 68: 239-251, 1983.
8. Hirakawa, K., Ueda, S., Nakagawa, Y., Suzuki, K., Fukuma, S. et al. Cancer 51: 1976-1981, 1983.
9. Nakagawa, Y., Hirakawa, K., Ueda, S., Suzuki, K., Fukuma, S. et al. Cancer Treat. Rep. 67: 833-835, 1983.
10. Obbens, E.AMT., Feun, L.G., Leavens, M.E., Savaraj, N., Stewart, D.J. and Gutterman, J.U. J. Neuro-Oncol. 3: 61-67, 1985.
11. Bogdahn, U., Fleischer, B., Hilfenhaus, J., Röthig, H.J., Krauseneck, P., Mertens, H.G. and Przunick H. J. Neuro-Oncol. 3: 125-130, 1985.
12. Mahaley, M.S., Urso, M.B., Whaley, R.A., Blue, M., Williams, T. E. et al. J. Neurosurg. 63: 719-725, 1985.
13. Duff, T.A., Borden, E., Bay, J., Peipmeier, J. and Sielaff, K. J. Neurosurg. 64: 408-413, 1986.
14. Nagai, M. J. Jap. Soc. Cancer Therapy 18: 60-68, 1983.
15. Takakura, K. Jpn. J. Cancer Chemother. 12: 913-920, 1985.
16. Furue, T. Rinsho Iyaku 1: 1103-1121, 1985.
17. Kitamura, K. in press
18. Nagai, M. and Arai, T. Shoni-Naika 18: 815-819, 1986.

19. Hanley, D.F., Wiranowska-Stewart, M. and Stewart,II, W.E. Int. J. Immunopharmac. 1: 219-226, 1979.
20. Billiau, A., de Somer, P., Edy, V.G., de Clercq, E. and Hermans H. Antimicrob. Agents Chemother. 16: 56-63, 1979.
21. Hayes, T.G. Arch. Virol. 67: 267-281. 1981.
22. Borden, E.C., Hogan T.F. and Voelkel, J.G. Cancer Res. 42: 4948-4953, 1982.
23. Nagai, M. and Arai, T. J. Neuro-Oncol. 4: 106, 1986.
24. Rohatiner, A.Z.S., Prior, P.F., Burton, A.C., Amith, A.T., Balkwill, F.R. et al. Br. J. Cancer 47: 419-422, 1983.
25. Nederman, T. and Benediktsson, G. Acta Radiol. Oncol. 21: 231-234, 1982.
26. Korosue, K., Tamaki, N., Matsumoto, S., Takeshita, I. and Fukui, M. Neurol. Med. Chir. (Tokyo) 24: 233-239, 1984.
27. Nagai, M. and Arai, T. Radiother. System Res. Suppl. 3: 24-25, 1986.

16

INTERFERON TREATMENT OF NEURO-ENDOCRINE TUMORS OF PANCREAS AND GUT

K. Öberg[x], B. Eriksson[x] and I. Norheim[xx]

Ludwig Inst for Cancer Research, Uppsala Branch[x] and Dept of Internal Medicine[xx], University Hospital, 751 85 Uppsala, Sweden

INTRODUCTION

Neuro-endocrine gut tumors are relatively slowly growing neoplasms, but due to increased secretion of biologically active hormones, they often provoke severe clinical symptoms such as diarrhoea, flushing, asthma, hypoglycemia, recurrent ulcers, necrolytic migratory erythema and right heart failure (1,2,3,4). The majority of the secretory products from the neuro-endocrine tumors are peptide hormones, but also other substances are released, such as glycoproteins (HCG subunits and chromogranin A), serotonin, bradykinins and prostaglandins. The origin of neuro-endocrine gut and pancreatic tumors has been discussed and is not yet elucidated but they are all sharing in common so called APUD- characteristics, which means uptake of amines, precursors, storage and release of peptides and hormones (5). Neuro-endocrine pancreatic tumors are suggested to develop from pancreatic stem-cells (nesidioblasts) with the potential capacity to differentiate into both exocrine- and endocrine-type of cells.

Clinical syndromes related to neuro endocrine pancreatic tumors are:

The Zollinger-Ellison syndrome, with recurrent gastric ulcers and diarrhoea due to excessive secretion of gastrin; hypoglycemic syndromes due to insulin excess; glucagonoma syndrome with diabetes, necrolytic migratory erythema, anemia and weight loss due to increased secretion of glugagon; the WDHA-syndrome including water diarrhoea, hypokalemia, achlorhydria due to increased production of primarily vasoactive intestinal polypeptide (VIP), but also PHM and calcitonin; somatostatinoma syndrome including gall-stones, obstipation, diarrhoea and diabetic glucose tolerance due to excessive production of somatostatin.

The majority of these tumors are potentially malignant and more than 50% present metastases at time of diagnosis. The only exception is insulin producing tumors, of which more than 90% are benign. About 30-40% of patients with malignant neuro-endocrine tumors of the pancreas present without any characteristic clinical features. These so called "non functioning tumors" are clearly neuro-endocrine, displaying positive staining with Grimelius' silver-staining technique and secretory granules on electron microscopy. A number of these tumors secrete substances that do not cause any specific clinical symptoms; e.g. pancreatic polypeptide (PP), HCG alpha and beta subunits and chromogranin A. Other explanations for the lack of clinical symptoms are down regulation of the hormone receptors or secretion of biologically inactive hormones, production of active hormones but no release and finally simultaneous secretion of inhibitory peptides such as somatostatin. Malignant neuro-endocrine tumors of the pancreas are also known to secrete ACTH or corticotropine releasing factor (CRF) causing Cushings' syndrome as well as growth hormone releasing factor (GHRF) causing acromegaly. Both these tumor types are very rare.

Neuro-endocrine tumors of the gut might be divided into three main categories, fore-gut, mid-gut and hind-gut carcinoids (6). The so called "fore-gut" carcinoid tumors might develop from neuro-endocrine cells of the bronchus, thymus, gastric or duodenal mucosa. These tumors produce substances such as 5 HTP, histamin, ACTH, GHRF, PP, HCG-alpha and beta subunits, gastrin, somatostatin and calcitonin and cause clinical symptoms related to the hormon production. The most common localization of gut neuro-endocrine tumors is that of the "mid-gut" region, the majority of tumors originating from the appendix. Almost all appendix carcinoids are benign and might have developed from neuronal elements (7).

The second most common localization of neuro-endocrine "mid-gut" tumors is the ileum and jejunum and these tumors originate from the enterochromaffine cells (EC-cells) of the intestinal mucosa. These cells have the

capacity to produce serotonin and peptide hormones such as tachykinins and chromogranin A. The majority of these tumors have a malignant potential and metastatize to regional lymphnodes and the liver. When liver metastases occur, the patient might develop the carcinoid syndrome including diarrhoea, flush, asthma and right heart failure (4). Ethiological agents include serotonin, tachykinins, bradykinins and prostaglandins (4,8).
The third group of neuro-endocrine gut tumors are the so called "hind-gut" carcinoids, which may develop from neuro-endocrine cells in the mucosa of the distal colon and rectum. The most common localization is rectum. These tumors do not present any certain clinical symptoms except for obstruction or bleeding. The tumor cells contain immunoreactive somatostatin, PP, PYY and HCG-alpha subunits.

TREATMENT

The primary treatment of neuro-endocrine gut and pancreatic tumors is surgery (1). However, as the majority of malignant tumors display metastases already at the time of diagnosis, other treatment procedures are mandatory. The most important aim of any treatment of neuro-endocrine neoplasm is to alleviate clinical symptoms due to hormone excess. The clinical syndromes such as severe diarrhoea, recurrent gastric ulcers, hypoglycemia are life-threatning because of severe metabolic disturbances.

As a majority of the tumors present clinical symptoms, it is as important to control these by inhibiting hormone production and/or release and blocking their peripheral biological effects as controlling the rather slow tumor growth. It has been suggested that some of the secretory products from these neuro-endocrine tumors might be autocrine growth factors for the tumors. It has already been demonstrated for small cell lung cancer which is closely related to carcinoids, that bombesin is an autocrine growth factor (9). Thus, by controlling the production and release of secretory products from the tumor itself, tumor growth might also be controlled.

During 1960th and -70th a specific cytotoxic agent streptozocin became available and has been used alone or in combination with other cytotoxics such as 5-FU or adriamycin (10, 11). Objective responses have been documented in 60-70% of patients with malignant endocrine pancreatic tumors, when the combination of streptozocin plus 5-FU or adriamycin has been used. However, the closely related carcinoids showed objective responses in only 10-30% of the patients (10). During the last years, analogues of natural somatostatin e.g. SMS 201-995 has been available (12). This drug has been very useful in controlling clinical symptoms of different neuro-endocrine gut and pancreatic tumors. It has been most valuable in the treatment of the severe diarrhoea in patients with WDHA-syndrome, but it also controls the diarrhoea as well as the flushing reaction in carcinoid patients. Neither cytotoxic treatment nor somatostatin analogues have demonstrated any permanent cure to patients with malignant neuro-endocrine gut and pancreatic tumors. The ultimate therapy for these tumors has not yet been found and there is a certain need of developing new therapeutic regimens.

INTERFERON

We have used both human leukocyte interferon and recombinant interferon alpha-2 (Intron-A. Scheering Corp. USA). The human leukocyte interferon has been produced according to the method of Cantell (13) at the Interferon Laboratory, Biomedical Center, Uppsala (G. Alm). The patients are started on subcutaneous or intramuscular daily doses of 3 MU of human leukocyte interferon for three days, followed by 6 MU/day thereafter. In the study with recombinant alpha-2 interferon (Intron-A) we have used an initial dose of 2 MU/m^2 subcutaneously for three days, then increasing up to 5 MU/m^2 three times a week. If the patient presented serious side-effects, the dose was reduced down to maximal tolerable dose (5-6 MU s.c. three times a week) for long-term treatment.

EVALUATION OF CLINICAL RESPONSES

OBJECTIVE RESPONSE:

The criteria for objective responses are those presented

in other studies with cytotoxic drugs or somatostatin analogues. A complete remission is obtained if the tumor markers normalize and all visuable tumor tissue disappear on computerized tomography (CT). An objective response is achieved if the patient shows a more than 50% reduction of the principal tumor marker and/or reduction of tumor size measured as the product of 2 perpendicular diameters on CT-scan. Stable disease is less than 50% reduction of tumor markers and/or tumor size without any new lesions seen on CT-scan. Progressive disease is noted when the tumor markers and/or tumor size increase by more than 25% or if new lesions appears.

SUBJECTIVE RESPONSE:

Before start of treatment the patient was asked about clinical symptoms such as flushing, diarrhoea and other symptoms related to these tumors. The number of attacks and the severity of these episodes were then followed during treatment and noticed in a questionnaire.

RESULTS

We are presenting data from four different studies on interferon treatment of 87 patients with neuro-endocrine tumors. In the first study, 22 patients with malignant endocrine pancreatic tumors have been treated with human leukocyte interferon. Seven patients presented the WDHA-syndrome, four the Zollinger-Ellison syndrome, one patient had an insulinproducing tumor and one a somatostatinproducing tumor. Nine patients had clinically "nonfunctioning" tumors. All these patients had previously been subjected to treatment with chemotherapy but demonstrated progressive disease. Seventeen out of 22 patients (77%) showed an objective response with a median duration of 8.5 months (range 2-36 + months). Fifteen out of 22 patients had decreased tumor markers. Six had significantly decreased tumor masses and two patients showed a complete remission. Stable disease was noted in one patient with a duration of 27 months. Progressive disease was noted in 4 of the 22 patients (14). The effects of interferon on circulating hormone levels in one of the patients is illustrated in Figure 1 och the effect on tumor masses is illustrated in Figure 2, a and b.

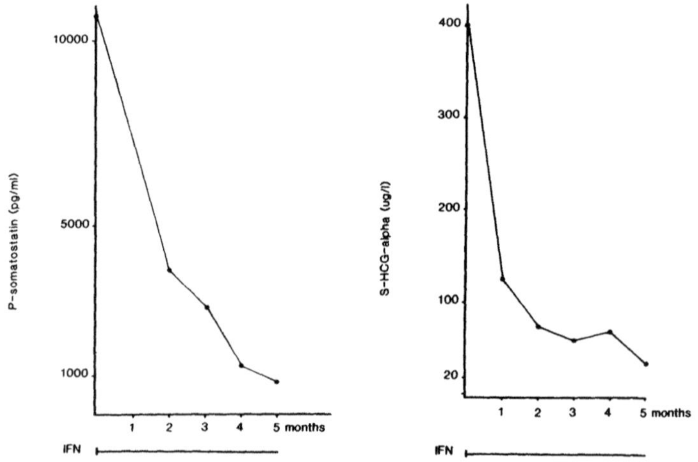

Fig 1. Plasma somatostatin and serum HCG-alpha levels during treatment with interferon in a patient with a malignant endocrine pancreatic tumor.

Fig 2A

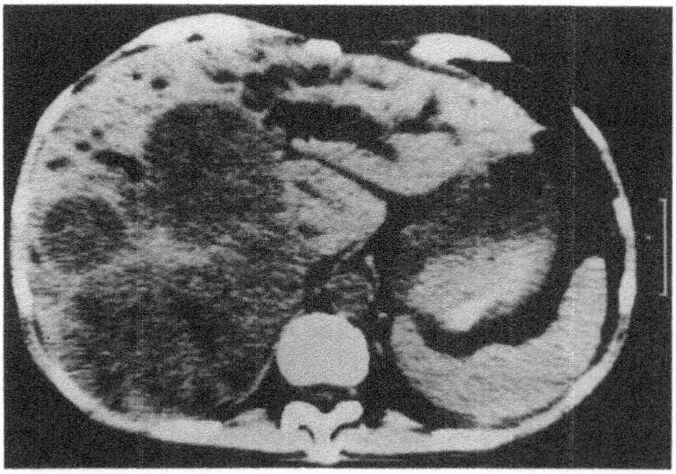

Fig 2B

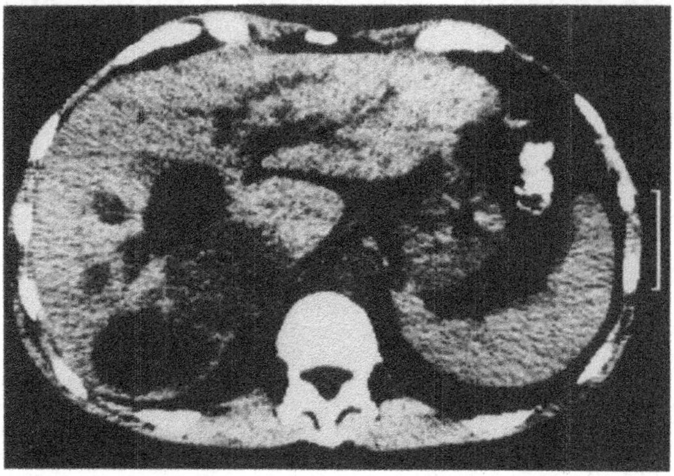

Fig 2A and B. Computed tomography of the liver in a patient with a malignant endocrine pancreatic tumor, gastrinoma, before interferon treatment (A) and after 3 months therapy (B). Note the significant reduction of liver metastases.

The <u>second study</u> is a long-term study in 36 patients with carcinoid tumors (29 patients had "mid-gut" carcinoids, four lung carcinoids and one ovarian, rectal and unlocalized primary tumor, respectively). Thirty-two out of 36 patients had liver metastases and 19/36 patients had received previous cytotoxic treatment. An objective tumor response was noticed in 17 of the 36 patients (47%), 14 of 29 patients with "mid-gut" carcinoids (48%) and 3 of the 4 patients with lung carcinoids (75%). A further 14 patients showed stable disease (39%). Five of the 36 patients (14%) showed progressive disease. Sixteen of the 17 patients with objective responses had a significant decrease of tumor markers and four of them also had a significant reduction of tumor size. Two patients showed a complete remission. The median duration of objective response was 34 months and the median duration of stable disease 25 months (15). The effects on the main tumor marker, urinary 5-HIAA, is illustrated in Figure 3.

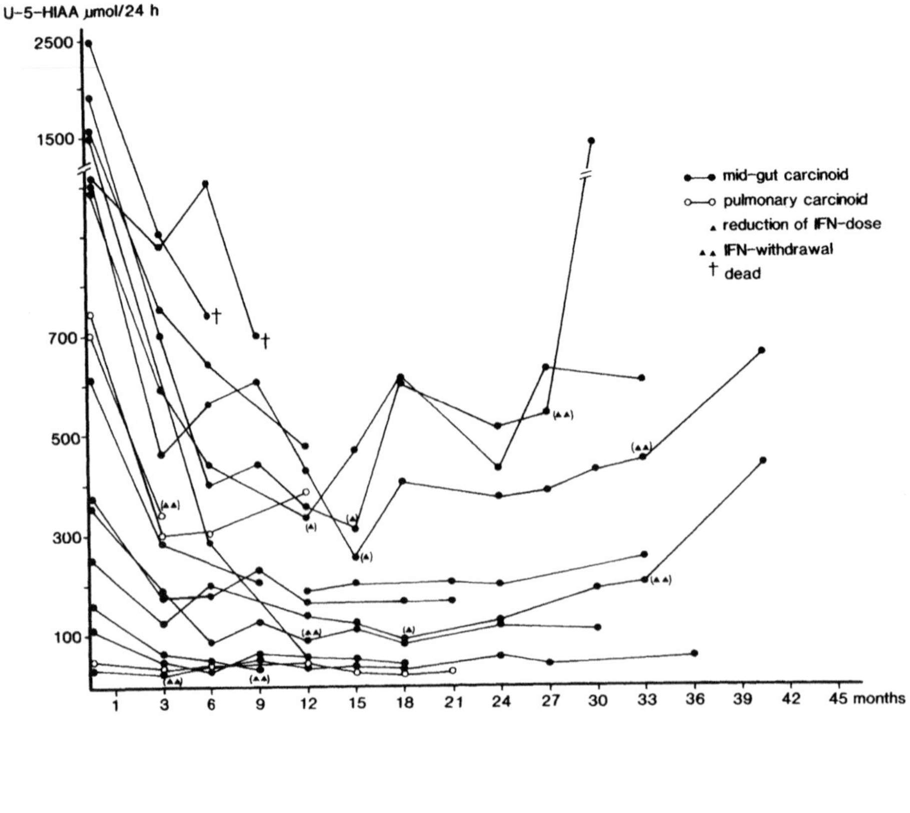

The third study is a randomized controlled study between human leukocyte interferon and streptozocin plus 5-FU in malignant "mid-gut" carcinoid tumors. Doses of streptozocin and 5-FU were those recommended by other investigators (19). Twenty patients with "mid-gut" carcinoid tumors and liver metastases were included in the study. Ten patients received interferon and ten patients chemotherapy. After six months of therapy an objective response was noted in 5/10 patients on interferon (50%) and in none of the patients on chemotherapy. Stable disease was noted in 50% of the patients on interferon as well as in 50% of the patients on chemotherapy. Progressive disease was not seen in any of the patients on interferon, whereas five of the patients (50%) on chemotherapy progressed. A chi-square analysis showed that interferon was significantly better than chemotherapy ($p=0.0067$).

In the fourth study recombinant interferon alpha-2 (Intron-A) was used in 19 patients with carcinoid tumors. Seventeen of the patients had malignant "mid-gut" carcinoids and two malignant lung carcinoids. Ten out of the 19 patients had earlier received interferon treatment, but this treatment was withdrawn 2 to 12 months prior to treatment with Intron-A. After 6 months therapy, ten out of 19 patients (53%) had an objective response, whereas 7 patients showed stable disease (37%) and 2/19 patients (10%) progressive disease.

In all the studies the number of patients with subjective responses was greater than the number with an objective response. About 70-80% of the patients experienced a reduction of the clinical symptoms and improved quality of life.

ADVERSE EFFECTS

The side-effects of interferon treatment are listed in Table 1. About 90% of the patients showed "flu-like" symptoms for an initial 3-6 days of treatment, which then gradually declined and in most patients disappeared when the treatment continued for a longer period of time. Sixty percent of the patients experienced mild to moderate fatigue throughout the whole treatment period. About 50-65% of the patients developed granulocytopenia and/or of trombocytopenia as well as decreasing haemoglobin content. However, only very few patients had a severe reduction of these blood components necessitating a dose reduction. About 1/3 of the patients showed slightly to moderately increased levels of liver enzymes. This adverse reaction was reversed when the dose was reduced. Increased serum trigycerides was noted in 1/3 of the patients, especially in patients receiving human leukocyte interferon. Liver steatosis developed in about 20% of the patients. Another adverse reaction, which has not been reported earlier, was the autoimmune phenomenon noticed in about 10-20% of the patients. A SLE-syndrome was seen in one patient, thyroid autoantibodies in five and parietal cell antibodies in one. We have also seen exacerbation of psoriasis in a few patients. Mental depression was noted in only one patient.

TABLE 1

ADVERSE EFFECTS (Long-term study, carcinoids)

"Flu-like" symptoms	(89%)
Decreased Hemoglobin content	(58%)
Anemia (<110 g/l)	(31%)
Decreased no of leucocytes	(67%)
Leucocytes (<2.0 x 10^9/l)	(3%)
Decreased no of platelets	(74%)
Platelets (<150 x 10^9/l)	(14%)
Increased liver enzymes	
s-ASAT	(31%)
s-ALAT	(31%)
s-Alkaline phosphatases	(22%)
s-Bilirubin	(8%)
s-LD	(6%)
Increased blood lipids	
(s-Triglycerids)	(32%)
Liver steatosis	(19%)
Thyroid autoantibodies	(15%)
Hypothyroidism	(6%)
Parietal cell antibodies	(3%)
SLE-syndrome	(3%)
Leucoplacia (Buccal)	(3%)
Mental depression	(3%)

DISCUSSION

Patients with malignant endocrine pancreatic tumors have demonstrated very good results with a combination therapy of streptozocin and 5-FU or adriamycin (10, 11). The objective response rates have been about 60 to 70%, which are comparable to our results with human leukocyte interferon. Until further studies including a greater number of patients confirm our response rates, we think that chemotherapy should be the first line treatment in malignant endocrine pancreatic tumors. However, if the patients fail to respond to chemotherapy or when they present a very advanced disease and can not tolerate chemotherapy, interferon treatment should be considered. When patients with malignant endocrine pancreatic tumors are followed for longer periods of time, the tumor might become resistant to the treatment and thus an alternation between chemotherapy and interferon might improve the results. Neither chemotherapy nor interferon has so far cured any patient with malignant endocrine pancreatic tumors, although complete remissions with varying duration have been obtained with both therapies.

Carcinoid tumors are closely related to malignant endocrine pancreatic tumors, but they show quite different sensitivity to chemotherapy. We have obtained short-lasting objective responses in only about 10% of the patients and the best result reported so far is about 30% (10). The new somatostatin analogue SMS 201-995 has also been suggested as the treatment of choice (12,21), but variable (30-72%) objective response rates have been published and the effects short-lasting in most of the patients. Furthermore, the dose must steadily be increased because of tachyphylaxis and no convincing data on tumor regression have been reported. We think that the somatostatin analogue is an excellent "acute" drug in patients with carcinoid crisis and/or severe clinical symptoms and might be an adjunct to other treatments.

Interferon treatment has so far showed the highest objective response rates of about 50%. We have obtained similar results in the three different studies reported, comprising 75 patients. Some of them are followed up to five years. Similar objective response rates have been obtained with both human leukocyte interferon and recombinant interferon alpha-2 (Intron-A). The majority of the responses are noticed in tumor markers with concomitant amelioration of the clinical symptoms. More than 70% of the patients experienced decreased carcinoid symptoms, such as flushing and diarrhoea. Only a smaller fraction of the patients (10-15%) showed a significant decrease of tumor size. However, both patients with objective responses and stable disease, totally about 80% of the patients, showed an arrest of the tumor growth for extended periods of time.

The mechanism of actions of interferon in patients with malignant neuro-endocrine tumors are not explained at the moment. Interferon is known to decrease protein synthesis and DNA replication, which might account for the decreased production of different peptide hormones (16,17). It has also been demonstrated that interferon interferes with oncogen expression. The diminished production and/or release noticed after interferon therapy of hormones and other secreted proteins from the tumors, which might act as autocrine growth factors, might account for the inhibited growth of the tumors. Furthermore, studies from our own group has demonstrated increased expression of beta-2 microglobulin on carcinoid tumor cells after treatment with interferon, which might stimulate the immunesystem and via an indirect mechanism arrest tumor growth (18).

The adverse effects of interferon treatment are those reported earlier and the "flu-like" symptoms and fatigue are the most dose limiting symptoms. All side-effects including haematologic, increased liver enzymes, liver steatosis and increased serum triglycerides are dose dependent and reversible. More serious side-effects are the autoimmune phenomena seen in some of the patients. We believed initially that those side-effects were due to contamination with small amounts of gamma-interferon in the interferon preparation (19). However, recent studies have demonstrated that no such contamination can be found in the human leukocyte interferon preparation (20) and furthermore, similar autoimmune effects have been noticed in patients on recombinant alpha-2 interferon. We thus believe that these autoimmune reactions might be effects of the alpha interferons themselves.

FUTURE ASPECTS

At the moment neither optimal doses nor dose intervals have been elucidated for interferon treatment. Furthermore, the route of administration of interferon has to be investigated further, perhaps intraarterial or intravenous infusions directly to the tumor might be more efficient than subcutaneous injections. The combination of interferon with cytotoxic drug has to be evaluated and also treatment with interferon inducers.

SUMMARY

At the moment interferon treatment seems to be the treatment of choice for malignant carcinoid tumors, more efficient than both chemotherapy and somatostatin analogues. In patients with malignant endocrine pancreatic tumor, interferon treatment seems to be as efficient as chemotherapy but we still consider chemotherapy as first line treatment. Somatostatin analogues are only adjuncts to chemotherapy or interferon to control clinical symptoms in patients with malignant neuroendocrine tumors.

REFERENCES

1. Welbourn, R.B. Ann. Surg. 185:1-12, 1977.
2. Modlin, I.M. Surg. Gyn. Obstretics 149:751-769, 1979.
3. öberg, K. Ups. J. Med. Sci. Suppl 39:107-130, 1984.
4. Grahame-Smith, D.F. Am. J. Cardiol 21:376-387, 1968.
5. Pearse, A.G.E. Clin. Endocrinol. Metab. 17:211-222, 1980.
6. Williams, E.D and Sandler, M. Lancet i:238-239, 1963.
7. Lundquist, M. and Wilander, E. Cancer. In press.
8. Norheim, I., Theodorsson-Norheim, E., öberg, K and Brodin, E. J. Clin. Endocrinol. Metab. 63:605-612, 1986.
9. Cutitta, F, Carney, D.N., Mulshine, J., Moody, T.W., Fedorko, J., Fishler, A. and Minna, J.D. Nature 316:823-826, 1985.
10. Moertel, C.G and Hanley, J.A. Cancer Clin. Trials 2:327-331, 1979.
11. Johansson, H., Boström, H., öberg, K. Ups. J. Med. Sci. Suppl 39:179-189, 1984.
12. Wood, S.M., Kraenzlin, M.E., Adrian, T.E. and Bloom, S.R. Gut 26:438-444, 1985.
13. Cantell, K. and Hirvonen S. J. Gen-Virol. 39:541-543, 1978.
14. Eriksson, B., öberg, K., Alm, G., Lundqvist, G., Magnusson, A., Wide, L. and Wilander, E. Lancet ii:1307-1309, 1986.
15. öberg, K., Norheim, I., Lind, E., Jonsdottir, B., Magnusson, A., Lundqvist, G., Wide, L. and Alm, G. Cancer Treat. Rep. 70:1297-1304, 1986.
16. Quesada, J.R. and Gultterman, J.U. Eur. J. Canc. Clin. Oncol. 24:1213-1215, 1981.
17. Inglot, A. Arch. Virol. 76:1-13, 1983.
18. Funa, K., Gazdar, A.F., Doyle, A., Mattson, K., Niivanen, A., Koivuniemi, A., öberg, K., Wilander, E., Minna, J. and Linnoila, R.J. Clin. Immunol. and Immunopathol. 41:159-164, 1986.
19. Burman, P., Tötterman, H.T., öberg, K. and Karlsson, F.A. J. Clin. Endocrinol. Metabol. 63:5:1080-1090, 1986.
20. Cantell, K. Immunobiol. 172:231-242, 1986.
21. Kvols, L.K., Moertel, C.G., O'Connell, M.J., Schutt, A.J., Rubin, J. and Hahn, R.G. N. Engl. J. Med. 315:663-666, 1986.

17

THE ROLE OF BETA INTERFERON IN THE TREATMENT OF CERVICAL DYSPLASIA

G. DE PALO*, B. STEFANON*, M. GHIONE**
* Istituto Nazionale per lo Studio e la Cura dei Tumori, via Venezian 1, Milan, Italy
** Istituto di Microbiologia Medica, Università di Milano, via Mangiagalli 31, Milan, Italy

ABSTRACT

Cervical dysplastic lesions, better defined as intraepithelial neoplasia frequently associated with papilloma virus infections, presently are treated by conservative procedures and tissue destruction methods are widely employed. Comparable or even better results in terms of complete remission are obtained with intra-perilesional injection of human fibroblast interferon. The efficacy of interferon is attributed to its antiviral, cytostatic and immunomodulatory properties.

INTRODUCTION

The term "cervical dysplasia" is not related to a defined morbid condition. Indeed, this rather vague expression of cytologic descent, originally intended to denote the presence of morphologically

M. Revel (ed.), CLINICAL ASPECTS OF INTERFERONS.

aberrant cells in cervical smears, has been adopted by general pratictioners to encompass a number of preneoplastic conditions of the uterine cervix that are supposed or proven to be somewhat at risk of invasive carcinomatous evolution.

In the late sixties, Richart (1) proposed to group the "dysplastic" cervical lesions under the heading of cervical intraepithelial neoplasia (CIN) and divided them according to the grade of severity (I, II, III).

The question is not of mere verbal value, but bears upon the interpretation of the natural history of, as well as the therapeutic approach to, human cervical cancer.

The natural history of cervical cancer is schematically given in Fig. 1.

As shown in the diagram, cervical "dysplastic" lesions can follow diverse evolutive patterns. Cytological studies performed in the past few years, have shown that about 50% of patients with various forms of untreated dysplasia progress to carcinoma in situ, 28% remain in the stage

Figure 1. Natural history of cervical cancer

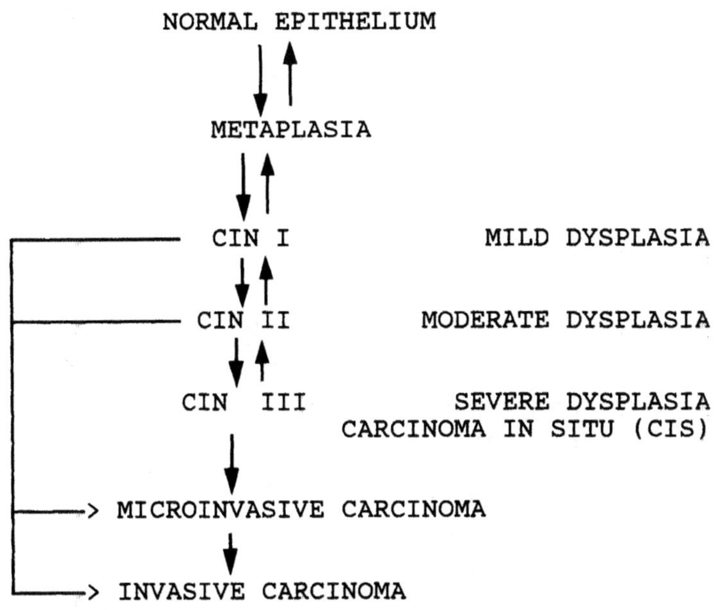

in which they were detected, 6% show spontaneous regression, and 16% progress to a higher grade of dysplasia (2).

The high trend towards malignant evolution makes practically mandatory the treatment of these morbid entities, once the pathology has been assessed. No serious objections have been raised against this policy, whereas the choice of treatment is still fraught with a number of problems. In fact the treatment of CIN is dependent more on the anatomic localization of the lesion(s) than the histological degree of severity. The discussion of the criteria for treatment selection is beyond the aim of this paper (see for more details ref. 3), but a summary of the indications is given in Fig. 2.

FIG. 2 - DIAGNOSIS AND TREATMENT OF CIN

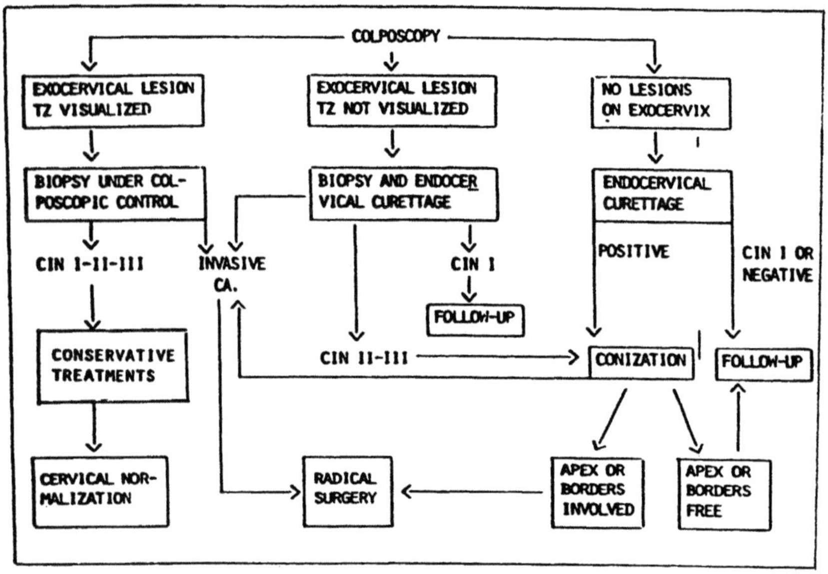

TZ = TRANSFORMATION ZONE

Fig. 2 shows that CIN confined to the exocervix and not extending into the canal (as confirmed by the colposcopic visualization of the transformation zone) can be treated conservatively, whereas lesions that extend into the canal, those in which endocervical curettage is positive, or those in which the transformation zone is not visualized during colposcopy should be treated by conization.

Conservative treatments (based on topical application of low or high temperature including cryotherapy, cold coagulation, electrodiathermy and, last but not least, CO_2 laser evaporization or resection) can eradicate localized lesions and, in contrast to conization, do not cause infertility or incompetence of the cervix and other complications of pregnancy and labor. Conservative methods are limited by the same factors that support their indication. Indeed, the tissue destruction area is kept, for obvious reasons, circumscribed as possible, and not only peripheral zones but also deeply located foci may be left partly or totally untouched. The relevance of the issue is still more evident in the light of recent results indicating that an increasing number of cases of CIN, and also its vulvar equivalent (VIN), and frequently in young women, are found to be associated with human papilloma virus infection (PVI) in the form of adjacent condylomata or superimposed viral cytopathic effect (see Table 1 and also references 4 and 5).

Table 1. Coexistence of PVI with dysplastic or neoplastic lesions in the cervical epithelium

Author		Coexistence of PVI %	
Purola and Savia (6)	Dysplasia	25	
Syrjanen (7)	Dysplasia in pts < 20 yrs	100	
Meisels and Morin (8)	All	25.6	cytology
	Dysplasia	29.9	
	CIS	21.8	
	Microinvasive ca.	13.7	
De Palo <u>et al</u> (9)	PV-associated CIN II-III	29.5	33,4
	PV-associated CIN II-III + PVI	16.5	
	CIN II-III + PVI	52.0	
	Invasice ca. +PVI	2.0	

Another point should be taken into consideration. Scattered viral microfoci have been histologically detected in apparently normal mucosa of patients affected by colposcopically evident and apparently single or multiple and discrete PV cervical lesions. Therefore PV should be considered a multifocal disease of the uterine cervix and, possibly, of the lower female genital tract (5). This being the case, it is obvious that the physical destruction of one or more macroscopic foci can influence the clinical course of the disease, but it can hardly be expected per se to eradicate the infection. Analogously to that known to occur in other cases of infectious pathology, it makes sense to ascribe the occurrence of clinical complete remission associated with focal physical destruction procedures to the efficient immune reaction unchained by the removal of the main pathologic burden. Therefore, these interventions do not induce complete remission when the immune reaction is suppressed or improperly modulated, as happens in a number of morbid circumstances.

The detection of virus footprints along the road followed by the evolution of cervical preneoplastic and neoplastic lesions led to testing of the therapeutic activity of antiviral agents. Experience with antiviral synthetic compounds (chiefly represented by analogs of nucleic acid components) is limited, and their mutagenic properties do not encourage their widespread use. This being the situation, the candidacy of interferon (IFN) for treatment of these lesions seems to be more than legitimate on the basis of its antiviral, cytostatic and immunomodulating properties (10).

MATERIALS AND METHODS

Research in this field has been carried out by our group by studying the efficacy of intra-perilesional injection of human fibroblast interferon (HF-IFN). The compound is well tolerated and is credited with a more protracted persistence in the tissues than human lymphocyte interferon (HL-IFN).

Informed and consenting patients affected by cervical PV lesions associated with CIN were submitted to intra-perilesional treatment with HF-IFN (lyophilized FRONE: Serono Pharmaceutical Institute, Rome, Italy, specific activity 10^7 U/mg protein). The administration was carried out under colposcopic control by injecting 2-3 x 10^6 IU of HF-IFN, dissolved by extemporaneous preparation (in two ml of saline solution) every day for 5 days/week for 2-3 cycles. Two patients with PV-associated CIN III and adjacent PVI were, after intra-perilesional injections, treated by topically applied cream (1.2 x 10^6 IU/day x 5 consecutive days). Application of the cream with vaginal pessary was discontinued after the first cases owing to leakage of the softened cream.

The characteristics of the patients are reported in Table 2.

Table 2. Characteristics of 19 patients with PV-associated CIN ± PVI or CIN + PVI treated with HF-IFN

- Median age	32 (range 23-48)
- No gravidity	4
- Contemporaneous vulvo-vaginal condylomatosis	1
- Synchronous or asynchronous condylomata in partner	1
- Contemporaneous venereal disease	1 (syphilis)
- Histology	
PV-associated CIN III	3
PV-associated CIN III+PVI	6
CIN III + PVI	1
PV-associated CIN III+ CIN III	1
PV-associated CIN II	1
PV-associated CIN II+PVI	6
CIN II + PVI	1
- PV-antigen positive	8/19

All patients had large lesions, and in 17 patients the lesion involved the entire cervix. The histological examination was carried out on small biopsy samples taken under colposcopic control. In all patients, the bulk of the visible cervical lesion was largely left untouched by the biopsy procedure. All patients were treated in the Out-patient Department and were instructed to prevent reinfection by using a condom.

Complete regression (CR) was defined either as disappearance of all lesions at colposcopy without any lesion left for biopsy, or histological regression of the intraepithelial neoplasia; partial regression (PR) was defined as reduction by at least 50% in the diameters of the lesions as seen at colposcopy without histological modification; no change (NC) was defined as disease not meeting either of the above two criteria. In a few patients in which CR of CIN occurred but PVI foci still persisted, a cautery treatment was applied. The type of response was evaluated within 15 days from the end of treatment. Its duration was calculated from the start of response. The patients were followed with colposcopic and cytological examination carried out every three months after the completion of treatment. The median duration of follow-up calculated from the start of treatment was 26 months.

RESULTS

The results are reported in Table 3.

Table 3. Response to intra-perilesional HF-IFN treatment in 19 patients with PV-associated CIN ± PVI or CIN + PVI

	No.of cases	Response to HF-IFN		
		CR	PR	NC
PV-associated CIN III+PVI	6	5 (46+,43+,31+, 13+*,5+*)	-	1
PV-associated CIN III+CIN III	1	-	1	-
PV-associated CIN III	3	1 (35+)	1	1
PV-associated CIN II + PVI	6	3 (29+*,23+*,3+*)	1	2
PV-associated CIN II	1	1 (12)	-	-
CIN III + PVI	1	-	-	1
CIN II + PVI	1	1 (25+)	-	-
Total	19	11 (57.9%)	3	5

* Cautery on residual PVI

In parenthesis, the duration of remission in months

A CR was obtained in 11 out of 19 patients (57.9%).

The median duration of CR was 25 months. With one exception, no relapse occurred in the CR patients during the follow up period. The results are analogous to those observed in other groups of patients presenting comparable lesions and submitted to other conservative therapeutic approaches.

DISCUSSION

The data can be accompained by a few words of conclusive comment aimed at calling the reader's attention to the fact that the choice of the conservative therapy of "dysplastic" cervical lesions is presently restricted to two types of treatment, namely CO_2 laser and IFN application.

By far broader appears to be the latitude of action granted by IFN treatment if one considers that a large and steadily increasing number of CIN are found to be associated with PVI. This conclusion is in agreement with literature data; indeed local (11) and systemic (12) administration of IFN have been successfully employed for the treatment not only of condylomatous lesions of the lower genital tract, but also other PVI such as laryngeal papillomatosis in children or adults and epidermodysplasia verruciformis (for a comprehensive review see ref. 9). The cytostatic, antiviral and immunomodulating properties of IFN undoubtedly concur to explain its activity in the treatment of morbid conditions such as cervical and vulvar intra-epithelial neoplasia, whose pathogenesis involves viral infection, neoplastic growth and immune derangement.

Local injection of HF-IFN results in a high concentration of the compound inside the lesion and around it, and exposes to the treatment a tissue area that is not only wider, but also deeper than the one reached by other methods.

No tissue destruction is associated with IFN treatment and therefore no cicatricial sequelae shall be feared. Furthermore IFN administration is accompanied by negligible side effects.

For these reasons IFN should have a place in the gynecologic therapeutic armamentarium.

REFERENCES

1. Richart, R.M. Clin. Obstet. Gynecol. 10: 748-784, 1967.

2. Richart, R.M., Barron, B.A. Am. J. Obstet. Gynecol. 105: 386-393, 1969.

3. Singer, A. In: Herpes and Papilloma Viruses. Their Role in the Carcinogenesis of the Lower Genital Tract (Eds. G. De Palo, F. Rilke, H. Zur Hausen), Serono Symposia Publications from Raven Press, vol. 31, New York 1986 pp. 277-290.

4. Pilotti, S., Rilke, F., Alasio, L., Fontanelli, R. Diagn. Gynecol. Obst., 4, 357-362, 1982.

5. Pilotti, S., Rilke, F., Stefanon, B. In: Herpes and Papilloma Viruses. Their Role in the Carcinogenesis of the Lower Genital Tract (Eds. G. De Palo, F. Rilke, H. Zur Hausen) Serono Symposia Publications from Raven Press, vol. 31, New York 986, pp. 209-220.

6. Purola, E., Savia, E. Acta Cytol. 21: 26-31, 1977.

7. Syrjanen, K.J. Surg. Gynecol. Obstet. 150: 372-376, 1980.

8. Meisels, A., Morin, C. Gynecol. Oncol. 12: S111-S123, 1981.

9. De Palo, G., Stefanon, B., Del Vecchio, M. In: Herpes and Papilloma Viruses. Their Role in the Carcinogenesis of the Lower Genital Tract. (Eds G. De Palo, F. Rilke, H. Zur Hausen) Serono Symposia Publications from Raven Press, vol. 31, New York 1986, pp. 305-327.

10. Stewart, W.E. The Interferon System. Springer Verlag Publ. Berlin, Heidelberg, New York 2nd ed.,1981 passim.

11. De Palo, G., Stefanon, B., Rilke, F., Pilotti, S., Ghione, M. J. Reprod. Med. 30: 404-408, 1985.

12. Schonfeld, A., Schatter,R A., Crespi, M., Levavi, H., Shoham, J., Nitke, S., Wallach, D., Hahn, T., Yarden, O., Doerner, T. The Lancet, i: 1038-1041, 1984.

18

INTERFERONS IN COMBINATION WITH CYTOTOXIC MODALITIES FOR CANCER TREATMENT*

S. M. Schmid and E. C. Borden

Department of Human Oncology
The University of Wisconsin
Clinical Cancer Center
Madison, Wisconsin 53792, U.S.A.

ABSTRACT:

The study of combination therapy with interferons (IFNs) is relatively new, with the largest body of data derived from *in vitro* systems. Animal studies and clinical trials have begun to probe treatment effectiveness and associated toxicities. While a variety of responses have been reported in differing systems, it was apparent that IFN employed with selected cytotoxic agents or radiation can be beneficial. The data suggest more preclinical and animal studies are required to define new and optimal ways for IFNs to complement or augment therapy for human malignancies.

Phase II clinical trials have established the anti-tumor effects of IFNs as single agents in more than a dozen malignancies. Single modalities, however, are not usually optimal treatment for any but early stage cancers. IFNs will not be an exception to this clinical

*Studies in our laboratory in this area supported by Triton Biosciences Inc. and the American Cancer Society. ECB is an American Cancer Society Professor of Clinical Oncology.

axiom. Optimum uses of IFNs in conjunction with other modalities must thus be defined.

A combination of treatments should yield results not achieved by either modality alone. This may occur either on a cellular level as a result of complementary biochemical effects or on an organismal level by elimination of cell clones refractory to a single therapeutic approach. On a biochemical level, IFNs regulate gene expression, modulate expression of proteins on the cell surface, and induce synthesis of new enzymes. These alterations result in modulation of levels of receptors for other cytokines, regulatory proteins on the surface of immune effector cells, and enzymes which modulate cell growth. On a cellular basis, these effects can translate into an alteration in the state of differentiation, rate of proliferation, and sometimes cell function. IFNs may be thus used to augment the effects of other molecules on endothelial, immune, hematopoeitic, or endocrine cells. They may also be used to potentiate the effects of cytotoxic drugs on malignant cell proliferation.

In the whole vertebrate, antitumor effectiveness of a treatment modality is generally inversely proportional to tumor burden. Thus modalities may complement the effects of each other by each eliminating a given quantum of the malignant cell population. For example, surgery may be curative of a local but not a systemic neoplasm. Alternatively, one drug may eliminate susceptible malignant clones while a biologic response modifier eliminates the chemoresistant clones by stimulation of immune effector cell function. In combination with other modalities, IFNs for malignant disease will probably enhance or complement effectiveness of cytotoxic agents on both cellular and organismal levels; our overview will provide examples of potential for use of IFNs on both levels.

The most extensive investigations have been with *in vitro* systems using human and murine cell lines. The species specificity of IFN has led to xenograft studies in nude mice which exclude immune modulation of tumor cell growth and can probe the antiproliferative effects *in vivo*. Limited availability of murine IFN and other rodent IFNs has slowed assessment of effects against transplantable murine tumors. IFNs may enhance efficacy of cytotoxic agents in treatment regimens without overlapping of toxicities. Although a few clinical trials have been conducted, assessment of IFN with cytotoxic modalities is still largely in a preclinical evaluation stage. Clinical application of IFNs with cytotoxic agents could be improved if synergistic combinations can be defined in preclinical models and underlying mechanisms dissected. This review will discuss IFN in combination with six different modalities: antimetabolites, alkaloids, alkylating agents, antibiotics, other cytotoxic drugs, and radiation.

Antimetabolites

Synergistic or additive antiproliferative activity has been reported with human IFN beta and 5-fluorouracil (5-FU) in human cell lines such as HeLa. Radiation transformed human WI38-CT1 were also synergistically inhibited, while the diploid parent cells WI38 demonstrated no increase in antiproliferative effect (1,2). Human breast carcinoma cells (MCF-7) were not potentiated. The use of the potent IFN inducer, poly I:C with 5-FU, was antagonistic in human colon carcinoma cells HT-29. Poly I:C induced reduced incorporation of

5-FU into RNA, and the metabolism and incorporation of FU and FUrd was inhibited (3).

Toxicity experiments in vivo with murine IFN and 5-FU indicate that IFN can ameliorate 5-FU lethal toxicity in mice (4). This diminished 5-FU toxicity also occurred when IFN was administered for two days following 5-FU treatment. Combined administration of 5-FU and the IFN inducer poly I:C also ameliorated toxicity and weight loss in mice. The dose of 5-FU could be increased to 350 mg/kg/week x 3 weeks in the presence of the potent IFN inducer poly I:C without significant mortality. Cell cycle analysis of bone marrow cells harvested from poly I:C treated mice demonstrated non-toxic accumulation of cells in the G1 phase of the cell cycle. The inhibition of cell cycle transit may allow 5-FU exposed cells time to repair DNA or generate RNA and thymidylate synthetase. Although reduced toxicity would be a novel action, correlation of changes in antitumor effects of 5-FU by poly I:C or IFN has not been fully studied.

The use of cytosine arabinoside (Ara-C) with IFNs has produced no more than additive antiproliferative effects in vitro. IFN alpha + Ara-C resulted in additive antiproliferative effects in KB cells (5). The use of IFN beta with Ara-C in HeLa cells produced sub-additive effects (2), while our studies with recombinant IFN $beta_{ser}$ with Ara-C in ACHN renal carcinoma cells suggest a pronounced antagonism of the antiproliferative effect of each agent.

Antiproliferative activity of combination methotrexate (MTX)-IFN alpha has been studied in human tumor cells (6). Osteosarcoma cell lines 393T, T_256, and lymphoblastoid cell line Daudi were all sensitive to IFN alpha and MTX, but synergistic antiproliferative effects were not observed when the drugs were used in a concomitant or sequential treatment schedule.

The combined use of murine IFN alpha/beta with MTX for the *in vivo* treatment of L1210 leukemia indicated that IFN augmented all MTX treatment regimens to a significant degree (7).

IFN gamma was shown to increase Class I expression even in insensitive variant cells expressing low levels of Class I molecules. In an innovative approach methotrexate (MTX) has been incorporated in unilamellar liposomes with staph protein A conjugated monoclonal antibodies (Abs) (Moab) to H-2 Class I Ags. Treatment of mouse thymoma cells with murine IFN gamma and MTX-MoAb liposomes produced increased cell killing compared to MTX-MoAb liposomes used alone (8).

The use of antimetabolites with IFNs has produced mixed results in various *in vitro* and *in vivo* models. Because antiproliferative effects of IFNs may alter antimetabolite activity, more studies will be required to produce evidence for the rational use of IFN in regimens with antimetabolites.

Vinca Alkaloids

Several investigators have shown additive and synergistic activity of IFNs and vinblastine. Synergistic antiproliferative activity with IFN alpha 2a was demonstrated in myeloma, MCF-7, and WiDR colon carcinoma cells (9). IFN $beta_{ser}$ and vinblastine produced synergistic effects in ACHN renal carcinoma cells (10). Interaction of IFN and vinblastine may be due to overlapping effects on the cytoskeletal system by both agents which produce lethal damage or compounded cytostatic effect. Cytoskeletal changes induced by IFN (11,12) and abrogation of antiviral effects of IFN in

EMC-infected L929 cells by vinblastine, cyclochalasin B, or colchicine (13) indicate that IFN effects are mediated partly through cytoskeletal organization. IFN alpha 2a was used in combination with vinzolidine to test antiproliferative effects on several human cell lines. The combination produced additive effects when used against myeloma 8226, breast MCF-7, and colon WiDR tumor cells (9).

Several phase I trials have been conducted with IFNs and vinblastine with renal cell carcinoma patients (14,15,16). One study evaluated 23 patients treated with 3 x 10^6 units im natural leukocyte IFN alpha x 5 days with 2 days off and vinblastine (0.15 mg/kg) on day 1 of each week. Vinblastine doses were lowered due to dose-limiting hematologic toxicity in 59% of the patients. In a clinical trial of IFN alpha-2a with vinblastine in 16 patients with renal cell carcinoma, patients received IFN alpha-2a (36 x 10^6 units im) 3 times a week and vinblastine (0.10-0.15 mg/kg iv) every 2 to 3 weeks. All patients had side effects of fever, fatigue, and weight loss. This relatively small study demonstrated an objective response rate of 33% (6 of 16 patients). We have conducted a clinical phase I trial of lymphoblastoid IFN alpha-N1 and vinblastine in 18 patients with advanced renal cell carcinoma (16). The treatment schedule consisted of IFN alpha-N1 at 3 x 10^6 units/m^2/day im on day 1, and escalated to 10 x 10^6 units/m^2/day on days 3-10 with vinblastine continuous infusion of 0.15 mg/m^2/day on days 1 through 5. The treatment cycle was repeated monthly. A partial response occurred in one patient who received 5 x 10^6 units/m^2/day at the end of the third cycle. Toxicities associated with IFN-vinblastine were manageable, but further study will be required to determine the maximum tolerated dose (MTD) of both agents used in combination and the effect on MTD when employing different preparations of natural or recombinant produced IFNs. Clinical activity of this treatment combination is still to be determined, although in vitro and in vivo therapeutic activity of this data suggest synergistic antitumor activity.

Antibiotics

Doxorubicin toxicity for HeLa cells was potentiated with IFN beta (2). Doxorubicin + IFN alpha 2a produced sub-additive effects in T348 colon carcinoma cells (17). The effects were dependent on cell cycling and growth phase of the cells in culture. Colony forming studies with IFN alpha 2b and doxorubicin were performed with human cell lines of ovarian carcinoma (BG-1), cervical carcinoma (ME-180, CASKI), melanoma (SKMEL-28), and renal carcinoma (CAKI-2) (18). Combinations of 1 hr exposures, 1 hr of one drug followed by longterm exposure of the second drug, and longterm exposure of both drugs were analyzed for additive and synergistic anti-colony forming activity by the multiplicative method of analysis. Combinations of IFN with doxorubicin in 1 hr sequences produced synergistic activity in the BG-1, CASKI, and SKMEL-28 cell lines. Primary human tumor cells from the colon, ovary, endometrium, cervix, and a melanoma were tested with IFN alpha 2b and doxorubicin. The results of 1 hr sequential treatment produced synergistic effects in cells of a cervical carcinoma, while additive and sub-additive inhibition of colony forming was seen in the other tumor samples. Longterm exposure of the tumor cells to both IFN at 10,000 units/ml and doxorubicin at 0.0015 μg/ml produced synergistic effects in a cervical carcinoma, while samples of colon, endometrium, and ovary treated with 1,000 units/ml yielded additive antiproliferative results.

Human breast xenografts implanted sc in nude mice have been treated with combination human IFN alpha-N1 and adriamycin (19,20). Therapy began after the implanted tumors reached diameters of 0.3-0.9 cm. Growth of the tumor was analyzed by measuring the tumor size. Daily treatment with IFN and weekly treatment with adriamycin resulted in synergistic inhibition of tumor growth.

Clinical phase I trials of IFN with doxorubicin employed IFN alpha 2b sc (10 x 10^6 units/m^2 TIW x 2 weeks) and doxorubicin iv (0-40 mg/m^2 on day 5 of each cycle) with one week off between cycles (21,22). Significant myelosuppression occurred in patients treated with 40 mg/m^2 doxorubicin but returned to baseline levels within 1 week after cessation of treatment. Partial responses were seen in 2 of 14 patients. Another study involved 17 renal cell carcinoma patients who were treated with IFN alpha 2b (10 x 10^6 units/m^2 sc and 20 x 10^6 units/m^2 iv for each patient) followed by doxorubicin (20 mg/m^2 iv over 2 hours) at weekly intervals (22). Toxicities were similar to the previous trial of IFN-doxorubicin, but no responses occurred in the patient group.

Bleomycin cytotoxicity was additive when combined with IFN beta _in vitro_ in HeLa cells. The biochemical events associated with IFN action, DNA synthesis inhibition, 2'-5'A synthetase, and protein kinase induction were unaffected by bleomycin treatment. The antiviral state against vesicular stomatitis virus (VSV) was also maintained during bleomycin treatment (23). High dose IFN alpha 2b (10,000 units/ml) with bleomycin 0.3 ug/ml used in 1 hr sequential treatments produced synergistic inhibition of BG-1 ovarian adenocarcinoma colony formation (18). Mitomycin C + IFN beta produced additive antiproliferative effects in HeLa cells which was accompanied by an increased antiviral against VSV (23). The combination of aclacinomycin with IFN beta produced synergistic antiproliferative effects in HeLa cells (2). Thus synergistic antiproliferative effects *in vitro* reported for human cell lines employing IFN with doxorubicin, bleomycin, and aclacinomycin warrant further investigation.

Alkylating Agents

Melphalan + IFN alpha 2a gave sub-additive to additive antiproliferative results in RPMI 8226 myeloma cells used in a colonogenic assay system (9). The use of IFN beta with melphalan in HeLa cells produced additive effects (2). IFN alpha A has been combined with BCNU (1,3-Bis(2-chloroethyl)-1-nitrosourea) in colony forming assays with supra-additive results in A375 melanoma and A101D melanoma while additive results were seen in A498 renal carcinoma cells (24). ACNU (3-[(4-amino-s-methyl-5-pyrimidinyl)methyl]-1-(2-chlorethyl)-1-nitrosourea hydrochloride) was combined with IFN beta and additive effects were seen in HeLa cells (2).

The treatment of LSTRA leukemia in CDF_1 mice has been studied using murine IFN alpha/beta with BCNU (25). Single dose BCNU 7 days after tumor inoculation sc followed by daily IFN treatment produced augmentation in mean survival time and percent cures over that demonstrated with single agent treatment.

IFN-alpha 2a combined with BCNU treatment was tested in a clinical phase I trial in 12 patients with advanced cancer (24). Patients were given BCNU iv (50 mg/m^2/day escalated to 150 mg/m^2/day on days 1 to 3) and IFN-A im (12 x 10^6 units/m^2 3 times a week (TIW)). Dose escalation of BCNU led to progressive leukopenia and thrombocytopenia in 12 patients. Treatment with BCNU was delayed at the second cycle in 6 of 9 patients due to myelosuppression. The weight loss due to treatment ranged from 0.19-8.5 kg. In 10/12 patients the IFN dose was reduced due to the toxicity. While patient numbers were low, IFN-BCNU combined treatment should be investigated further to confirm safe dose levels in treatment.

The combination of murine IFN alpha/beta with cyclophosphamide for spontaneous lymphoma in AKR mice resulted in a significant increase in the mean time of survival over either single agent therapy (26). Another study demonstrated that murine IFN alpha/beta combined with cyclophosphamide was synergistic in prolonging survival of A/J mice inoculated with C1300 murine neuroblastoma cells (27). Human IFN alpha-N1 has been tested in combination with cyclophosphamide with human breast tumor cells sc xenografts in nude mice (19, 20). Treatment began after tumor implants reached a size of 0.3-0.9 cm. The daily treatment with IFN combined with weekly treatment with cyclophosphamide produced tumor growth inhibition that was augmented over single agent therapy to a significant degree.

Cyclophosphamide has been combined with IFN in other rodent tumor model systems. The use of rat IFN alpha/beta with cyclophosphamide in the treatment of Ls 175 liposarcoma implanted sc in BN female rats provided no augmentation of mean survival time (28).

Human IFN alpha 2a diminished cyclophosphamide efficacy for TBD 932 lymphosarcoma in vivo (29).

Diminished survival of hamsters occured when treated concomitantly with cyclophosphamide and IFN during the first three days after tumor inoculation. This antagonistic behavior was altered to synergism when IFN was administered daily and the cyclophosphamide dose lowered. Toxicity studies revealed that normal hamsters suffered no adverse effects from the combined therapy ruling out a lethal interaction between IFN and cyclophosphamide. IFN alpha 2a did not suppress cytochrome P-450 metabolism which is required for conversion of cyclophosphamide to an active phosphamide mustard. The need for understanding the complex interaction of this combination is of clinical importance.

The antitumor effects on MBT-2 transitional cell carcinoma in mice was augmented significantly by the IFN inducer poly I:C when given on succeeding alternate days after cyclophophamide treatment (30). The enhanced effect of poly I:C (2.5 mg/kg) was seen with 25,50 and 100 mg/kg doses of cyclophosphamide. IFN induction was clearly documented at 6 and 24 hr post poly I:C treatment in mice.

Other Cytotoxic Agents

IFN alpha 2a combined with VP-16 in RPMI 8226 myeloma cells produced sub-additive to additive effects in a clonogenic assay system (9). Cis-platinum (cis-DDP) combined with IFN alpha 2a produced synergistic antiproliferative effects in RPMI 8226 cells when the IFN exposure was continuous (9). The use of IFN beta + cis-DDP with a 1 hr treatment produced synergistic activity in HeLa cells and additive effects were demonstrated in MCF-7 cells (9). Synergistic inhibition of colony formation was seen in BG-1 ovarian adenocarcinoma cells treated continuously with IFN alpha 2b (100 or 10,000 units/ml) and cis-DDP (0.25 µg/ml) (18). Our studies of cis-DDP + IFN $beta_{ser}$ in ACHN renal carcinoma cells demonstrate synergistic effects which can be significantly enhanced with the addition of IFN gamma. The use of hydroxyurea with IFN beta in HeLa cells resulted in additive inhibiton of cell proliferation (2).

Murine IFN alpha/beta has been combined with cis-DDP for P388 leukemia in CDF_1 mice. Single dose cis-DDP with daily treatments of IFN resulted in significant augmentation of mean survival time compared to single agent treatment (31).

Radiation and IFN

IFN *in vitro* has sensitized cells to radiation cytotoxicity. Radiation sensitization of mouse 3T3 cells resulted from mouse IFN (32). In both mouse and human cells, the shoulder of the radiation survival curve was decreased without alteration of the slope of the exponential portion (33). Interferons may thus interfere with enzymes important in radiation repair or induce more effective radiation damage with less sublethal cumulative damage. In the limited number of cell lines evaluated, radiation sensitization has been more prominent with IFN beta (33, 34). IFN alpha 1 has, however, also proved effective (35).

IFN has served as a whole body radioprotector *in vivo* (36). When given 24 hrs after lethal radiation, mice receiving mouse IFN as a single injection, had median survival prolonged from 5.5 to 9 days. Phase I trials of the combination have been initiated. The critical preclinical data suggests need for further evaluation of this promising combination.

CONCLUSION

IFNs are the first biological, prepared by recombinant DNA technology, to impact on treatment of human neoplastic disease. Clinical use of IFN in conjunction with other therapeutic modalities has just begun to be probed. Such studies involve an almost infinite number of variables of schedule, route, and timing of IFN administration. It will be possible to evaluate but a few of these in humans. Optimal therapeutic use will therefore critically depend on judicious preclinical studies *in vitro* and *in vivo*. The principles derived from use of IFNs in combination with other modalities will thus be a model for the next decade of research on therapeutic use of human cytokines for malignant disease.

REFERENCES

1. Miyoshi, T., Ogawa, S., Kanamori, T., Nobuhara, M., and Namba, M. Cancer Lett. 17:239-247, 1983.
2. Namba, M. Yamamoto, S., Tanaka, H., Kanamori, T., Nobuhara, M., and Kimoto, T. Cancer 54:2262-2267, 1984.
3. Ligo, M., and Glazer, R.I. Cancer Res. 45:1953-1957, 1985.
4. Stolfi, R.L., Martin, D.S., Sawyer, R.C., and Spiegelman, S. Cancer Res. 43:561-566, 1983.
5. Oku, T., Imanshi, J., and Kishida, T. GANN 73:667-674, 1985.
6. Bostrom, L.A. Cancer Lett. 10:83-89, 1980.
7. Slater, L.M., Wetzel, M.W., and Cesario, T. Cancer 48:5-9, 1981.
8. Machy, P., Arnold, B., Alino, S., and Leserman, L.D. J. Immunol. 136:3110-3115, 1986.
9. Aapro, M.S., Alberts, D.S., and Salmon, S.E. Cancer Chemother. Pharmacol. 10:161-166, 1983.
10. Cummings, K.B., Schmid, S.M., Bryan, G.T., and Borden, E.C. World J. Urol. 3:230-233, 1986.
11. Pfeffer, L.M., Wang, E., and Tamm, I. J. Cell Biol. 85:9-17, 1980.
12. Gerfaux, J., Rousset, S., Chany-Fournier, F., and Chany, C. Cancer Res. 41:3629-3634, 1981.
13. Bourgeade, M.F. and Chany, C. Exp. Biol. Med. 153:501-504, 1976.
14. Figlin, R.A., deKernion, J.B., Maldazys, J., and Sarna, G. Cancer Treat. Rep. 69:263-267, 1985.
15. Fossa, S.D., deGaris, S.T., Heie, M.S., Flokkmann, A., Lien, H.H., Salverson, A., and Moe, B. Cancer 57:1700-1704, 1986.
16. Ravdin, P.M., Borden, E.C., Magers, C.F., Whisant, J.K., and Trump, D.L. Submitted for publication, 1987.
17. van der Bosch, J. and Zirvi, K.A. Cancer Res. 42:3789-3792, 1982.
18. Welander, C.E., Morgan, T.M., Homesley, H.D., Trotta, P.P., and Spiegel, R.J. Int. J. Cancer 35:721-729, 1985.
19. Balkwill, F.R. and Moodie, E.M. Cancer Res. 44:904-908, 1984.
20. Balkwill, F.R., Mowshowitz, S., Seilman, S.S., Moodie, E.M., Griffen, D.B., Fantes, K.H., and Wolf, C.R. Cancer Res. 44:5249-5255, 1984.
21. Sarosy, G.A., Brown, T.D., Von Hoff, D.D., Spiegel, R.J., Golando, J.P., Beougher, K.L., Kuhn, J.G., and Kisner, D.L. Cancer Res. 46:5368-5371, 1986.

22. Muss, H.B., Welander, C., Caponera, M., Reavis, K., Cruz, J.M., Cooper, M.R., Jackson, D.V. Jr., Richards, F. II, Stuart, J.J., Spurr, C.L., White, D.R., Zekan, P.J., and Capizzi, R.L. Cancer Treat. Rep. 69:721-722, 1985.
23. Suhadolnik, R.J., Sawada, Y., Flick, M.B., Reichenbach, N.L., and Mosca, J.D. Cancer Res. 43:5462-5466, 1983.
24. Creagan, E.T., Kovach, J.S., Long, H.J., and Richardson, R.L. J. Clin. Oncol. 4:408-413, 1986.
25. Chirigos, M.A. and Pearson, J.W. J. Natl. Can. Inst. (USA) 51:1367-1368, 1973.
26. Gresser, I., Maury, C., and Tovey, M. Eur. J. Cancer 14:97-99, 1978.
27. Kangas, L., Cantell, K., Gronroos, M., Maenpaa, J., and Perila, M. Ann. Chir. Gynecol. 199:60-63, 1985.
28. Marquet, R.L., Schellekens, H., Westbrook, D.L., and Jeekel, J. Int. J. Cancer 31:223-226, 1983.
29. Lee, S.H., Chiu, H., Renton, K.W., and Stebbins, N. Biochem. Pharmacol. 33:3439-3443, 1984.
30. Borden, E.C., Sidky, Y.A., Groveman, D.S., and Bryan, G.T. Cancer Res. 45:45-50, 1985.
31. Mowshowitz, S.L., Chin-Bow, and Smith, G.D. J. IFN Res. 2:587-591, 1985.
32. Dritschilo, A., Mossman, K., Gray, M., Sreevalsan, T. Am. J. of Clin. Oncol. 5:79-82, 1982.
33. Gould, M.N., Kakria, R.C., Olson, S., and Borden, E.C. J. Interferon Res. 4:123-128, 1984.
34. Nederman, T. and Benedikeson, G. ACTA Radiol. Oncol. 21:231-234, 1982.
35. Chang A.Y.C., and Keng, P.C. J. Interferon Res. 3:379-385, 1983.
36. Ortaldo, J.R. and McCoy, J.L. Rad. Res. 81:262-266, 1980.

III

Biological properties and experimental systems

19

TOXIC EFFECTS

G. M. S. SCOTT
Clinical Microbiology, University College Hospital, London, WC1E 6AU, U.K.

ABSTRACT

All interferons consistently cause a variety of symptoms and signs in man which are interpreted as "toxic" effects because they are unpleasant, but may reflect the normal pathophysiologic effects of high levels of endogenous interferon produced acutely during virus infections or chronically during active autoimmune processes. Many of the effects appear to be non-specific and are probably in part mediated by inflammatory prostaglandin synthesis. Toxic effects may be augmented by impurities in interferons derived from any source. They may be reduced by giving prostaglandin antagonists or steroids or by giving interferons in the evening. Both subjectively and objectively, acute febrile responses and associated symptoms are often reduced with repeated dosing, but tend to be replaced by chronic fatigue and malaise. Idiosyncratic responses such as glomerulonephritis, thrombocytopenia or thyroid disease may involve autoimmune processes.

INTRODUCTION

No mention was made of any adverse effect of exogenous interferon until 1966, when a crude preparation of leucocyte interferon (IFN) given intravenously to patients with a variety of illnesses caused shivering, fever and malaise[1]. There is no doubt that this preparation would have contained a variety of lymphokines and cytokines in addition to the interferon. Early, partially-purified preparations of leucocyte interferon[2] may well have contained cytotoxins (eg tumour necrosis factor) which also cause fever. IFNs were purified to a high degree in the 1970's and

M. Revel (ed.), CLINICAL ASPECTS OF INTERFERONS. Copyright 1988.

although each purification step may have reduced the degree of inflammatory and toxic effects to some degree[3,4], these properties have been observed with every natural preparation tested in man[5,6]. The advent of gene cloning led to single gene products for tests in man, and although the first preparations tested may not have been entirely free of pyrogenic bacterial products, with the manufacture and testing of purer and purer products it has become increasingly clear that human IFNs are indeed endowed with toxic properties[7,8,9,10].

It remains to be seen whether the toxic effects of IFNs are in fact "unwanted". They cannot be regarded as desirable in a pharmaceutical product and are most certainly unwanted by the patients, yet they may be important in the therapeutic activity of the IFN. For example, it could be argued that the febrile response to IFN is an important host defense factor in infection and interferons have greater antiviral activity at higher temperature[11]; prostaglandins induced by IFN may also influence the response of cells to the antiviral effects[12]; steroids induced by IFN may be therapeutic and may also modify responses to subsequent doses[13]. However, the properties of IFN have created serious problems for pharmaceutical companies and regulatory authorities, being additional to the difficulties in developing and licensing a new generation of gene products for therapeutic (rather than simply for hormonal replacement) indications. In the final analysis, however, interferons appear to have important therapeutic properties and their value in each indication must take into account the acceptability of the toxic effects.

Several interferons are under development for clinical use in man. Most information is available about natural leucocyte and lymphoblastoid interferons; these are probably very similar mixtures of alfa species with $alfa_2$ predominating. Recently tested preparations of lymphoblastoid IFN have been purified to a very high degree[5]. Natural fibroblast interferon[14] and a modified cloned beta interferon[9] continue in trials. The most extensively tested cloned interferons are $rIFN\text{-}alpha_{2a}$[8] and $alfa_{2b}$[7]; the genes from which these are derived give rise to products which differ by one amino-acid, but perhaps also in other structural ways as indicated by their relative immunogenicity (see below). Clinical trials with rIFN-gamma[10] have only recently begun. Considering how different these preparations are, it is remarkable how similar their toxic properties appear to be.

All the reports of administration of interferons to man mention certain side-effects of treatment which are quite consistently related to dose. In addition, rare anecdotal cases are reported of effects which appear to be idiosyncratic. These may be more likely in particular patients or with particular preparations of IFN but this is not yet clear. Thirdly, a further unwanted effect is the induction of IFN neutralising factors by treatment. These toxic effects are listed and discussed below.

Table 1. Effects of interferons in man

1. Predictable effects of a single dose

a. Febrile reaction complex: headache, fever, malaise, myalgia. (sometimes with chills and rigors, nausea and vomiting, anorexia, confusion, dysaesthesiae, arthralgia, stiff neck, dysuria, etc.)
   Objectively pyrexia, tachycardia, variations in blood pressure.
b. Rapid changes in peripheral white cell counts.
c. Changes in cortisol levels, blood glucose and trace elements.
d. Local inflammation after intradermal injection.

2. Effects of longer courses of interferon

a. Persistence of any or all of the symptoms of the febrile reaction complex: eg continuing or remitting-relapsing fever or headache.
b. Fatigue, somnolence, minor mental changes, chronic headache, orthostatic hypotension.
c. Bone marrow suppression.
d. Loss of hair, aesthenia and anorexia.
e. Loss of weight (adults) or failure to gain weight (children).
f. Hepatocellular damage (rises in serum transaminases and alkaline phosphatase).
g. Hormonal changes, effects on lipid metabolism.
h. Local inflammation after intranasal administration.

3. Particular effects of very high doses of interferon

a. Cerebral impairment with EEG changes. Fits.

b. Metabolic changes, particularly in calcium.

4. Idiosyncratic effects

a. Cardiotoxicity (dysrhythmias, infarction).

b. Renal transplant rejection (vasculitis).

c. Acute glomerulonephritis.

d. Hepatitis.

e. Induction of autoantibodies.

f. Induction of interferon neutralising activity.

g. Epididymitis and parotitis.

h. Reactivation of herpes simplex labialis.

FEBRILE REACTION COMPLEX

Febrile reactions to a single dose of interferon begin two to four hours after injection depending to a certain extent on the dose, the IFN preparation and the route of administration. Bolus intravenous injections are discouraged because rapid, unpredictable changes in blood pressure and pulse rate can occur[2]. The delay is longest after intradermal injections, as might be expected, and it seems that a slightly higher dose is necessary to consistently cause the reaction than if interferon is given subcutaneously or intramuscularly. Leucocyte and fibroblast IFNs given by the lumbar intrathecal route cause fever but after a delay very similar to that seen with intramuscular injection.

There is a threshold dose above which interferon becomes detectable in the serum when the febrile complex is almost inevitable. This is around 2Mu and is determined by the approximate equivalent antiviral activity, rather than by weight of different interferons. Thus the threshold for gamma interferon is ten to one hundred times that for alfa to give the same sort of reaction. However, there has not been a well controlled comparison between any two preparations in normal volunteers to be absolutely certain of this point. Fewer reactions were seen with $alfa_1$ than $alfa_2$ IFNs when they were compared in patients on a weight for weight basis[15], and this may be explained by the lower specific activity of the former.

The first symptom of a reaction is usually low backache, followed by headache and then chills or fever. The pulse rate and temperature rise as the symptoms begin, reach a peak around eight hours and the syndrome completely resolves around 15 hours after injection. Blood pressure may rise or fall in the order of 20-30 mm Hg but inconsistently and unpredictably. (Cardiac dysrhythmias are a rare complication but not sufficiently documented[16]. They may occur by a direct effect on myocardial cells[17] or by changes in calcium metabolism[18]). At the height of the febrile reaction, the subject may experience any of a number of diverse symptoms, including nausea and vomiting, dysuria and mental impairment. Paraesthesiae and dysaesthesiae can occur and make venepuncture uncomfortable. Some subjects report euphoria after recovering from the reaction and during prolonged treatment. Reactivation of herpes labialis may occur one to two days after IFN injection.

This febrile reaction is suppressed by prostaglandin synthetase inhibitors including aspirin, acetaminophen or indomethacin and by corticosteroids. In controlled experiments in normal volunteers using 3Mu IFN-alfa$_2$, symptoms were much less but fever was incompletely abolished and tachycardia not much affected by high doses of indomethacin[13]. One volunteer had a recurrence of herpes labialis despite indomethacin. It has been shown that interferons may induce the formation of inflammatory prostaglandins in vitro[19,20] but do not induce the formation of pyrogenic mediators, in particular interleukin 1, in macrophages[21]. Thus theoretically the fever may be due to the direct induction of prostaglandins in the circulation or in the temperature regulatory centre of the hypothalamus. This may explain why reactions do not occur when the dose of interferon is insufficient to give rise to detectable blood levels, why intramuscular IFN-beta is less reactogenic than alfa and why there is a delay in the onset of the reaction even in those receiving intrathecal interferon[14]. In general, diseases characterised by detectable circulating endogenous interferon (such as influenza) are more likely to be associated with general symptoms than, for example, the common cold. The reaction is also subjectively less when the interferon is given in the evening[23]. It has not been established whether there is any physiologic basis for this observation.

Rapid changes in circulating white blood cell counts (WBC) are seen in association with the reaction. For example, the granulocyte count will rise during the early phase (maximal around 8 hours) and be subnormal at 24 hours[6,13]. In cancer patients, the granulocyte count often falls during this phase suggesting that there is no marginated pool[5]. In normals, the lymphocyte count tends to fall as the granulocyte count rises, so the total white count may not vary significantly. However, total white cell counts will be below that of controls at 24 hr after injection in normals and in patients[5,6,13].

Serum total 11-hydroxycorticosteroids rise to a peak 8 hours after administration of a single dose of dose of interferon alfa (leucocyte-derived or rDNA IFN-alfa$_2$) sufficient to cause a febrile reaction[13]. The magnitude of the rise is similar to that seen with a standard test dose of 250 mcg of tetracosactrin (synthetic adrenocorticotrophic hormone) and hyperglycaemia may be observed. Similarly, there are changes in plasma trace elements (zinc and copper), which would infer the induction of "leukocyte endogenous mediator" (sometimes considered to be interleukin 1), but may be a closely related molecule. Neither changes in steroids or in trace elements are affected by indomethacin pretreatment of volunteers. Patients with underlying tumours may not show the same adrenal response as healthy volunteers[24].

One strange property of the febrile complex is that it seems to be universal with all human interferons in man, chimpanzees and rabbits, but human alfa interferons do not cause fever in rhesus monkeys even at high dose[25], and may do so only if given intraventricularly (but not intravenously) in the cat[26]. Human IFN also has direct effects on explanted cat neuronal tissue[27]. So far as has been ascertained, mouse interferons do not seem to cause fever in mice. Fever-inducing properties in rabbits can be separated from antiviral activity in crude rabbit cell interferon[28].

Therefore interferons which have strong cross-species specificity also have different physiological effects in each host species.

EFFECTS OF PROLONGED INTERFERON TREATMENT

Local administration

Perhaps the most experience has been with high concentrations of interferon given intranasally by drops or spray in the prevention of common colds. It is an unfortunate observation that the minimum dose any IFN-alfa preparation needed to effectively prevent rhinovirus infection (about 2 Mu three times per day) causes nasal discomfort with bloody discharge in about half of volunteers after two weeks of treatment[29]. There is an intense submucosal lymphocyte infiltrate with mucosal ulceration. IFN-beta(ser) may be less inflammatory in these doses but it remains to be seen whether it is as effective[30].

Interferons given intrathecally in the treatment of multiple sclerosis, intermittently over a period of months causes a persistant pleomorphic cellular infiltrate in the cerebrospinal fluid[14]. This is in addition to the usual effects of systemic administration.

Systemic administration

Most studies have been done in patients with chronic virus infections (such as with hepatitis B) or with tumours, and it is difficult to be sure that the underlying disease (such as disseminated malignancy) or additional therapy (such as toxic antimetabolites) does not modify the reaction to interferon. It is clear that the acute symptoms of the febrile reaction complex are reduced with successive injections providing they are given more often than once every week. Some patients continue to have fever but the temporal relation to a specific dose is less clear than for a single dose. The reason for tachyphylaxis is not known. Two alternative hypotheses have been suggested. First, essential precursors for pyrogenic intermediates such as arachidonic acid may be consumed. Secondly, one part of the response may cause a negative feedback on subsequent reactions; an obvious candidate would be the corticosteroids induced by interferon. Neither hypothesis has been tested. The acute symptoms of headache, fever and myalgia tend to reduce, to be replaced by chronic lassitude and fatigue. Occasionally, these are severe enough to warrant reducing or stopping treatment, especially in outpatients continuing to work. The cause of the fatigue has not been identified.

Prolonged interferon appears to suppress cell growth in man. This is manifest as a general bone marrow depression. Leucopenia, thrombocytopenia and anaemia are common. Occasionally the dose must be reduced on empiric grounds but there does not seem to be a risk of bleeding or particular susceptibility to opportunistic infections during treatment. (Occasionally a bleeding diathesis may occur but this is apparently due neither to thrombocytopenia nor to hepatic damage[31].) Peripheral cell counts plateau and may rise after some weeks of treatment indicating a reduced responsiveness of the bone marrow to prolonged treatment. The effects of IFNs on bone marrow precursors have been studied in detail. They suppress both myeloid and erythroid precursors, the latter only in high concentration[32]. It has been suggested that different interferons exert different degrees of suppression on different lines according to their cell of origin[33], but this remains controversial.

Occasionally, there are other signs of cell growth inhibition such as hair loss. It is worth considering hypothyroidism as a possible cause. Patients may lose or fail to gain weight, but this may in part be due to an effect on appetite.

Hepatitis

Many patients undergoing prolonged interferon treatment have reversible changes in liver cell-associated enzymes. Interferon suppresses the hepatic enzyme cytochrome P450[34], possibly interfering with the metabolism of other agents. In normals, minor rises in serum gamma glutamyl transpeptidase and alkaline phosphatase occur but rarely with IFN-gamma in moderate dosage. In patients with chronic active hepatitis, therapeutic dose regimens cause a sharp rise in serum aspartate transaminase, but this is often delayed for up to three months and indicates a good outcome of treatment. It is unreasonable to consider this a direct toxic effect of interferon. However, very large doses (>50Mu/day) used in cancer cause consistent rises in alkaline phosphatase and transaminases[18] and the mechanism is not known. Hepatic enzymes will usually return to normal within two weeks of stopping treatment. Although fulminant hepatic necrosis occurred in newborn mice given mouse interferon[35], it seems an exceedingly rare complication in man which may well be due to combination with other factors.

Other metabolic and hormonal complications

Very high doses of interferon may cause hypocalcaemia[17]. Hyperglycaemia may accompany the toxic reaction[6], presumably associated with high corticosteroid levels[13]. Minor falls in progesterone and oestradiol have been seen in normal volunteers receiving interferon through one menstrual cycle[36]. There may also be falls in circulating high density lipoprotein cholesterol levels and marked rises in serum triglycerides[37]. The metabolic pathways for each of these disturbances are totally obscure.

Cerebral and central nervous system toxicity

The acute febrile reaction often precipitates mild confusion and difficulty in concentration. Progressive changes occur with continued treatment and may supervene as the febrile reaction becomes less prominent. Continued high doses lead to confusion, stupor, coma and occasionally death[38]. Electroencephalographic abnormalities are common[18,38]. Major seizures are not uncommon at very high doses[39] but occasionally occur at lower doses (eg 10Mu intramuscularly or less intrathecally), presumably in those with a predisposition.

Renal toxicity

In a trial of leucocyte interferon in the prevention of virus reactivation after renal transplant, a significant delay in cytomegaloviruria was observed without unexpected untoward effect[40]. When a slightly higher dose of IFN-alfa$_{2a}$ was tested in the same model, all the patients rejected their donated kidneys and histologically a vasculitis was apparent[41]. Occasional reports of nephrotic syndrome and acute glomerulitis have emerged[42,43]. These often occur in patients with diseases such as hairy cell leukaemia which are prone to such complications of the untreated disease, so the effect of interferon may be irrelevant or to accelerate natural autoimmune processes.

ANTIBODY FORMATION

Autoantibody formation

Long term therapy with leucocyte interferon precipitates thyroid microsomal or thyroglobulin antibodies and rarely either hypo- or hyper-thyroidism have been observed[44]. In one group of patients with carcinoid, these findings were unexpectedly frequent and *in vitro* tests suggested that they could have been due to traces of IFN-gamma contaminating the natural leucocyte-derived preparation[45]. However, it seems likely that alfa IFNs have this property which is dependent to a large extent on the susceptibility of the patient. Note that many of the non-specific symptoms of interferon are those of autoimmune diseases, which are often characterised by circulating interferon. One patient developed parotitis and epidydimitis clearly related to IFN-alfa treatment of hairy cell leukaemia[46], and one patient developed typical peripheral vasculitis[47].

Interferon antibodies

Interferon neutralising factors in serum have been described with many preparations tested[48,49,50,51]. They may be present before treatment is started particularly in those with neoplasia. However, treatment often seems to precipitate such antibody formation. There has been concern about the relative frequency of antibody formation with different preparations. There is a contrast in this respect between two preparations of IFN-$alfa_2$ (Table 2) which supposedly vary by only one amino-acid but in practice probably differ more substantially because of their methods of preparation and purification.

Table 2 Interferon neutralising factors in the serum*
after treatment with

Indication	INTRON-A alfa$_{2b}$		ROFERON alfa$_{2a}$	
Breast cancer	1/30	3	1/17	6
Small cell, bronchus	1/7	14		
Lung cancer			4/14	29
Lymphoma	1/47	2	13/60	22
Multiple myeloma	0/43		2/12	17
Kaposi's sarcoma	1/24	4	22/80	37
Other sarcoma			4/14	29
Colon/rectal carcinoma	1/37	3		
Gastrointestinal tumours			2/14	14
Renal carcinoma	3/114	3	85/185	46
Malignant melanoma	1/11	9	18/108	17
Bladder cancer	0/18			
Leukaemias			3/65	5
Hodgkin's disease	0/3			
Others	1/14	7	0/10	
TOTALS	10/348	3	154/579	27

*data collated from ref 7 and other published clinical trials.

It is by no means certain that all neutralising factors are indeed antibodies. Occasionally development of neutralising antibody in high titre is associated with a sudden abrogation of response to treatment and of side-effects.

DISCUSSION

Interferon treatment is not innocuous. Simple toxic effects such as nasal inflammation preclude its prolonged use in the prevention of colds. The febrile reaction complex may be sufficient to prevent a patient working or concentrating. There are ways around the febrile reaction (eg with acetaminophen) but it is not possible to prevent the insidious onset of fatigue and lethargy which follow prolonged treatment at therapeutic doses.

In oncology it is not unusual to expose patients to unpleasant therapies but these are only acceptable when there are clear potential benefits. Most publications on clinical trials of interferon state that the side-effects were mild and well-tolerated. However, several patients in most published trials are withdrawn because of intolerable symptoms or unacceptable changes in the peripheral white cell or platelet counts.

Not all human interferons are the same. *In vitro* tests have shown that alpha, beta and gamma interferons differ structurally, in their relative antiviral, antitumour and immunomodulating powers and the different alfa subtypes to vary in their affinity for cells from different species. Furthermore, cloned interferons are clearly different to natural preparations. It may seem that rDNA $alfa_2$ interferon which is the prevalent subtype in a mixture of natural leucocyte or lymphoblastoid interferons, is likely to have the same properties as the natural mixture. In practice, there are major differences in the final products and differences in biological properties. Cloned interferons from *Escherichia coli* are not glycosylated and are likely to contain a proportion of abnormal molecular forms. They may be contaminated with bacterial products in the same way that natural interferons may contain other cytokines. Lymphoblastoid interferon is the only natural preparation available which has been purified to near homogeneity. Both this preparation and cloned interferons possess the same toxic effects as crude natural preparations. We must therefore conclude that these effects are part of the physiologic effects of endogenous interferons, but that they may be enhanced by impurities. Results of further studies using high doses of gamma and combinations of human interferons are awaited with interest. More work is needed to elucidate mechanisms for some of the observations made. The role of chronic interferonaemia in autoimmune diseases and Kaposi's sarcoma is not known. It is tempting to speculate that interferon itself acts as endogenous pyrogen distinct from interleukin 1.

REFERENCES

1. Falcoff, E., Falcoff, R., Fournier, F. and Chany, C. Ann. Inst. Pasteur 111: 562-584, 1966.
2. Strander, H., Cantell, K., Carlstrom, G. and Jakobsson, P.J. Natn. Cancer Inst. 51: 733-742, 1973.
3. Ingimarrson, S., Cantell, K. and Strander, H. J. Infect. Dis. 140: 560-563, 1979.
4. Scott, G., Stewart, W.E.11, Tyrrell, D.A.J., Cantell, K. and Edy, V. J. Interferon Res. 1: 79-86, 1980.
5. Priestman, T.J. Lancet 2: 113-118, 1980.
6. Scott, G.M., Secher, D.S., Flowers, D., Bate, J., Cantell, K. and Tyrrell, D.A.J. Br. Med. J. 282: 1345-1348, 1981.
7. Spiegel, R.J. Semin. Oncol. 13: 89-101, 1986.
8. Jones, G.J. and Itri, L.M. Cancer 57: 1709-1715, 1986.
9. Sama, G., Pertcheck, M., Figlin, R. and Ardalan, B. Cancer Treat. Rep. 70: 1365-1372, 1986.
10. Kurzrock, R., Quesada, J.R., Rosenblum M.G., Sherwin, S.A. and Gutterman, J.U. Cancer Treat. Rep. 70: 1357-1364, 1986.
11. Heron, I. and Berg, K. Nature 274: 508-510, 1978.
12. Stringfellow, D.A. Science 201: 376-378, 1978.
13. Scott, G.M., Ward, R.J., Wright, D.J., Robinson, J.A., Onwubalili, J.K. and Gauci, C.L. Antimicrob. Agents Chemother. 23: 589-592, 1983.
14. Jacobs, L., Salazar, A.M., Herndon, R., Reese, P.A., Freeman, A., Josefowicz, R., Cuetter, A., Husain, F., Smith, W.A., Ekes, R. and O'Malley, J.A. Lancet 2: 1411-1413, 1986.
15. Hawkins, M.J., Borden, E.S. and Edwards B.S. J. Clin. Oncol. 2: 221-226, 1984.
16. Oldham, R.R. Science 219: 902, 1982.
17. Blalock, J.E. and Stanton G.J. Nature 283: 406, 1980
18. Rohatiner, A.Z.S., Prior, P.F., Burton, A.C., Smith, A.T., Balkwill, F.R. and Lister,T.A. Br. J. Cancer 47: 419-422, 1983
19. Yaron, M., Yaron, I., Gurari-Rotman, D., Revel, M., Lindner, H. and Zor, V. Nature 267: 457-459, 1977.
20. Fuse, A., Mahmud, I., and Kuwata, T. Cancer Res. 42: 3209-3214, 1982.
21. Dinarello, C.A., Bernheim, H.A., Duff, G.W., Hung, V.L., Nagabushan, T.C., Hamilton, N.C. and Coceani, F. J. Clin. Invest. 74: 906-913, 1984.
22. Lucero, M.A., Magdelenat, H., Fridman, W.H., Pouillart, P., Billiau, A., Cantell, K. and Falcoff, E. Eur. J. Cancer Clin. Oncol. 18: 243-251, 1982.
23. Abrams, P.G., McClamrock, E. and Foon, K.A. N. Engl. J. Med. 312: 443-444, 1985.
24. Merigan, T.C., Sikora, K., Breeden, J., Levy, R., and Rosenberg, S.A. N. Engl. J. Med. 299: 1449-1453, 1978.
25. Finter, N.B., Woodroffe, J. and Priestman, T. Nature 298: 301, 1982.
26. Feldberg, W. and Scott, G. J. Physiol. 355:28P, 1984
27. Calvet, M-C. and Gresser, I. Nature 278:558-560, 1979
28. Elfenbeim, D.S., Murphy, P.A., Chesney, P.J., Seto, D.S. and Carver, D.H. J. gen. Virol. 42: 185-187, 1979
29. Hayden, F.G., Mills, S.E. and Johns, M.E. J. Infect. Dis. 148: 914-921, 1983.

30. Hayden, F.G., Innes, D.J., Mills, S.E. and Levine, P.A. In: The Biology of the Interferon System 1986 (Eds. K. Cantell and H. Schellekens), Martinus Nijhoff, Dordrecht, 1987, pp.449-456.
31. Mirro, J., Kalwinsky, D., Whisnant, J., Weck, P. Chesney, C. and Murphy, S. Cancer Treat. Rep. 69: 315-317, 1985.
32. Broxmeyer, H., Lu, L., Platzer, E., Feit. C., Juliano L. and Rubin, Y. J. Immunol. 131: 1300-1313, 1983
33. Kataoka, T., Ohhashi, F., Sukurai, Y., and Ida, N. Cancer Chemother. Pharmacol. 9: 75-80, 1982
34. Mannering, G.J., Renton, K.W., El Azhary, R., and Deloria L.B. Ann. N.Y. Acad. Sci. 350: 314-331, 1980.
35. Gresser, I., Tovey, M.G., Bandu, M.T., Maury, C. and Brouty-Boye, D. Nature 258: 76-78, 1975.
36. Kauppila, A., Cantell, K., Janne, O., Kokko, E. and Vikho, R. Int. J. Cancer 29: 291-294, 1982.
37. Cantell, K., Enholm, C., Mattila, K. and Kostianen, E. N. Engl. J. Med. 302: 1032-1033, 1980
38. Smedley, H., Katrak, M., Sikora, K. and Wheeler, T. Br. Med. J. 286: 262-264, 1983.34.
39. Rohatiner A.Z., Balkwill, F.R., Griffin, D.B., Malpas, J.S. and Lister, T.A. Cancer Chemother. Pharmacol. 9: 97-102, 1982.
40. Cheeseman, S.H., Rubin, R., Stewart, J., Tolkoff-Rubin, N., Cosimi, A., Cantell, K., Gilbert, J., Winkle, S., Herrin, J., Black, P., Russell, P. and Hirsch, M. N. Engl. J. Med. 300: 1345-1349,1979.
41. Kramer, P., Ten Kate, F., Bijnen, A. and Weimar, W. Lancet 1: 989-990, 1984.
42. Averbuch, S.D., Austin, H.A., Sherwin, S.A., Antonovych, T., Bunn, P.A. and Longo, D.L. N. Engl. J. Med. 300:32-35,1984.
43. Herrman, J. and Gabriel, F. N. Engl. J. Med. 316: 112-113, 1987
44. Fentiman, I.S., Thomas, B.S., Balkwill, F.R., Rubens, R.D. and Hayward, J.L. Lancet 1: 1166, 1985.
45. Burman, P., Totterman, T.H., Oberg, K. and Karlsson, F.A. J. Clin. Endocrinol. Metab. 63: 1086-1090, 1986.
46. Bevan, P.C. Lancet 2: 561, 1985.
47. Sangster, G., Kaye, S.B., Calman, K.C. and Toy, J.L. Eur. J. Clin. Oncol. 19: 1647-1649, 1983.
48. Mogensen, K.E., Daubas, P., Gresser, I., Sereni, D. and Varet, B. Lancet 2: 1227-1228, 1981.
49. Vallbracht, A., Treuner, J., Manncke, K.H. and Niethammer, D. J. Interferon Res. 2: 107-110, 1982.
50. Krown, S.E., Real, F., Cunningham-Rundles, S., Mysowski, P.L., Koziner, B., Fein, S., Mittleman, A., Oettgen, H.F. and Safai, B. N. Engl. J. Med. 308: 1071-1076, 1983.
51. Quesada, J.R. and Gutterman, J.U. J. Natl. Cancer Inst. 70: 1041-1046, 1983.

20

PHARMACOKINETICS OF INTERFERONS AND PRACTICAL IMPLICATIONS

V. BOCCI

Institute of General Physiology, University of Siena, Via Laterina 8, 53100 Siena, Italy

ABSTRACT

The pharmacokinetics of IFN's α, β and γ are reviewed pointing out their short half-lives, their distribution and major catabolic sites. After several years of IFN clinical trials based on empirical approaches it is now clear that IFNs are pharmacodynamically neither comparable to antibiotics nor to cytotoxic agents and it may be useful to test unconventional routes of administration which may improve their therapeutic index.

INTRODUCTION

Natural interferons (IFNs) are proteins (class α) and glycoproteins (classes β and γ(which, while they present some amino acid homology (about 29% between IFNs α and β), show significant physicochemical differences such as hydrophobicity (IFN $\beta > \gamma \geqslant \alpha$) and the carbohydrate content which are in part responsible for their different distribution and fate (1). Thus, the metabolic heterogeneity associated to their potential dual action as immunomodulators makes these compounds somewhat difficult to use as drugs. A thorough knowledge of their pharmacokinetics, biological activities and mechanisms of action is becoming indispensable in order to use them on a rational basis and to take advantage of their efficacy while reducing to a minimum their side effects. Moreover, the present availability of recombinant (R) IFNs produced by E. coli complicates the problem because bacterially produced R IFNs β and γ are not glycosylated and have different pharmacodynamic characteristics compared to their natural counterparts (2). However, this

M. Revel (ed.), CLINICAL ASPECTS OF INTERFERONS. Copyright 1988. Kluwer Academic Publishers, Boston. All rights reserved.

problem is going to be eased in the future when most of the R IFNs will be produced by eukariotic cells and, particularly in the case of glycoproteins, will be glycosylated.

PHARMACOKINETICS OF IFNs

First of all I will consider the metabolic behaviour that is common to all IFNs. When they are administered via the intravenous (IV) route and mix in the plasma, several processes occur simultaneously: some IFN passes from the plasma pool into the interstitial fluids of various organs, some is either filtrated by the kidneys or taken up by the liver and a minimal amount becomes bound to circulating leukocytes or to endothelial cells (3). Transcapillary passage is regulated by blood flow and capillary permeability that is almost total in the case of the liver, spleen and bone marrow, partially reduced in the case of the kidneys, intestines and skin and almost negligible for muscles, lungs, subcutis, bone tissue and central nervous system (CNS). Capillary permeability of different tumors varies considerably (4) and therefore the amount of IFN actually reaching tumor cells is neither predictable, nor has it been evaluated so far in animal model systems. It is obvious that this complex and heterogeneous pattern of distribution is conditioning the efficacy and the toxicity of the drug, particularly when one is dealing with chronic viral diseases or tumors secluded in anatomical sanctuaries.

Disappearance of IFNs from plasma occurs in a triphasic (5) and more frequently biphasic manner depending upon the amount of IFN injected and the accuracy of analysis. IFN distribution has usually been assessed using a two-compartment open model described by the equation $C_t = Ae^{-\alpha t} + Be^{-\beta t}$ where C_t is the concentration of IFN at time 5, A and B are concentration values and α and β are constants estimated by using standard methods (6). Serum half-lives ($t\frac{1}{2}$) of a protein during the initial (fast) distribution phase and the successive (slow) metabolic phase are calculated as $0.693/\alpha$ and $0.693/\beta$, respectively (6). It is useful to emphasize that half-life values of IFNs (when used as an immunomodulator) and of other biological response modifiers (BRMs) should be taken only as an indication that these drugs remain in the circulation for a brief time, while in fact biological effects may last for a longer time depending upon the lifespan of the stimulated cells or the "memory" of biochemical mechanisms. Moreover among published data there is a considerable variability of half-life values largely due to various dosages and usually there is a trend towards a longer half-life with increasing dosage.

After IV administration as a bolus, half-lives (slow phase) of leukocyte IFN and human R IFNα_2 are of 1.5 hr (7) and of 0.75-2 hrs (8), respectively. Metabolic heterogeneity of IFN is evident by observing that after IV administration half-lives (slow phase) of natural IFNs β and γ are of only 13 mins (9) and of 2.6-31 mins (10), respectively. Human R IFNβ (serine) and R IFNγ (E. coli) display similarly half lives of 69.8 mins (11) and of 30 mins (12), respectively.

IFNs have been most frequently administered via the intramuscular (IM) and subcutaneous (SC) routes. Under such circumstances, IFN is absorbed from the site of injection (usually one or two sites in the buttock or deltoid muscles) and reaches the plasma pool so that IFN plasma levels increase progressively during the first 5-9 hrs (t max) and then decline because disappearance of IFN from the plasma prevails over the IFN either being absorbed or returning via the lymphatic or blood vessels. The IFN peak level (C max) is obviously much lower than that achieved immediately after IV administration but IFN plasma levels are sustained for a longer time and in general are practically undetectable 24 hrs thereafter. Both IM and SC administration of IFN dissolved in saline yield similar plasma curves although the IFN peak level may be slightly delayed after SC injection.

After IM administration ofleukocyte IFN and R IFNα_2 half-lives of about 7 hrs (13) and of 5-8.2 hrs (8,13), respectively, have been measured. On the other hand both natural IFNs β and γ after IM administration are hardly detectable in the plasma (10,14) and most of the clinicians prefer to administer these IFNs by slow IV infusion. However, R IFN serine and R IFN (E. coli) after IM administration yield a plasma curve with half-lives of 70 mins (11) and of 227-462 mins (12), respectively, although some of the circulating R IFN (measured by Elisa-assay) is biologically inactive.

The conclusion that natural human IFNβ after IM administration is biounavailable because it remains fixed or is catabolized at the site of injection (1) is somewhat unprecise because in the rabbit model we have shown (Bocci et al., manuscript in preparation) that at least some of IFNβ is absorbed via lymphatics. This result clarifies why IFNβ, in spite of its undetectability in plasma, causes an increase of the (2'-5') oligoadenylate synthetase in leukocytes (15), enhancement of the natural killer (NK) activity (16) and increased expression of hormonal receptors (17). During the transit through the lymph and lymph nodes, IFNβ comes into contact with a great number of effector cells and, as it slowly emerges into the plasma, it undergoes a very rapid turnover and therefore does not become detectable.

Unless we want to use IFN as an antiviral drug in acute viral diseases or as a cytotoxic drug there is no real advantage in maintaining high and constant IFN levels in plasma as it can be achieved by continuous IV infusion, because renal and hepatic catabolism are related to the IFN concentration (3) and therefore there is a considerable loss of material. In practical terms the rapid IFN elimination is now interpreted as a useful detoxification process because the higher the IFN plasma levels, the worse the toxicity (18). From an antiproliferative point of view IFN is not a selective drug and while it may or may not inhibit tumor growth (depending upon tumor cells being IFN sensitive or rresistant) it certainly inhibits proliferation of normal cells, particularly of the hemopoietic lineages.

Learning how and where IFN molecules are produced and function in physiological conditions (19), how they move around in the body fluids (3), how and where they are eliminated (3) and why they cause toxic effects (20) has profoundly influenced my thinking and this led to the conclusion that, with the few exceptions mentioned before, IFNs, as well as other BRMs, (particularly interleukin 2) are not physiologically supposed to be circulatory proteins such as albumin and immunoglobulins (2,18) and therefore when we use them we should bear in mind that physiologically they behave as paracrine hormones.

When IFN and other BRMs are mainly used as immunoenhancers in the therapy of tumors and chronic viral diseases, we should aim ideally to stimulate the highest number of effecto cells (NK, macrophages, etc.) minimizing toxic effects. Moreover, the phenomena now shown in vitro and in vivo of receptor down-regulation and cell unresponsiveness (21, 22) should be borne in mind in deciding the frequency of IFN administration.

NEW SUGGESTIONS

In order to improve the utilization of IFN, most of it should be shifted into the lymph pool thereby increasing the lymph/plasma ratio several fold (usually below or near 1). To this end I proposed (23) administering IFN subcutaneously by way of multiple (at least 10) small volume (0.1-0.3 ml) injections of IFN dissolved in a hyperoncotic (12.5-15%) human albumin solution. Different areas of the extremities', trunk's and neck's subcutis can be injected once or at most twice weekly. This method (indirect lymphatic administration) tries to overcome the practical difficulties with the direct intralymphatic administration and is based upon the fact that albumin acts as an interstitial fluid expander which, by increasing locally the colloid osmotic pressure of the interstitial fluid, favours lymphatic absorption. A more extensive description and pharmacokinetic data have been reported (18,24) and I would be glad to give assistance to any clinician willing to try this route. The method already tried in a few patients has not shown any significant drawback and so far the number of injections have been accepted with the best patient compliance. Needless to say, the lymphatic administration method is also being advocated for other BRMs such as interleukin 2, thymic hormones, transfer factor, monoclonal antibodies etc.

In a book dedicated to clinical aspects of IFN, I cannot omit a brief discussion on the foreseeable improvement of IFN pharmacodynamics when administered by particular routes.

Pros and cons of the intraperitoneal (IP) route have been previously discussed (25) and Berek et al. (26) have reported that peritoneal fluid has, at least for 24 h following IP administration, IFN levels 30-100 fold higher than blood levels. IFN levels in the peritoneal cavity remain detectable for a further 5 days so that it appears that the cavity acts as a reservoir and IFN is mostly absorbed through subdiaphragmatic lymphatics (25). Further delayed absorption can be envisaged by administering IFN entrapped in controlled-release polymers (27), or by simply keeping the patient at rest during the day (thus reducing the excursions of the diaphragm) sitting in bed, i.e. pelvis-down position. Slow peritoneal clearance of IFN may mean a prolonged IFN action at cystostatic levels and an even longer immunoadjuvan effect.

The possible persistence of viruses in the CNS of immunocompetent hosts (28) and the fact that most of the human gliomas remain incurable (29) compel the improvement of the delivery of IFN and other BRMs to the CNS to attempt an effective antiviral or antitumoral therapy. With the exceptions of small anatomical areas (infundibular recess, area postrema and median eminence) and of some tumors (30,31) the blood-brain barrier (BBB), constituted by tightly closed capillaries with a continuous basement membrane, is not leaky and as is shown by the presence of only a trace of IFN in the cerebro-spinal fluid (csf), allows I believe, only a very limited passage of IFN from plasma to the CNS (3) via the choroid plexuses.

How can we overcome this impasse? Theoretically two approaches are possible: the first is to deliver the drug via carotid or vertebral circulation (regional intra-arterial therapy) while the BBB has been made transiently permeable by either dehydrocholate (32), or mannitol (33) or etoposide (34). The second is to administer IFN into the CNS more or less directly. To this end the intralumbar route is frequently used but allows a partial perfusion of CNS while the intraventricular route via Ommaya's reservoir (35) favours a wider distribution of the drug. Moreover, and wherever possible, intra or peritumoral (glioblastomas) injection of IFN has been performed and intralesional administration of human IFNβ appears more effective than IFNα (36, 37). In all cases disappearance of IFN from cfs is much slower than from plasma and, owing to the slow drainage of cfs into the dural sinuses, IFN can be detected for days (37). On the other hand, particularly using the second approach, IFN plasma levels are very low and consequently bone marrow depression and the flu-like syndrome are rare or mild. The main issue here is to foresee the most effective approach and the one which leads to minimal CNS toxicity. This is fearsome syndrome and while it has been noted after continuous IV infusion with doses as high as 100 megaU/m^2/daily (38), it has been barely noticed while using the second type of approach in spite of the fact that 10^3-10^4 U/ml are measurable in the cfs (36, 37).

A likely explanation of this striking difference is that CNS toxicity is caused by the interaction of IFN with CNS endothelium, pericytes and glial cell surrounding cerebral microvessels. Vasoactive substances such as noradrenaline, serotinin, histamine and neurotransmitters such as γ-amino-butyric acid and acetylcholine released by endothelial cells (39) may be responsible for vasogenic and cellular edema. If this interpretation is correct the first approach aiming at reversibly opening the BBB should be considered with great caution as the advantage of an increased IFN delivery to the CNS may be outweighed by the insurgence of an unacceptable toxicity, possibly due to the combined action of IFN and the permeabilizing agent. Thus after tumor debulking, IFN application using the tumor cyst device (35) may be at the moment the best option for treating malignant gliomas. Furthermore, the small amount of IFN lleaking slowly into the general circulation may deploy a useful immunoadjuvant effect.

CONCLUSIONS

It appears by now that selection of the IFN type, dosages, routes and schedules of administration should be dictated by the type of disease, the pharmacokinetic pattern and the prevalent biological activity to be exploited for obtaining the therapeutic effect. There are no golden rules and present knowledge suggests the following conclusions:

a) continuous IV infusion of IFN seems useful in the treatment of fulminant viral diseases (40) and in the case of CNS infections it could be implemented by intrathecal administration (37).

b) Conventional IM or SC administration of IFNα appears useful for prolonged treatment of chronic viral diseases (41). It appears satisfactory in inducing remission of hairy cell leukemia (42) probably because IFNα attains sufficient plasma levels and, owing to the discountinuous type of the bone marrow's capillaries, reaches this organ easily and it allows either cell differentiation or cytostasis. The role of reactivation of NK cells in this disease remains uncertain (43). IFNγ is ineffective and it would be interesting to evaluate whether IFNβ would be equally effective when administered intramuscularly instead of IV infusion. Other myelopoietic diseases such as chronic myelogenous leukemia (44) and essential thrombocythaemia (45) are positively influenced by this type of IFN treatment.

c) Tumors such as low-grade non-Hodgkin's lymphoma (46), cutaneous T-cell lymphoma (47), Kaposi's sarcoma (48) and multiple myeloma (49) need high IFNα levels obtainable by IM administration: when these tumors are IFN sensitive, antiproliferative effect develops and significant improvement has been observed. In some cases the maximum antiproliferative effect may be below the maximum tolerated dose.

d) On the other hand, either leukemias in the acute phase or visceral neoplasms and metastases are most likely IFN insensitive because of lack of IFN receptors and/or transducing signals and corrective mechanisms. Wherever possible, after debulking, IFN activation of the cellular immune response may be advantageous. In order to consolidate the remission, minimal residual malilgnant disease should be treated with immunotherapy. In such cases the use of BBMs via the indirect lymphatic route (18) may be advantageous because it simulates the physiological distribution and the mode of action of these hormone-like compounds (19). The sequential or combined treatment with BBMs (thymic hormones, interlelukin 2, IFNs, etc.) is likely to be more effective than using a single drug. This route may also be proficiently used for the treatment of chronic viral diseases.

e) Intracavitary or topical administration (25) of IFNs either in localized tumors, or in cutaneous, or mucosal viral diseases may be useful when diffusion of IFNs through anatomical barriers is impeded, or when a localized high IFN concentration is desirable.

f) Finally observance of circadian rhythms of these hormone-like drugs (50) may improve their therapeutic effect and minimize side-effects.

ACKNOWLEDGEMENTS

This work was supported by Progetti finalizzati 1986 "Oncologia" and "Controllo delle malattie de infezione" CNR, Roma, Italy.

REFERENCES

1. Bocci, V. Trends Biochem. Sci. 8: 432-434, 1983
2. Bocci, V., Pessina, G.P., Pacini, A., Paulesu, L., Muscettola, M., Naldani, A. and Maioli, E. In: The Interferon System (Eds. F. Dianzani and G.B. Rossi), Serono Symposia Vol. 24, Raven Press, New Yor, 1985, pp. 129-135.
3. Bocci, V. In: Interferon. Vol. 4: In Vivo and Clinical Studies (Eds. N.B. Finter and R.K. Oldham), Elsevier Science Publishers B.V., Amsterdam, 1985, pp. 47-72.
4. Warren, B.A. In: Tumour Blood Circulation: Angiogenesis, Vascular Morphology and Blood Flow of Experimental Tumours (Ed. H.I. Peterson), CRC Press Inc., Boca Raton, Florida, 1979, pp. 20-35.
5. Bocci, V., Pessina, G.P., Pacini, A., Paulesu, L., Muscettola, M., Naldini, A. and Lunghetti, G. Gen. Pharmac. 16: 277-279, 1985.
6. Greenblatt, D.J. and Koch-Weser, J. New England J. Med. 293: 702-705, 1975.
7. Emodi, G., Just, M., Hernandez, R. and Hirt, H.R. J. Natl. Cancer Inst. 54: 1045-1049, 1975.
8. Bornemann, L.D., Spiegel, H.E., Dziewanowska, Z.E., Krown,, S.E. and Colburn, W.A. Eur. J. Clin. Pharmacol. 28: 469-471, 1985.
9. Treuner, J., Dannecker, G., Joester, K.-E., Hettinger, A. and Niethammer, D. J. Interferon Res. 1: 373-380, 1981.
10. Gutterman, J.U., Rosenblum, M.G., Rios, A., Fritsche, H.A. and Quesada, J.R. Cancer Res. 44: 4164-4171, 1984.
11. Hawkins, M., Horning, S., Konrad,M., Anderson, S., Sielaff, K. Rosno, S., Schiesel, J., Davis, T., DeMets, D., Merigan, T. and Borden, E. Cancer Res. 45: 5914-5920, 1985.
12. Kurzrock, R., Rosenblum, M.G., Sherwin, S.A., Rios, A., Talpaz, M., Quesada, J.R. and Gutterman, J.U. Cancer Res. 45: 2866-2872, 1985.
13. Gutterman, J.U., Fine, S., Quesada, J., Horning, S.J., Levine, J.F., Alexanian, R., Bernhardt, L., Kramer, M., Spiegel, H., Colburn, W., Trown, P., Merigan, T. and Dziewanowski, Z. Ann. Int. Med. 96: 549-556, 1982.
14. Quesada, J.R., Gutterman, J.U. and Hersh, E.M. J. Interferon Res. 2: 593-599, 1982.
15. Schonfeld, A., Nitke, S., Schattner, A., Wallach, D., Crespi, M., Hahn, T., Levavi, H., Yarden, O., Shoham, J., Doerner, T. Revel, M. The Lancet i: 1038-1042, 1984.
16. Lucero, M.A., Magdelenat, H., Fridman, W.H., Pouillart, P., Billardon, C., Billiau, A., Cantell, K. and Falcoff, E. Eur. J. Cancer Clin. Oncol. 18: 243-251, 1982.
17. Pouillart, P., Palangie, T., Jouve, M., Garcia-Giralt, E., Fridman, W.H., Magdelenat, H., Falcoff, E. and Billiau, A. Eur.

J. Cancer Clin. Oncol. 18 929-935, 1982.
18. Bocci, V. J. Biol. Resp. Mod. 4: 340-352, 1985.
19. Bocci, V. Immunol. Today 6: 7-9, 1985.
20. Scott, G.M. In: Interferon (Ed. I Gresser), Vol. 5, Academic Press, New York, 1983, pp. 85-114.
21. Billard, C., Sigaux, F., Castaigne, S., Valensi, F., Flandrin, G., Degos, L., Falcoff, E. and Aguet, M. Blood 67: 821-826, 1986.
22. Wietzerbin, J., Gaudelet, C., Aguet, M. and Falcoff, E. J. Immunol. 136: 2451-2455, 1986.
23. Bocci, V. In: Interferons (Ed. T. Kishida), Japan Convention Services Inc., Osaka, 1983, pp. 38-44.
24. Bocci, V., Muscettola, M., Naldini, A., Bianchi, E. and Segre, G. Gen. Pharmac. 17: 93-96, 1986.
25. Bocci, V. Cancer Drug Delivery 1: 337-351, 1984.
26. Berek, J.S., Hacker, N.F., Lichtenstein, A., Jung, T., Spina C., Knox, R.M., Brady, J., Greene, T., Ettinger, L.M., Lagasse, L.D., Bonnem, E.M., Spiegel, R.J. and Zighelboim, J. Cancer Res. 45: 4447-4453, 1985.
27. Langer, R. Pharmac. Ther. 21: 35-51, 1983.
28. Johnson, R.T. In: Neuroimmunology (Ed. P. Behan and F. Spreafico), Raven Press, New York, 1984, pp. 229-236.
29. Bullard, D.E., Gillespie, G.Y., Mahaley, M.S. and Bigner, D.D. Seminars Oncol. 13: 94-109, 1986.
30. Blasberg, R.G. and Groothuis, D.R. Seminars Oncol. 13: 70-82, 1986.
31. Hasegawa, H., Ushio, Y., Hayakawa, T., Yamada, K. and Mogami, H. J. Neurosurg. 59: 304-310, 1983.
32. Spigelman, M.K., Zappulla, R.A., Malis, L.I., Holland, J.F., Goldsmith, S.J. and Goldberg, J.D. Neurosurgery 12: 606-612, 1983.
33. Neuwelt, E.A. and Rapoport, S.I. Fed. Proc. 43: 214-219, 1984.
34. Spigelman, M.K., Zappulla, R.A., Strauchen, J.A., Feuer, E.J., Johnson, J., Goldsmith, S.J., Malis, L.I. and Holland, J.F. Cancer Res. 46: 1453-1457, 1986.
35. Omaya, A.K. Cancer Drug Delivery 1: 169-179, 1984.
36. Koyama, Y. In: Interferons (Ed. T. Kishida), Japan Convention Services Inc., Osaka, 1983, pp. 189-195.
37. Nagai, M. and Arai, T. Neurosurg. Rev. 7: 55-64, 1984.
38. Rohatiner, A.Z.S., Prior, P., Burton, A., Balkwill, F. and Lister, T.A. In: Progress in Experimental Tumor Research (Ed. F. Homburger), Vol. 29, S. Karger, Basel, 1985, pp. 197-202.
39. Parnavelas, J.G., Kelly, W. and Burnstock, G. Nature 316: 724-725, 1985.
40. Levin, S. and Hahn, T. The Lancet i: 592-594,1982.
41. Omata, M., Imazeki, F., Yokosuka, O., Ito, Y., Uchiumi, K., Mori, J. and Okuda, K. Gastroenterology 88: 870-880, 1985.
42. Ehmann, W.C. and Silber, R. Am. J. Med. 80: 1111-1114, 1986.
43. Ratain, M.J., Golomb, H.M., Vardiman, J.W., Vokes, E.E., Jacobs, R.H. and Daly, K. Blood 65: 644-648, 1985.
44. Talpaz, M., Kantarjian, H.M., McCredie, K., Trujillo, J.M., Keating, M.J. and Gutterman, J.U. New England J. Med. 314: 1065-1069, 1986.
45. Velu, T., Delwiche, F., Gangji, D., Monsieur, R., Flament, J.,

Stryckmans, P., Wybran, J. and Bellens, R. Oncology 42: suppl. 1, 10-14,, 1985.
46. Foon, K.A., Sherwin, S.A., Abrams, P.G., Longo, D.L., Fer, M.F., Stevenson, H.D., Ochs, J.J., Bottino, G.C., Schoenberger, C.S., Zeffren, J., Jaffe, E.S. and Oldham, R.K. NewEngland J. Med. 311: 1148-1152, 1984.
47. Bunn, P.A., Foon, K.A., Ihde, D.C., Longo,, D.L., Eddy, J., Winkler, C.F., Veach, S.R., Zeffren, J., Sherwin, S. and Oldham, R. Ann. Intern. Med. 101: 484-487, 1984.
48. Volberding, P.A. and Mitsuyasu, R. Seminars Oncol. 12: suppl. 5, 2-6, 1985.
49. Quesada, J.R., Alexanian, R., Hawkins, M., Barlogie, B., Borden, E., Itri, L. and Gutterman, J.U. Blood 67: 275-278, 1986.
50. Bocci, V. Cancer Drug Delivery 2: 313-318, 1985.

21

MONITORING OF INTERFERON THERAPY BY ASSAY OF INTERFERON-INDUCED (2'-5')OLIGO A SYNTHETASE IN HUMAN PERIPHERAL WHITE BLOOD CELLS

A. SCHATTNER and M. REVEL

Department of Virology, Weizmann Institute of Science, Rehovot, Israel

ABSTRACT

A simple and rapid assay was developed to measure levels of the interferon (IFN)-induced enzyme 2'-5' oligo adenylate synthetase in peripheral blood mononuclear cells (PBM cells) and granulocytes. We used this assay to monitor therapy with either IFN-alpha or beta administered by various routes. Results indicate that serial OASE determinations are a valuable addition to clinical trials with IFN that may identify responders and help to resolve problems of the optimal dose, route and schedule of IFN administration.

INTRODUCTION

The number of patients undergoing interferon (IFN) therapy for viral or malignant diseases is constantly increasing. However, even today the precise indications, timing, route of administration and dosage of IFN have not been positively determined and in many cases IFN-therapy is still empirical and based on initial, often incomplete, observations. A true breakthrough in the therapeutic use of IFN may be achieved with further understanding of the mechanism of IFN-action at the molecular level (1), but more practically, two conditions would have to be fulfilled. The first, is the actual availability of 'unlimited' amounts of pure IFN for controlled experimentation. This goal has already been achieved with the recent availability of IFN's produced by recombinant DNA technology (2). The second, is the use of a unified monitoring system to determine in real-time the effects of IFN-administration which would enable the treating physician to adjust the route, dose or frequency of IFN injections necessary to obtain an optimal response in the individual treated.

Several approaches have been used to monitor IFN-therapy (Table 1), but a rapid sensitive and convenient assay was still needed. Direct assay of IFN activity in the serum is possible (6,7,13) but it involves lengthy biological procedures which are often imprecise because of the low levels and short half-life of IFN activity usually observed. Furthermore, IFN-β is not readily detected in serum after i.m. or s.c. injection (10). As an alternative to the IFN assay in the serum we suggested (12) the measurement of the variations in an enzyme, 2'-5' oligo A synthetase, which is induced in cells exposed to IFN's. Significant activity of 2'-5' A synthetase can be detected in a wide range of mammalian tissue extracts (14), and it increases sharply within hours of treating the cells with IFN (15). The 2'-5' A synthetase is a dsRNA-dependent enzyme which polymerizes ATP to form a series of oligonucleotides (dimers, trimers and longer chains of adenylic residues) linked by a 2'-5' phosphodiester bond (1,16,17). These oligoadenylates, at nanomolar concentrations activate the latent ribonuclease F (or L) and inhibit mRNA translation. Thus, viral protein synthesis and replication are impaired through the action of the IFN-induced 2'-5' A synthetase. Several forms of the enzyme are found in human cells and were cloned and sequenced (18,19). Our work shows that 2'5' A can serve as a marker not only for the _in vitro_ response of cells to IFN, but also for a sensitive monitoring of the IFN system _in vivo_, both in patients undergoing IFN therapy (12,24-28) and in patients with various diseases (20-23, 25) and the results obtained will be reviewed here.

Table 1. Different approaches used to monitor IFN-therapy

A. Monitoring by toxicity (3) (e.g. symptomatic, hematologic, etc.) or by clinical effects (4-5) (e.g. decrease in spleen size and a number of pathological cells in hairy-cell leukemia; decrease in viral antigens in hepatitis B, etc.).

B. Monitoring by serum IFN levels and the appearance of serum neutralizing factors (6,7).

C. Monitoring by following the immunomodulatory effects of IFN (8-10) [e.g. effect on natural killer cells (NK) cells, major histocompatibility antigens (β_2 microglobulin) etc.)].

D. Monitoring by assessing the antiviral state of the peripheral blood mononuclear cells (11).

E. Monitoring by measuring changes in IFN-induced enzymes in the peripheral blood or other cells (12).

MATERIALS AND METHODS

The assay procedure is summarized in Fig. 1. Two ml of heparinized venous blood are diluted with 2 ml phosphate-buffered saline (PBS), layered on 3 ml of Ficoll-Hypaque (Pharmacia Fine Chemicals), and centrifuged for 30 min at 400 g. The interphase cells (mononuclear cells) are collected, washed twice in 5 ml PBS, and counted. An aliquot of 10^6 cells is pelleted in an Eppendorf microtest tube, and 0.1 ml lysis buffer (20 mM Hepes buffer pH 7.5, 5 mM $MgCl_2$, 120 mM KCl, 7 mM dithiothreitol, 10% glycerol, 0.5% Nonidet-P40) is added at 4^oC. After a few minutes, the extract is clarified by centrifugation for 6 min at 8,000 g and can then be frozen at -70^oC. Extracts from granulocytes can be prepared from the same sample as described (25).

For enzyme assay, 0.01 ml thawed extract is added to 0.025 ml poly (rI): (rC)-agarose beads (PL Biochemicals; the beads have been previously washed with 1 ml of buffer C: 10 mM Hepes buffer, pH 7.5, 50 mM KCl, 50 mM $MgCl_2$, 7 mM dithiothreitol, 20% glycerol). After gentle mixing, the beads suspended in the extract are incubated for 15 min at 30^oC; at this step, the (2'-5') oligo A synthetase binds to the beads. Tubes with cell extracts from 3 healthy donors ant tubes without cell extract are routinely included in addition to the extracts to be tested. The beads are pelleted in the Eppendorf microfuge, washed with 1 ml buffer C, and all liquid is carefully removed. A reaction mixture of 0.01 ml is added which contains 10 mM Hepes buffer, pH 7.5, 5 mM $MgCl_2$, 7 mM dithiothreitol, 10% glycerol, 2.5 mM [^{32}P]-α-ATP (0.1-0.3 Ci/mmol), 3 mg/ml creatine kinase, 10 mM creatine phosphate, and 40 μg/ml poly (rI)(rC). Incubation is carried out for 14-20 hours at 30^oC. Then, 1 unit of bacterial alkaline phosphatase in 0.01 ml of 140 mM Tris-base is added. After 1 hour at 37^oC, 0.02 ml water is added, and the beads are removed by centrifugation. From the supernatant, 0.01 ml is applied to a 0.3 ml column of alumina (acid alumina WA1 from Sigma) equilibrated in 1 M glycine-HCl buffer, pH 2. Three ml of the same buffer are applied to the column and directly collected in scintillation vials which are then counted in the ^{3}H-channel of a Tricarb scintillation counter (Packard) by Cerenkov radiation. This procedure measures the $(A2'p)_nA$ nucleotides formed (29).

The results can be expressed in cpm or pmol ATP incorporated per hour and per μg protein, or per 10^5 cells. The assay itself uses only

2.5×10^4 cells and usually gives $6\text{-}8\times10^3$ cpm. We routinely express the results in percent of the mean enzyme activity found for the samples from healthy donors.

RESULTS

Enzyme level in PBM cells from healthy donors

In a group of 63 samples from healthy blood donors, the mean (2'-5') oligo A synthetase activity was 130 pmoles $(A2'p)_nA$ per hour and 10^5 cells (about 5μg protein). The standard deviation was ± 40%, and this is taken as the statistical range of normal values. The (2'-5') oligo A synthetase level of mononuclear cells in the normal human population appears, therefore, high and constant. We did find, in this group of 63, 3 high values outside the statistical deviation, but these were later diagnosed as pathologic cases. We also found enzyme activity in the granulocytes, but this amounted to only 0.1-0.3 fold the activity of mononuclear cells. Erythrocytes, thrombocytes, and serum had no significant enzyme activity in our assays. Incubation of mononuclear cells and granulocytes for 20 hours *in vitro* with 100 U/ml human HuIFN-α led to at least 5-fold increases in the enzyme level.

Enzyme increases in patients receiving IFN therapy

IFN-α

The first study was done on a group of patients receiving intramuscular (I.M.) injections of HuIFN-α prepared from fresh human leukocytes by the method of Cantell and Hirvonen (30). (This IFN was produced at the Israel Biological Institute, Ness Ziona, by Dr. Hagai Rosenberg and T. Bino.) The dose was $1\text{-}3 \times 10^6$ units, according to the weight of the patient ($0.8\text{-}1\times10^5$ U/kg).

Figure 2 shows the enzyme activity in the PBM cells of some of these patients relative to the mean enzyme value of the healthy population (solid horizontal line). The standard deviation of the normal value is shown by the two broken lines. Blood samples were taken immediately before the first injection (day 0) and then every 24 hours just prior to the next injection of IFN. For 2 of the patients (curves 1 and 2), there was a rapid 6-8 fold increase in enzyme level during the first 2 days following the onset of therapy. Curve 1 indicates the enzyme changes in a critically ill 2-year-old

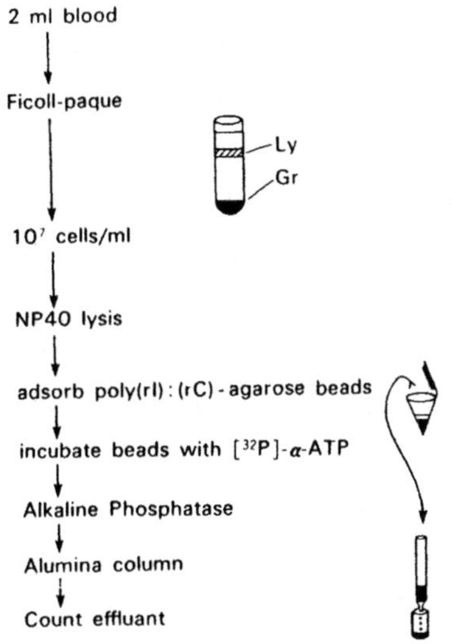

Fig. 1. Assay of (2'-5') oligo A synthetase in human peripheral blood lymphocytes (mononuclear cells) and granulocytes (polymorphonuclear cells).

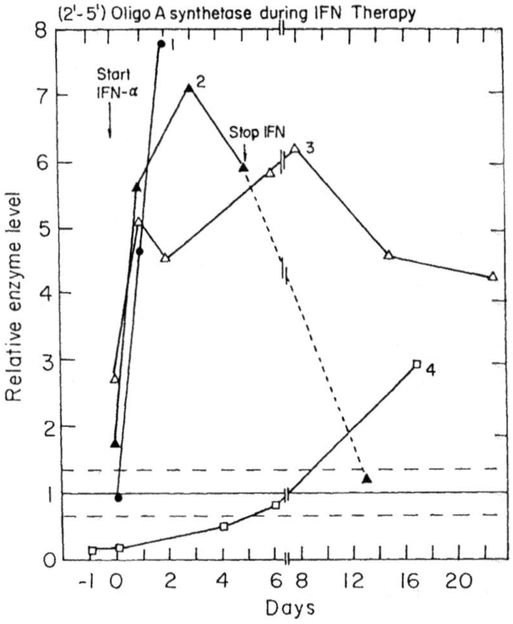

Fig. 2. (2'-5') oligo A synthetase levels in peripheral blood mononuclear cells of 4 patients undergoing IFN-α therapy, relative to healthy donors (see text for details). (From ref. 12).

boy with fulminant hepatitis who received 10^6 U of IFN daily. The enzyme activity increased from 6,800 cpm on day 0, to 33,200 on day 1 and 56,200 cpm on day 2. Despite treatments, however, death occurred. Curve 2 relates to a case of severe viral encephalitis in a 14-month-old infant who received 1.2×10^6 U daily. The symptoms disappeared after 5 days of treatment and IFN injections were discontinued. During treatment the enzyme level was high, but returned to normal one week after the last injection (broken line). Curve 3 represents the case of a young girl with laryngeal papillomatosis. The starting enzyme level on day 0 was higher than normal, as observed in most viral diseases (see below). The injection schedule in this patient was 1.5×10^6 U every second day for several months. Over the 3-week period we followed this case, the enzyme level was stable at 4-5 fold the normal values. The increased enzyme level thus appears to correlate well with IFN treatment since continued injections result in a relatively constant elevation in enzyme, while interruption of the treatment results in a decrease to the normal values.

Curve 4 of Fig. 2 illustrates the importance of monitoring the patient's response to IFN. This was a case of lymphoproliferative disease in a 21-year-old woman with a septic fever and a severe combined immunodeficiency (after immunosuppressive treatment). When the enzyme was first measured, the values obtained were much lower than normal, as is often observed in such patients (see below). Daily injections of 3×10^6 U of HuIFN-α were started on day 0, but the increase in enzyme was very low. It took 6 days of daily injections to attain the normal level, and after 17 days she reached a level 3 times higher than the normal, a 15-fold increase over her starting level. The abnormal enzyme response may be related to her disease or to previous treatment, indicating some deficiency in the IFN system. Such information on the response of the patient should be very precious to the physician. There was actually no visible clinical improvement during IFN therapy in the patient.

In addition to PBM cells we have examined the 2'-5' A synthetase levels in the granulocytes of patients treated with IFN-α. We have found that the enzyme level in granulocytes (polymorphonuclear cells, (PMNL) increases significantly following _in vitro_ incubation with IFN

(Table 2) and proceeded to examine serially the levels in the PML of two of the patients previously described (No. 1 and 3 on Fig. 2). The results (Table 2) show that 2'-5' A synthetase in granulocytes can also be used to monitor therapy with IFN, however, the considerable variability of of enzyme levels in granulocytes of healthy individuals (25) as well as the shorter survival time of granulocytes compared to PBM cells (31) make the latter more suited to 2-5A monitoring.

HuIFN-β

The use of HuIFN-β by systemic injections has been criticized because, after I.M. injections, IFN activity cannot be detected in the serum (10,32). Better results are obtained by intravenous (I.V.) infusions, but the half-life of the activity is very short and it disappears from the serum 2 hours after injection (33). We have measured the changes in (2'-5') oligo A synthetase activity following intramuscular or subcutaneous (perineal) injections of 3×10^6 U of HuIFN-β (InterYeda Ltd., Rehovot, Israel) (27).

An intramuscular injection of 3×10^6 units of IFN-α produces a 6-8 fold increase in the enzyme level of peripheral blood mononuclear (PBM) cells after 24 hrs (12). We could show that IFN-β given intramuscularly or subcutaneously (Fig. 3) also increased 2'-5' A synthetase levels in PBM cells although no IFN-β was found in serum. In our trials, with 3×10^6 units IFN-β intramuscularly, PBM cells enzyme levels rose 2-4 fold after the second injection, whereas with 9×10^6 units it rose 6-8 fold and more rapidly (Fig. 3).

The effect of IFN-β on PBM cells 2'-5' A synthetase levels seemed slower and more sustained than that reported for IFN-α (12) and a "booster" effect of repeated IFN-β injections is often seen (Fig. 3). Topical application of IFN-β ('Frone', InterYeda Ltd., Israel) to the vulval area had no significant effect on the 2'-5' A synthetase levels of PBM cells, except in one patient with cervical dysplasia in whom the cream was applied to the cervix, which suggests some uptake through this route (27).

The changes in 2'-5' A synthetase levels were accompanied by a 3-5 log reduction of VSV growth, occurring after the second IFN-β injection and lasting for about 1 week in short-term cultures of PBM cells (27). Natural killer cell activity (8) also rose after intramuscular or subcutaneous IFN-β injections (27) as has been shown by others (10).

Table 2. (2'5') oligo A synthetase in polymorphonuclear (PMN) and mononuclear cells (PBM) before and after IFN-α treatment.

		PMN cells			PBM cells		
IFN treatment	Samples	Control	IFN	IFN/ control	Control	IFN	IFN/ control
In vitro[†]	I	65[*]	455	7	90	570	6.3
	II	40	505	12.6	115	440	3.8
	III	125	1220	9.8	130	360	2.8
	IV	15	290	19.3	95	535	5.6
In vivo[‡]	V	25	515	21	125	600	4.8
	VI	170	435-495	2.6-2.9	350	585-750	1.7-2.2

[*]pmoles ATP incorporated in 1 h into $(A2'p)_nA$ by extract from 10^5 cells.

[†]Cells of four healthy individuals were incubated for 18 h with or without 1,000 units IFN-β/ml.

[‡]Patients undergoing therapy with IFN. Activities right before (control) and after IFN treatment are given. Sample V: 24 h after the second injection of 1 x 10^6 units IFN-α. Sample VI: range of activities observed in six repeated determinations during 1 month of treatment by repeated injections of 1.5 x 10^6 units IFN-α.

A moderate, transient leucopenia and fever often followed the IFN-β injection. These other monitoring methods are, however, more elaborate, lengthier, and less precise than the (2'-5') oligo A synthetase assay which we feel is best suited for monitoring IFN therapy. A double-blind trial recently demonstrated that the (2'-5') oligo A synthetase level in PBMC was the only parameter correlated with IFN treatment and not with placebo (Levin, S., et al, private communication).

DISCUSSION

Since our first reports, our assay as well as other assays for 2'-5' A synthetase determinations have been quite widely used to monitor therapy with IFN or IFN-inducers in vivo. For example, the action of IFN in mice treated with polyadenylic-polyuridylic acid [poly(A)·poly(U)] was assessed by 2'-5' A synthetase determinations. Increased enzyme levels were found in the spleen, liver and even lung tissue extracts and the efficacy of different routes of administration in inducing IFN could be compared (34). When the same IFN-inducer was later used in humans with operable breast cancer, considerable enhancement in the level of

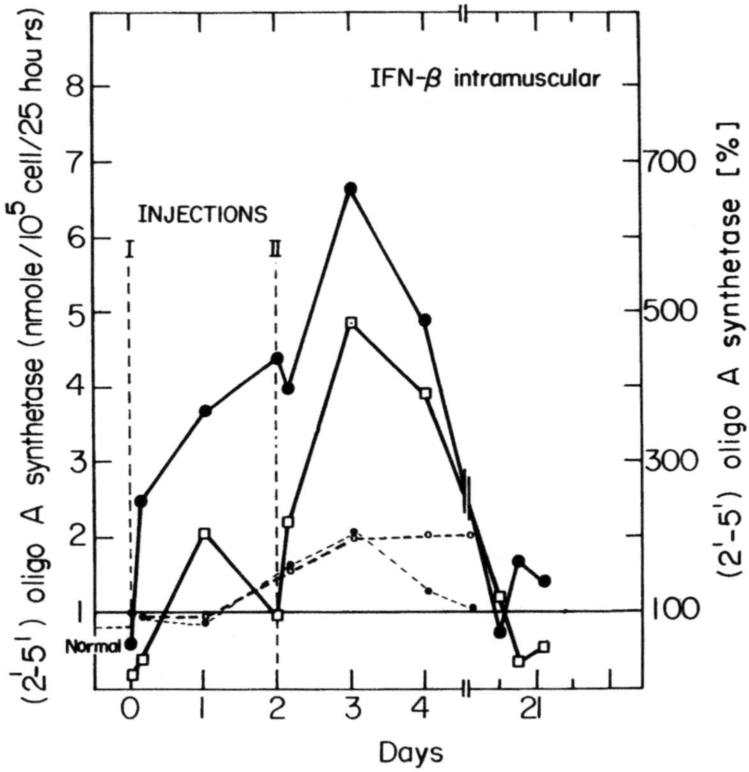

Fig. 3. (2'-5') oligo A synthetase levels in peripheral blood mononuclear cells of two patients receiving IFN-β intramuscularly, 9 x 10^6 units (solid lines) or 3 x 10^6 units (broken lines) at days 0 and 2, relative to healthy donors. (From ref. 27).

lymphocytic 2'-5' A synthetase was observed in 80% of 48 patients treated, even though no IFN could be detected in the serum (35). There was a concordance for most patients examined between 2-5A levels and increased NK cell activity, an expected beneficial result of IFN therapy for both malignant and viral diseases (36). Thus a subgroup of 'non-responder' patients may be delineated and further studies should clarify whether such patients need higher doses of IFN or IFN-inducer or whether other treatment modalities should be attempted. The clinical significance of 2'-5' A synthetase determinations during IFN therapy is supported by a recent communication from Japan. Eighteen patients with HBeAg-positive chronic hepatitis B were treated either with IFN-α or IFN-β. The level of 2-5A was determined in PBM cells and significant though variable increases were found by the third treatment day. Moreover, an inverse correlation was found in the HBV-specific DNA polymerase activity: Patients with large increases in 2'-5' A synthetase had large decreases in the polymerase, and often HBeAg disappeared from the serum (37). In other studies, 2'-5' A synthetase measurements helped to determine the dose and route of administration of IFN in patients with laryngeal papillomatosis, cutaneous T lymphoma or other neoplastic diseases (38-39), but were not sufficient to predict good clinical response in homosexual men with Kaposi's sarcoma (40). However, interpretation of the response in these patients is considerably more complicated because most of them have circulating acid-labile IFN-α (41) and the level of 2'-5' A synthetase in PBM cells is elevated prior to therapy with 'exogenous' IFN (40). In leukemias and lymphomas a good correlation was recently reported between induction of 2'-5' A synthetase and response to IFN. One of the mechanisms for the beneficial effect of IFN in hematological malignancies is considered to be the induction of differentiation of the pathological cells. It was found that in vitro IFN induced differentiation in 19 out of 29 patients with chronic lymphocytic leukemia (CLL), showing a close correlation between induction of differentiation and the induction of 2'-5' A synthetase in CLL cells (42). Thus, differentiation occurred in clones expressing enhanced levels of enzyme but not in those showing no increase in the enzyme, possibly indicating that resistance of the cells of some CLL patients to IFN was associated with the absence of differentiation. Additional experiments performed on freshly separated tumor cells from surgical specimens of patients with solid tumors also confirm that an in vitro 2'-5' A synthetase induction assay is a rapid and reproducible method for detecting sensitivity or resistance of primary malignant cells to IFN (43). The group from M.D. Anderson followed fourteen patients with chronic myelogenous leukemia (CML) for their response to IFN-α therapy and for their 2'-5' A synthetase levels in PBM cells. They classify their patients as being sensitive or resistant to IFN and report that 2'-5' A synthetase was induced to about a 100-fold increase in the sensitive patients as compared to a very mild increase in enzyme levels (less than 7-fold), which occurred in the IFN-resistant group (44).

The most important result of this monitoring assay concerns the possibility to use IFN-β by the i.m. or s.c. route, with successful clinical outcome (27). Other methods of monitoring such as β_2 microglobulin increase and Neopterin excretion (45) have recently confirmed these results (Ares-Serono, on file). These other methods are also most useful for monitoring IFN-γ therapy (45).

Immunoassay of 2'-5' oligo A synthetase

Recent work from J. Chebath et al. in this laboratory, may further extend the use of 2'-5' A synthetase in IFN monitoring. Cloning of 2'-5' synthetase A cDNAs and sequences of two forms of the enzyme (18) allowed to synthesise a common small peptide and produce polyclonal anti 2'-5' A synthetase antibodies in rabbits (19). Initial results show that the anti-2'-5' A synthetase can be used for an immunoassay of the enzyme and this approach may be adapted in the future for an improved method of monitoring IFN-therapy.

In summary, measurement of (2'-5') oligo A synthetase levels is a valuable addition to clinical IFN studies and may help to resolve problems of the optimal dose, route and schedule of IFN administration.

REFERENCES

1. Revel, M. In: Interferon 1 (Ed. I. Gresser), Academic Press, London, 1979, pp. 101-163.
2. Pestka, S. Sem. Hematol. 23: 27-37, 1986.
3. Quesada, Jr., Talpaz, M., Rios, A., Kurzrock, R. and Gutterman, J.U. J. Clin. Oncol. 4: 234-243, 1986.
4. Quesada, J.R., Reuben, J., Manning, J., Hersh, E.M. and Gutterman, J.U. N. Engl. J. Med. 310: 15-18, 1984.
5. Greenberg, H.B., Pollard, R.B., Lutwick, L.I., Gregory, P.B., Robinson, W.S. and Merigan, T.C. N. Engl. J. Med. 295: 517-522, 1976.
6. Gutterman, J.U., Fine, S., Quesada, J., Horning, S.J. et al. Annals Internal. Med. 96: 549-556, 1982.
7. Foon, K.A., Bottino, G.C., Abrams, P.G., Fer. M.F., Longo, D.L., Schonberger, C.S. and Oldham, R.K. Am. J. Med. 78: 216-220, 1985.
8. Huddlestone, J.R., Merigan, T.C. Jr. and Oldstone, M.B.A. Nature 272: 417-419, 1979.
9. Heron, I., Hokland, M., Hokland, P. and Berg, K. Ann. NY Acad. Sci. USA 350: 112-120, 1980.
10. Lucero, M.A., Magdelenat, H., Fridman, W.H., Pouillart, P., Billardon, C., Billiau, A., Cantell, K. and Falcoff, E. Eur. J. Cancer Clin. Oncol. 18: 243-251, 1982.
11. Levin, S. and Hahn T. Clin. Exp. Immunol. 46: 475-483, 1981.
12. Schattner, A., Merlin, G., Wallach, D., Rosenberg, H., Bino, T., Hahn, T., Levin, S. and Revel, M. J. Interferon Res. 1: 587-594, 1981.

13. Parry, R.P. and Parry, J.V. Lancet 1: 506-507, 1981.
14. Stark, G.R., Dower, W.J., Schimke, R.T., Brown, R.E. and Kerr, I.M. Nature 278: 471-473, 1979.
15. Kimchi, A., Shulman, L., Schmidt, A., Chernajovsky, Y., Fradin, A. and Revel, M. Proc. Natl. Acad. Sci. USA 76: 3208-3212, 1979.
16. Hovanessian, A.G., Brown, R.E. and Kerr, I.M. Nature 268: 537-540, 1977.
17. Schmidt, A., Zilberstein, A., Shulman, L., Federman, P., Berissi, H. and Revel, M. FEBS Letters 95: 257-264, 1978.
18. Benech, P., Mory, Y., Revel, M. and Chebath, J. EMBO J. 4: 2249-2256, 1985.
19. Chebath, J., Benech, P., Hovanessian, A., Galabru, J. and Revel, M. J. Biol. Chem. 1987, in press.
20. Schattner A., Wallach, D., Merlin, G., Hahn, T., Levin, S. and Revel, M. Lancet 2: 497-500, 1981.
21. Morag, A., Tobi, M., Ravid, Z., Revel, M. and Schattner, A. Lancet 1: 744, 1982.
22. Schattner A. Lancet 1: 171, 1984.
23. Revel, M., Schattner, A., Wallach, D., Merlin, G., Levavi, H., Hahn, T. and Levin, S. In: The Clinical Potential of Interferons (Eds. R. Kono and J. Vilcek) University of Tokyo Press, 1982, pp. 353-367.
24. Schattner, A., Merlin, G., Shapira, A., Revel, M. and Wallach, D. J. Interferon Res. 2: 285-289, 1982.
25. Schattner, A., Merlin, G., Bregman, V., Hahn, T., Levin, S., Revel, M. and Wallach, D. Clin. Exp. Immunol. 57: 265-270, 1984.
26. Wallach, D., Schattner, A., Merlin, G., Kimchi, A., Fellous, M. and Revel, M. In: Interferons (Eds. T.C. Merigan, R.M. Friedman, F.C. Fox), Academic Press, New York, 1982, pp. 449-463.
27. Schoenfeld, A., Nitke, S., Schattner, A., Wallach, D., Crespi, M., Hahn, T., Levavi, H., Ovadia, J., Shoham, J., Doerner, T. and Revel, M. Lancet 1: 1038-1042, 1984.
28. Schoenfeld, A., Ovadia, J., Levavi, C., Nitke, S., Wallach, D., Schattner, A. and Revel, M. Early Human Development 10: 279-286, 1985.
29. Merlin, G., Revel, M. and Wallach, D. Anal. Biochem. 110: 190-196, 1981.
30. Cantell, K. and Hirvonen, S. Tex. Rep. Biol. Med. 35: 138-144, 1977.
31. Ottenson, J. Acta Physiol. Scand. 32: 75-79, 1954.
32. Edy, F.G., Billiau, A. and De Somer, P. Lancet 1: 451-452, 1978.
33. Treuner, J., Niethammer, D., Dannecker, G., Hagmann, R., Neef, V. and Hofschneider, P.H. Lancet 1: 817-818, 1980.
34. Young, J.K., Hovanessian, A.G., Riviere, Y., Hue, G. and Lacour, F. Cell Immunol. 79: 298-308, 1983.
35. Hovanessian, A.G., Young, J.K., Buffet-Janvresse, C., Riviere, Y., Michelson, M., Lacour, J. and Lacour, F. Cancer 55: 357-362, 1986.
36. Schattner, A. and Duggan, D. Amer. J. Hematol. 18: 435-443, 1985.
37. Kuroki, T., Nishiguchi, S., Shimizu, Y. and Yamamoto, S. TNO-ISIR Meeting, Clearwater, 1985. Abstracts p. 87.
38. Merritt, J.A., Borden, E.C. and Ball, L.A. J. Interferon Res. 5: 191-198, 1985.
39. Lodemann, E., Nitsche, E.M., Lang, M.H., Gerein, V., Altmeyer, P., Holzmann, H., and Kornhuber, B. J. Interferon Res. 5: 621-628, 1985.

40. Preble, O.T., Rook, A.H., Steis, R., Silverman, R.H., Kraus, D., Quinnan, G.V., Masur, H., Jacob, J., Lonyo, D. and Gelmann, E.P. J. Infect. Dis. 152: 457-465, 1985.
41. DeStefano, E., Friedman, R.M., Friedman-Kien, A.E., Goedert, J.J., Henriksen, D., Preble, O.T., Sonnnabend, J.A. and Vilcek, J. J. Infect. Dis. 146: 451-455, 1982.
42. Ostlund, L., Einhorn, S., Robert, K.H., Juliusson, G. and Biberfeld, P. Blood 67: 152-159, 1986.
43. Einhorn, S., Ostlund, L., Vanky, F., Grander, D., Robert K-H. and Willems, J. J. Interferon Res. 6: Suppl. 1, p. 128, 1986.
44. Rosenblum, M.G., Maxwell, B.L., Talpaz, M., Kelleher, P.H., McCredie, K.B. and Gutterman, J.U. TNO-ISIR Meeting, Clearwater, USA, 1985. Abstracts p. 77.
45. Huber, C., Batchelor, J.R., Fuchs, D., Hausen, A., Lang, A., Niederweiser, D., Reibnegger, G., Swetly, P., Troppmair, J. and Wachter, H. J. Exp. Med. 160: 310-316, 1984.

22

INTERFERON STUDIES IN NON-HUMAN PRIMATES

H. SCHELLEKENS, P.H. VAN DER MEIDE AND K. NOOTER*
Primate Center TNO and *Radiobiological Institute TNO, PO Box 5815, 2280 HV Rijswijk, The Netherlands

INTRODUCTION

Biological response modifiers have multiple biological effects. Therefore, results obtained in vitro cannot be directly extrapolated to the therapeutic efficacy in humans. So animal studies are an essential prerequisite for the successful application of these agents in man. Interferons show a high degree of species specificity. Human interferon preparations are not active in "standard" laboratory animals as rats and mice. For studies in these species interferons from rat and mouse origin have to be produced. Initially the rodent interferons were produced from induced cell culture systems, which did not provide enough material for all the necessary studies. Recombinant DNA technology did not only solve the problem of production of human interferon, but also a number of animal interferons can now be produced in unlimited amounts (1).

Apart from animal studies with rodent interferons, essential in vivo data can be obtained from studies with human interferons in higher non-human primates. Studies in non-human primates, however, have their own ethical, economical and technical limitations. For example: there are no tumor models available. Inbred strains of non-human primates do not exist, which exclude transplantable tumors. Chemical induction of tumors in primates is unpredictable and needs an incubation period of years and is therefore not feasible.

We will discuss here the studies, that have been performed in non-human primates concerning the antimicrobial activity, the toxicity and pharmacokinetics of human interferons.

THE ANTIMICROBIAL ACTIVITY OF HUMAN INTERFERONS IN NON-HUMAN PRIMATES

In Table 1 all studies with human interferon in microbial infections in non-human primates are listed. The study of Andrews (2) has been included for historical reasons, although he did not

Table 1. Infections in non-human primates in which human interferons proved effective.

Viral infections	Species and references
vaccinia virus	cynomolgus monkeys* (2) baboons (3) rhesus monkeys (4-11, 26) marmosets (8)
yellow fever virus	rhesus monkeys (12)
rabies virus	cynomolgus monkeys (13, 14) rhesus monkeys (15)
varicella-zoster like viruses	rhesus monkeys (6) patas monkeys (16) african green monkeys (27)
hepatitis B virus	chimpanzees (17)
encephalomyo-carditis virus	squirrel monkeys (18-21) baboons (21)
herpes hominis type I	owl monkeys (22) African green monkey (5, 23, 24)
non-viral infections	
Plasmodium vivax	chimpanzees (25)

*Interferons prepared in monkey cells, chicken cells and rabbit cells were used in this study.

use human interferons but interferon produced from monkey, chicken and rabbit cells. Except for the study of malaria in chimpanzees reported by Ferreira (25), all studies concern a variety of viral infections.

Amazingly all studies published have obtained positive results. It could be that negative results have been obtained but were never published. However, that is unlikely considering their relevance and the costs of experiments in these expensive animals. The virus-species combinations that were most intensively studied were intradermal vaccinia infection in rhesus monkeys, systemic encephalomyocarditis virus infections in squirrel monkeys and herpes keratitis in African Green monkeys.

Intradermal vaccinia infection in rhesus monkeys.

This has been the most popular model to study the antiviral effect of human interferons. The model has a number of advantages. Vaccinia virus only gives a mild self-limiting skin lesion and only results in a minor inconvenience for the animals. The procedure is economical and the animals can be used afterwards for other purposes. The antiviral effect can be easily monitored by mere observation and because of the relation between interferon dose and lesion size suboptimal effects can easily be determined. In the era that human interferon was scarce the comparative small amounts of interferon were an argument in favour. The outlines and the results of the studies performed are listed in Tables 2a and 2b. Except in the study of Pinto (4) in all these studies lesion size and appearance were used as parameter for the antiviral effect. Pinto titrated the virus inoculum by infecting the skin of the animals with serial dilutions of his challenge virus. The vaccina infection is extremely sensitive. Pinto et al. (4) reported that 800 units/kg had a significant effect. Schellekens et al. reported a significant protection by 50,000 U/kg (28). This sensitivity of vaccinia virus in vivo sharply contrasts with the relative resistance and even lack of sensitivity of vaccinia virus in vitro (7). This may indicate that host defence mechanisms play an important role in the interferon effect in this model.

Table 2a. Human interferons in rhesus monkeys intradermally infected with vaccinia virus

Reference	Type of interferon	Dosages and route of injection	Treatments relative to infection	Results
Pinto et al., 1970 (4)	n-HuIFNα	80,000, 800 and 8 U/kg i.v.	-24, 0 and +24 hr -24 and 0 hr -24 or +24 hr	Effects with 80,000 and 800 U/kg when given -24, 0, +24 or -24 and 0 relative to infection
Neumann-Haefelin et al., 1976 (6)	n-HuIFNα	$5x10^5$ U/animal i.m.	daily from day -1 until day +3 or for 7 days starting the day lesions appear	No effect on primary lesions. Prevention of development of secondary lesions in immunosuppressed animals
Schellekens, 1979 (7)	n-HuIFNα	$5x10^5$ U/kg i.m.	day -1 until day +6	complete inhibition of lesion development
Weimar, 1980 (10)	n-HuIFNα n-HuIFNβ	$5x10^5$ U/kg i.m. and i.v. $1.25x10^5$ U/kg i.m. $0.5x10^5$ U/kg i.m.	day -1 until day +6	$5x10^5$ U/kg HuIFNα complete protection, lower doses significant but partial protection. HuIFN-β less active than HuIFN-α
Schellekens, 1981a (8)	n-HuIFNα	$5x10^5$ U/kg i.m. $2x10^6$ U/kg i.m.	day -1 and 0, day -5, day -10 and-9, day +1, 2 and 3, day +3, 4 and 5	Only effect of treatment schedules starting before infection except for treatment at day -10 and -9. No effect when treatment started after infection, not even in a dose of $2x10^6$ U/kg

Table 2b. Human interferons in rhesus monkeys intradermally infected with vaccinia virus

Reference	Type of interferon	Dosages and route of injection	Treatments relative to infection	Results
Schellekens et al., 1981b (9)	n-HuIFNα r-HuIFNα2	$5x10^5$ U/kg i.m. $1.25x10^5$ U/kg i.m. $5x10^4$ U/kg i.m.	day -1 until day +6	Dose dependent effect. No difference between recombinant and natural HuIFNα
Schellekens et al., 1982 (26)	n-HuIFNα r-HuIFNαA r-HuIFNβ	$5x10^5$ U/kg i.m.	day -1 until day +6	No difference between effect of recombinant and natural HuIFNα. r-HuIFNβ less active than HuIFNα's
v.d. Meide et al., 1985 (11)	n-HuIFNα n-HuIFNβ r-HuIFNαCon1 n-HuIFNγ r-HuIFNγ	10^6-$5x10^6$ U/kg i.m.	day -1 until day +6	n-HuIFNα serum buffy coats is more effective than n-IFNα from lymphoblastoid cells, r-IFNα Con_1 or n-IFNβ. n-IFNα is as effective as n-IFNγ or r-IFNγ

Abbreviations:
i.v. = in vitro
i.m. = intramuscular
i.c. = intracutaneous
n-HuIFN-α = natural
r-HuIFN-α = recombinant

Neumann-Haefelin et al. (6) were the only ones that failed to observe an effect on the development of primary lesions. Only an effect on the development of secondary lesions was observed. The rhesus monkeys however, were immunosuppressed which may have interfered with the host mediated defence mechanisms.

Timing of the interferon injection is of great importance. Only when given before infection systemic interferon treatment has an effect. Even a single injection 5 days before infection is sufficient to induce an effect (8). Interferon treatment started after infection has no effect, even if the doses applied exceed by far the doses that have a preventive effect (8). Other conclusions that can be drawn from these studies: alpha interferon has a better effect than beta interferon, independent of the route of application (10, 26); there is no difference between natural interferons and their recombinant DNA derived counterparts (9, 26); and that gamma interferon has at least the same antiviral activity in vivo as alpha interferons (11).

EMC infections in squirrel monkeys

In Table 3 the studies performed in squirrel monkeys infected with encephalomyocarditis virus are summarized. Also in this model recombinant DNA interferons have the same antiviral activity as their natural counterparts (18). Also this infection is quite sensitive and a significant effect can be obtained with 30,000 units/kg (20). Of interest is the observation that there is difference in antiviral efficacy of different α subtypes (20).

Herpes keratitis in experimental animals

This model has been exclusively studied by Neumann-Haefelin and his coworkers (Table 4). The special characteristics of this model are that it mimicks an important viral infection in humans and that the effects of local administration of interferon can be studied. In this model alpha interferon has the same efficacy as beta interferon (23) suggesting that the lower activity of systemically administered beta interferon is caused by its pharmacokinetic behaviour and not by its intrinsic properties. If beta interferon can reach the target organ it is as active as alpha

Table 3. Human interferons in squirrel monkeys infected with E.M.C. virus

Reference	Type of interferon	Dosages and route of injection	Treatments relative to infection	Results
Goeddel et al., 1980 (18)	r-HuIFNα2 n-HuIFNα	10^6 U/animal i.v.	-4 hr, +2 hr, 23, 29, 48, 72, 168 and 240 hr	Complete protection
Weck et al., 1983 (19)	n-HuIFNβ r-HuIFNβ	10^6, $3x10^4$ and 10^3 U/kg i.v.; 10^6 U/kg i.m.	6 doses over 4 days starting at -4 hr	Comparable significant effect of r-huIFNβ and n-HuIFNβ except from 10^3 U/kg
Stebbing et al., 1983 (20)	r-IFNαA r-IFNαD r-IFNαA/D	10^6, $3x10^4$ and 10^3 U/kg i.v.	8 doses during 4 days starting at -4 hr	r-IFNαD and r-IFNαA/D effective in all doses. r-IFNαA only active at highest dose.

Table 4. Human interferons in experimental herpes keratitis in African green monkeys

Reference	Type of interferon	Dosages and route of injection	Treatments relative to infection	Results
Neumann-Haefelin et al., 1975 (5)	n-HuIFNα	$2x10^4$ U per treatment	-5 to 72 hr 1 time daily; -5 to 20 hr 2 times daily; -½ to 90 hr 2 times daily; +20 to 72 hr 1 time daily; -6 to 96 hr 2 times daily	Inhibition of keratitis by prophylactic and simultaneous administration. Moderation of infection by therapeutic schedules
Neumann-Haefelin, 1977 (23)	n-HuIFNα n-HuIFNβ	10^5, 10^4, 10^3 and 10^2 U per treatment	-13, 0 and +24 hr	Both interferons effective except 10^2 U
Neumann-Haefelin, 1985 (24)	r-HuIFNγ r-HuIFNα2	r-HuIFNγ $3x10^6$, $3x10^5$, $3x10^4$, $3x10^3$ U per treatment. Combinations of $3x10^3$ and $3x10^2$ r-IFNγ with 500 and 50 U IFN-γ	-15 hr, 0 and +24 hr	r-HuIFNγ active except for $3x10^3$ units. Synergy of r-HuIFNγ and r-HuIFNα

interferon. Also in the herpes keratitis model gamma interferon proved to be as effective as alpha interferon (24).

Other viral infections in non-human primates

In the studies with yellow fever virus in rhesus monkeys reported by Finter (12) the effective dose of systemically administered interferon is also in the range of 30×10^3 U/kg. In the rabies infections in cynomolgus and rhesus monkeys reported by Baer et al. (15) and Hilfenhaus et al. (13, 14) there is no effect if the treatment is restricted to the preinfection period. Treatment schedules starting the day after infection and continuing for 9 to 13 days increased survival.

The studies in varicella-zoster-like infections in patas monkeys and African green monkeys (16, 27) also show a significant preventive effect of interferon. In these studies no curative schedules were tried. Levine (21) reported that baboons infected with EMC virus interferon given within 24 h after infection had a significant effect.

THE PHARMACOKINETIC BEHAVIOUR OF HUMAN INTERFERONS IN NON-HUMAN PRIMATES

It is obvious that a good understanding of the pharmacokinetic behaviour of IFNs in experimental animals and man will help to define the best route of administration and dosage schedules for the therapeutic use of these molecules in man. Preclinical and clinical pharmacokinetic studies with IFNs have been carried out indeed. Although there are some discrepancies in the results reported, there yet is a consensus with regard to the *in vivo* pharmacokinetics of IFNs. We here will start with a general description of how the IFNs behave upon different routes of administration. Thereafter, we will discuss the specific experiments carried out with human IFNs in non-human primates reported in the literature.

The interferons have been administered by the well-known routes (i.v. bolus and infusion, i.m. and s.c.) in a variety of animal species. In the majority of the preclinical pharmacokinetic studies the presented data obtained after an i.v. bolus dose of IFN could be fitted by equations describing an open, linear two-compartment model with drug excretions from the central compartment only. For all three IFN types the initial phase of the serum decay curve is rather steep and is followed by a longer lasting elimination phase. Kidneys and liver are thought to be of importance as catabolic organs for IFN. Only small amounts of IFN are found in the urine. With an i.m. or s.c. administration a depot of IFN is created of which the IFN is absorbed into the systemic circulation. The α-IFNs show a good systemic absorption irrespective of the route of administration: the systemic bioavailability can be upto 100% after an i.m. or s.c. injection. However, for the other types of IFNs (β and γ) lower or even undetectable serum levels are observed after i.m. or s.c. administration.

Pharmacokinetics of HuIFN-α in non-human primates

A specific problem that has been encountered by the evaluation of the experimental animal data is the fact that most animal studies have been done with human IFN types and not with species specific IFNs. This prompted the need for pharmacokinetic studies with human IFNs in non-human primates. In the mean time it is clear from the literature that no significant differences exist between the pharmacokinetic data obtained from lower animal (e.g., mice, rats and rabbits) and primates, including man.

Another problem that originally was thought to be of importance is the fact that in general recombinant-derived IFN molecules are not glycosylated. However, up to now no significant differences in pharmacokinetics between naturally- and recombinant-derived IFN preparations have been observed.

In Table 5 the most pertinent pharmacokinetic studies with HuIFN-α in non-human primates are summarized. In all studies a two- compartment distribution was found. A rapid distribution phase was followed by a slower elimination phase with a half-life value ranging from 1.8 to 7.1 hr (28,29,30). After i.m. or s.c. injection longer lasting serum titers were detected as compared to an i.v. bolus (28-32). The systemic bioavailability following i.m. injection ranged from 50% to 150%, mean of 93% in one study

(29), indicating near complete absorption of intact IFN-α from muscle and from 19-103% in another study (30).

Wills et al. (29) administered IFN per os and as could be expected no IFN was found in the circulation.

Bino et al. (33) determined IFN levels in different kinds of tissues and it was found that in the kidneys the levels were at least 7-fold higher than in the liver, spleen, lungs and heart. No IFN could be detected in brain and muscles. Subcellular fractionation of kidney revealed that the mitochondrial-lysosomal fraction had a high IFN content. It was found that HuIFN-α was rapidly degraded by lysosomal proteinases. No urine excretion of IFN was detected and all together these findings suggest that the kidney is a major site for IFN catabolism.

With regard to the question whether IFN can pass the blood-brain barrier the studies of Habif et al. (28) and Collins et al. (30) clearly demonstrated in macacas that there is a barrier to the penetration of IFN-α from serum to CSF and vice versa indeed, but that some IFN can pass this barrier in both directions. Comparable results were obtained for HuIFN-β in cynomolgus monkeys (34).

Pharmacokinetics of HuIFN-β in non-human primates

The most pertinent studies on the pharmacokinetics of HuIFN-β in non-human primates are simmarized in Table 6. Just like IFN-α, IFN-β shows a biphasic serum disappearance pattern (32,34,35). In man, IFN therapy will most likely always be given repeatedly over a prolonged period of time. Yamazaki et al. (32) and Hilfenhaus et al. (24) studied the pharmacokinetics of repeated HuIFN-β administrations. Only after repeated i.m. injections there was a slight increase in the 24 hr serum trough values.

The systemic bioavailability after i.m. administration of HuIFN-β was found to be 40% maximally (35). The pharmacokinetics of naturally-derived and recombinant HuIFN-β preparations were compared within one experimental set up (35). After an i.v. push injection no substantial differences were observed for most pharmacokinetic parameters except for a shorter distribution phase of

Table 5a. Pharmacokinetic studies with human alpha interferons in non-human primates.

reference	type of IFN, purity, spec. activity	species	route of administration, dosage	results
Skreko et al., 1973 (31)	n-HuIFN-α - -	gibbon	i.v. bolus $1x10^{6.5}$ U/animal	two-compartment distribution (rapid distribution, slower elimination).
			i.m. $1x10^{6.5}$ U/animal	maximum absorption at 6 hr, longer serum titers as compared to i.v. bolus.
			s.c. $1x10^{6.5}$ U/animal	much longer lasting serum titers as compared to i.v. bolus.
Habif et al., 1975 (28)	n-HuIFN-alpha - 2×10^4 U/mg	stump-tailed monkey	i.v. bolus $30x10^6$ U/animal	two-compartment distribution (elimination half-life = 7.1 hr), low titers in CSF.
	or 7×10^5 U/mg		i.m. $30x10^6$ U/animal	relatively high titers for at least 24 hr, higher titers in CSF as compared to i.v. bolus.
			intrathecally 1×10^6 U/animal to $10x\ 10^6$ U/animal	CSF clearance resembles serum clearance, low titers in serum.

Table 5b. Pharmacokinetic studies with human alpha interferons in non-human primates.

reference	type of IFN, purity, spec. activity	species	route of administration, dosage	results
Billiau et al. 1981 (42)	n-HuIFN-α - 1×10^6 U/mg	African green monkey	i.m. 1×10^6 U/animal	relatively high serum titers.
Yamazaki et al., 1981 (32)	n-HuIFN-α - 3×10^6 U/mg	Cynomolgus monkey	i.v. bolus 14×10^6 U/kg i.m. 12×10^6 U/kg	two-compartment distribution. maximum absorption at about 1 hr.
Bino et al., 1982 (33)	n-HuIFN-α - 3×10^6 U/mg	Cynomolgus monkey	i.v. bolus 2×10^6 U/kg	concentrations in kidneys about 7-fold higher than in liver, spleen, lungs, heart and muscle, no measurable urine titers.
Soike et al., 1983 (27)	r-HuIFN-α A	African green monkey	i.m. 3×10^6 U/kg	maximum absorption at 1-2 hr, detectable serum titers upto 12 hrs.
Wills et al., 1984 (29)	r-HuIFN-α A - 18×10^6 U/mg	African green monkey	i.v. bolus 3×10^6 U/kg	AUC_0 = 100 ng/ml.hr (ELISA) two compartment distribution (elimination half-life = 1.8 to 4.8 hr).
			i.v. infusion for 1 hr 3×10^6 U/kg	AUC_0 = 160 ng/ml.hr.

Table 5c. Pharmacokinetic studies with human alpha interferons in non-human primates.

reference	type of IFN, purity, spec. activity	species	route of administration, dosage	results
Wills et al., 1984 (29)	r-HuIFN-α A - 18 x 10^6 U/mg		i.m. 3 x 10^6 U/kg	AUC_0 = 120 ng/ml.hr, maximum absorption 1 to 4 hr, longer lasting serum titers as compared to i.v. administration, systemic bio-availability 93% compared to i.v. infusion.
			oral 6 x 10^6 U/kg	no measurable serum concentrations.
Collins et al., 1985	r-HuIFN-α A 95% 2-3 x 10^8 U/kg	rhesus monkey	i.v. bolus 2.5 x 10^6 U/kg	two-compartment distribution ($t_{½α}$ = 15-33 min, $t_{½β}$ = 1.7-4.6 hr), very low titers in CSF.
			i.m. 2.5 x 10^6 U/kg	systemic bioavailability 19-103%, maximum absorption at 3-6 hr, longer lasting serum titers as compared to i.v. bolus, no detectable titers in CSF.
			intraventricular 5 x 10^3 U/kg or 12 x 10^4 U/kg	no detectable serum titers.

Table 6a. Pharmacokinetic studies with human beta interferons in non-human primates.

reference	type of IFN, purity, spec. activity	species	route of administration, dosage	results
Billiau et al., 1981 (42)	n-HuIFN-β - 1×10^6 U/mg	African green monkey	i.m. 1×10^6 U/animal	relatively high serum titers.
			intrathecally 1×10^6 U/animal or 3×10^6 U/animal	no detectable or very low serum titers, hardly any penetration into brain tissues.
Hilfenhaus et al., 1981 (34)	n-HuIFN-β - 1×10^4 U/mg or 1×10^7 U/mg	Cynomolgus monkey	i.v. bolus 2.6×10^5 U/kg 2.2×10^5 U/kg (4 x q 24 hr)	two-compartment distribution, no accumulation in serum.
			i.v. infusion for 3 hr 2.0×10^5 U/kg	after one hr a constant level for two hr.
			i.m. 2.5×10^5 U/kg	very low titers.
			intrathecally 3×10^5 U/animal	very low serum titers.

Table 6b. Pharmacokinetic studies with human beta interferons in non-human primates.

reference	type of IFN, purity, spec. activity	species	route of administration, dosage	results
Yamazaki et al., 1981 (32)	n-HuIFN-β - 1×10^7 U/mg	Cynomolgus monkey	i.v. bolus 6×10^6 U/kg (7 x q 24 hr)	two-compartment distribution, no accumulation in serum.
			i.m. 6×10^6 U/kg (7 x q 24 hr)	accumulation in serum.
Gomi et al., 1984 (35)	r-HuIFN-β 99% 5×10^7 U/mg	crab eating monkey	i.v. bolus 1×10^6 U/kg 1×10^7 U/kg	two-compartment distribution ($t_{1/2}\,\alpha = 0.2$ hr, $t_{1/2}\,\beta = 3.97$ hr), linear pharmacokinetics.
			i.m. 1×10^7 U/kg	systemic bioavailability 42%.
	n-HuIFN-β - 1×10^7 U/mg		i.v. bolus 1×10^6 U/kg	no significant differences with recombinant prep., except for $t_{1/2}\,\alpha = 0.4$ hr.

Table 7. Pharmacokinetic studies with human gamma interferons in non-human primates

reference	type of IFN, purity, spec. activity	species	route of administration, dosage	results
Cantell et al., 1983 (40)	n-HuIFN-γ - 4×10^2 U/mg	rhesus monkey	i.v. bolus 3×10^5 U/animal	two-compartment distribution, detectable titers up to 4 hr.
			i.m. 3×10^5 U/animal	low but stable titers up to 7 hr.
			s.c. 3×10^5 U/animal	no serum titers.
Dawson et al., 1983 (41)	r-HuIFN-γ 98% 2×10^7 U/mg	rhesus monkey	i.v. bolus 0.5×10^6 U/kg	two-compartment distribution ($t_{1/2}\,\alpha$ = 0.02 hr, $t_{1/2}\,\beta$ = 1.3 hr).
			i.m. 0.5×10^6 U/kg	low titers for ½ to 6 hr.
			s.c. 0.5×10^6 U/kg	no levels

the recombinant preparation. However, in rabbits injected i.m. with either naturally-derived or recombinant HuIFN-β they found a much lower serum AUC (area under the plasma curve) with the recombinant preparation. These findings suggest that the recombinant IFN molecules might be distributed to the tissues more rapidly in the case of an i.v. bolus dose and that after i.m. administration the recombinant material is transferred less rapidly from the muscle tissue to the circulation. Both phenomena could be explained by a higher tissue affinity of the recombinant molecules.

In clinical studies it has been found that the systemic bioavailability of i.m. administered HuIFN-β is much worse than that of IFN-α (36). Comparable results were obtained in rabbits (37,38). In two studies in monkeys (32,37) the serum titers after i.m. administration were about equally high for both types of HuIFN. However, Gomi et al. (35) did not find any significant difference in pharmacokinetic behaviour of r-HuIFN-β upon i.m. administration in rabbits and monkeys.

Little information is available on a possible saturation kinetics of IFN at high dose administration. In one study (35) it was found that a ten-fold increase in HuIFN-β dose from 1×10^6 U/kg to 1×10^7 U/kg i.v. led to a proportional increase in AUC. For HuIFN-α in man also linear kinetics was observed for dosages ranging from 3×10^6 U to 72×10^6 U i.m. per person (39).

Very low serum titers were observed after intrathecal administration of HuIFN-β to monkeys (34,42) and despite high concentrations in the CSF, IFN remained undetectable in deeper layers of the brain (37).

Pharmacokinetics of HuIFN-γ in non-human primates

To our knowledge only two studies with HuIFN-γ in monkeys have been published (Table 7). From these studies (40,41) it is evident that HuIFN-γ also has a biphasic decay profile after i.v. push injection. However, Kurzrock et al. (43) reported a monophasic disappearance of r-HuIFN-γ in man with a half-life of elimination of around 30 min.

Intramuscular or s.c. administration of HuIFN-γ in rhesus monkeys led to very low or even undetectable titers, respectively (40,41). A comparable study, in which IFN-γ was determined by ELISA as well as by antiviral assay was carried out in man (43) and the results suggest that i.m. administration of IFN-γ may have resulted in absorbance into the serum of ELISA-positive material which had lost some of its antiviral activity. The authors reasoned that the partial loss of antiviral activity of IFN-γ may be explained on the basis of evidence suggesting that the functional antiviral unit of IFN-γ is a tetramer (44). It is possible that upon i.m. injection disaggregation of the tetramer takes place *in situ*, leading to low levels of antiviral activity in the serum.

Toxicity of human interferons in non-human primates

The toxic effects of human interferons in non-human primates have been a matter of dispute (45,46). There have been reports that human interferon preparations induce fever and leukopenia in rhesus monkeys and cynomolgus monkeys (9,26,47). Others, however, reported on the lack of side effects in patas monkey, cynomolgus monkeys and rhesus monkeys (45,48) Even the injection of 240 megaunits HuIFNα2 in a single bolus injection did not induce any change in body temperature and haematological and biochemical parameters. This amount of interferon can be compared to a 20,000 megaunits injection in a human or 50 to 100 times the dose that causes extreme toxicity as coma in man.

The only species, apart from man, that has been reported to be sensitive to the side effects of human interferons, is the chimpanzee (49). The chimpanzee of course is a special experimental animal and not all necessary toxicity testing can be performed in this species. But at least it offers the possibility for initial screening of toxicity, which is especially warranted for hybrid human interferons and other second generation molecules. Toxicity testing of human interferon in other non-human primates than chimpanzees which is obligatory for the registration of drugs in a number of countries, is useless and is a waste of animal life.

REFERENCES

1. Schellekens, H. and Meide, P.H. van der. In: The Interferon System, a current review (Baron et al., eds.), The University of Texas Press, 1987, in press.
2. Andrews, R.D. Brit. Med. J. 1728: 1728-1730, 1961.
3. Falcoff, R., Falcoff, E., Duboche, P. et Chany, C. C.R. Acad. Sc. Paris, t. 266, Série D: 297-300, 1968.
4. Pinto, C.A., Haff, R.F., Valenta, J.R., Pagano, J.F., DiCuollo, C.J. and Ferlauto, R.J. In: Symp. Series Immunobiol. Standard., vol. 14: 105-112, Karger, Basel/New York, 1970.
5. Neumann-Haefelin, D., Sundmacher, R., Sauter, B., Karges, H.E. and Manthey, K.F. Infection and Immunity 12, no. 1: 148-155, 1975.
6. Neumann-Haefelin, D., Shrestha, B. and Manthey, K.F. J. Infect. Dis. 133: A211-A216, 1976.
7. Schellekens, H., Weimar, W., Cantell, K. and Stitz, L. Nature 278: 742, 1979.
8. Schellekens, H., Stitz, L.W., Reus, A. de, Weimar, W. and Cantell, K. Antiviral Res. 1: 281-285, 1981a.
9. Schellekens, H., Reus, A. de, Bolhuis, R., Fountoulakis, M., Schein, C., Ecsödi, J., Nagata, S. and Weissmann, C. Nature 292:775-776, 1981b.
10. Weimar, W., Stitz, L., Billiau, A., Cantell, K. and Schellekens, H. J. gen. Virol. 48: 25-30, 1980.
11. Meide, P.H. van der, Harteveld, N., Stebbing, N. and Schellekens, H. Antiviral Res., suppl. 1: 199-207, 1985.
12. Finter, N.B. Interferons and interferon inducers. Elsevier Publishing company, Amsterdam, London, New York, 1973.
13. Hilfenhaus, J., Karges, H.E., Weinmann, E. and Barth, R. Infect. Imm. 11, no. 5: 1156-1158, 1975.
14. Hilfenhaus, J., Weinmann, E., Majer, M., Barth, R. and Jaeger, O. J. Infect. Dis. 135, no. 5: 846-849, 1977.
15. Bear, G.M., Shaddock, J.H., Moore, S.A., Yager, P.A., Baron, S.S., Levy, H.B. J. Infect. Dis. 136, 286-291, 1977.
16. Arvin, A.M., Martin, D.P., Gard, E.A. and Merigan, T.C. J. Infect. Dis. 147, no. 1: 149-154, 1983.
17. Desmyter, J., Groote, J. de, Desmet, V.J., Ray, M.B., Bradburne, A.F., Edy, V.G., Somer, P. de and Mortelmans, J. Lancet 645-647, 1976.
18. Goeddel, D.V., Yelverton, E., Ullrich, A., Heyneker, H.L., Miozzari, G., Holmes, W., Seeburg, P.H., Dull, T., May, L., Stebbing, N., Crea, R., Maeda, S., McCandliss, R., Sloma, A., Tabor, J.M., Gross, M., Familletti, P.C. and Pestka, S. Nature 287: 411-416, 1980.
19. Weck, P.K., Harkins, R.N. and Stebbing, N. J. gen. Virol. 64: 415-419, 1983.
20. Stebbing, N., Weck, P.K., Fenno, J.T., Estell, D.A. and Rinderknecht, E. Arch. Virol. 76: 365-372, 1983.
21. Levin, R., Yaari, A., Gitelman, J., Bino, T., Rosenberg, H. and Kohn, A. Antiviral Res. 6: 277-283, 1986.
22. Sugar, J., Kaufman, H.E. and Varnell, E.D. Invest. Ophthalmol. 12: 378380, 1973.

23. Neumann-Haefelin, D., Sundmacher, R., Skoda, R. and Cantell, K. (1977) Infect. Imm. 17, no. 2: 468-470, 1977.
24. Neumann-Haefelin, D., Sundmacher, R., Frey, H. and Merk, W. Med. Microbiol. Immunol. 174: 81-86, 1985.
25. Ferreira, A., Schofield, L., Enea, V., Schellekens, H., Meide, P., Collins, W.E., Nussenzweig, R.S. and Nussenzweig, V. Science 232: 881-884, 1986.
26. Schellekens, H., Eerd, P.M.C.A. van, Reus, A. de, Weck, P.K. and Stebbing, N. Antiviral Res. 2: 313-318, 1982.
27. Soike, K.F., Kramer, M.J. and Gerone, P.J. J. Infect. Dis. 147, no. 5: 933-938, 1983.
28. Habif, D.V., Lipton, R. and Cantell, K. Proc. Soc. Exp. Biol. Med. 149, 287-289, 1975.
29. Wills, R.J., Spiegel, H.E. and Soike, K.F. J. Interferon Research 4, 399-409, 1984.
30. Collins, J.M., Riccardi, R., Trown, P., O'Neill, D. and Poplack, D.G. Cancer Drug Delivery 2, 247-253, 1985.
31. Skreko, F., Zajac, I., Bahnsen, H.P., Haff, R.F. and Cantell, K. Proc. Soc. Exper. Biol. Med. 142, 946-947, 1973.
32. Yamazaki, S., Kohase, M., Tatsumi, N., Onishi, E. et al. In: The clinical potential of interferons. Eds. R. Kono and J. Vilček. Univ. Tokyo Press, pp. 281- 293, 1981.
33. Bino, T., Edery, H., Gertler, A. and Rosenberg, H. J. gen. Virol. 59, 39-45, 1982.
34. Hilfenhaus, J., Damm, H., Hofstaetter, T., Mauler, R. et al. J. Interferon Res. 1, 427-436, 1981.
35. Gomi, K., Morimoto, M., Inoue, A., Kobayashi, H. et al. Gann 75, 292-300, 1984.
36. Edy, V.G., Billiau, A. and De Somer, P. Lancet i, 451-452, 1978.
37. Billiau, A., De Somer, P., Edy, V.G., De Clercq, E. and Heremans, H. Antimicr. Agents Chemother. 16, 56-63, 1979.
38. Vilček, J., Sulea, I.T., Zerebreckyj, I.L. and Yip, Y.K. J. Clin. Microbiol. 11, 102-105, 1980.
39. Gutterman, J.U., Fine, S., Quesada, J., Horning, S.J. et al. Ann. Internal Med. 96, 549-556, 1982.
40. Cantell, K., Hirvonen, S., Pyhälä, L., De Reus, A. and Schellekens, H. J. gen. Virol. 64, 1823-1826, 1983.
41. Dawson, A., Hirsch, M., Otto, B., De Reus, A. and Schellekens, A. In: The biology of the interferon system. E. de Maeyer and H. Schellekens, eds., pp. 449-452, 1983.
42. Billiau, A., Heremans, H., Ververken, D., Van Damme, J., Carton, H. and De Somer, P. Archives of Virol. 68, 19-25, 1981.
43. Kurzrock, R., Rosenblum, M.G., Sherwin, S.A., Rios, A. et al. Cancer Res. 45, 2866-2872, 1985.
44. Pestka, S., Kelder, B., Familletti, P.C., Moschera, J.A. et al. J. Biol. Chem. 258, 9706-9709, 1983.
45. Finter, N.B., Woodrooffe, J. and Priestman, T.J.: Nature 298: 301, 1982.
46. Weissmann, C. and Schellekens, H.: Nature 298: 301, 1982.
47. Yamada, H., Nishiuchi, M., Sato, K., Kobayashi, F., Harada, Y., Bannai, H., Tatsumi, M. and Yamazaki, S. In: Interferons. Proc. of the Int. Symp. on Interferons (Kishida, T., ed.)278-285, 1984.

48. Schellekens, H.: In: Interferons, UCLA Symposia on Mol. and Cell. Biol. XXV (Merigan, T.C. et al., eds.), Academic Press, 1982, pp. 387-392.
49. Schellekens, H., Reus, A. de and Meide, P.H. v.d. J. Med. Primatol. 13: 235-245, 1984.

23

ANIMAL MODEL SYSTEMS TO EVALUATE THE ANTITUMOR ACTIVITY OF INTERFERONS

F.R. BALKWILL,
Head, Biological Therapy Laboratory,
Imperial Cancer Research Fund,
Lincoln's Inn Fields,
London, WC2A 3PX.

INTRODUCTION

Interferons, IFNs, have complex regulatory actions on many cells of the body. It would be impossible to understand and optimise their anticancer activity solely from in vitro data or clinical experience in man. Experiments in animals can provide important information and have also been catalytic in promoting clinical trials.

In this chapter I shall outline animal models in which useful information concerning therapeutic potential and mechanisms of actions of the IFNs can be obtained, and review the information they have provided.

The models can be divided into 2 main groups: 1) animal tumors growing in syngeneic hosts and 2) human tumors growing in immunodeficient mice. Each type of model has advantages and disadvantages: for instance, animal tumors are often produced by injecting cells from highly selected lines that are maintained in vitro; their sensitivity to therapy frequently differs from the same histologic type in man, and tumors often grow at inappropriate sites. Moreover transplantable animal tumors generally kill the host within a few weeks; the cells may be highly immunogenic, and because IFNs are generally species specific it is not possible to test human IFNs in these animal models (1).

Human IFNs can be tested on human tumors growing as xenografts in nude mice and in these models we can study tumors that accurately represent the metabolic characteristics of human cancer (2). Moreover xenografts of a particular tumor type are sensitive to the chemotherapeutic agents clinically active in these diseases (3).

M. Revel (ed.), CLINICAL ASPECTS OF INTERFERONS.

However, complex biologic agents such as IFNs have important effects on the tumor bearing host and may alter host response to the tumor. Human IFNs do not have any measurable effects on the nude mouse host and therefore we can only study the direct effects of these IFNs on the human tumors. Other disadvantages of the xenograft models are that the tumors often grow at inappropriate sites, and do not usually metastasize (2).

THE THERAPEUTIC POTENTIAL OF IFNS IN ANIMAL TUMOR MODELS

Effects on primary transplantable tumors

Various preparations of mouse IFN, differing in their degree of purity, have increased the survival time or reduced tumor size in mice implanted with many murine tumor cell lines, for instance Rc19, EL4, Lewis Lung, L1210 and B16 melanoma (for review see 4).

Efficacy of treatment depended on low tumor load and frequent dosing (4). In none of the experiments did IFN therapy alone effect a cure, but more recent work by Gresser et al showed complete regression of subcutaneous tumors of Friend leukaemia cells when treated with highly purified IFN (5). The majority of studies describe the use of murine IFN-α or β or combinations. Early studies with impure murine IFN-γ suggested that this may have greater activity but preparations probably contained other lymphokines such as tumor necrosis factor.

Production of IFN by the host may be important in the development of transplantable mouse tumors. Gresser et al (6) found that anti-IFN globulin injections enhanced the transplantability of a variety of mouse tumors in immunocompetent mice.

If we are to extrapolate from these animal studies to clinical situations it is important to relate the doses used in mice to those which are tolerated in man. However, it must be remembered that murine IFNs generally have higher specific activities than their human counterparts and are assayed on different cells, so that such comparisons can only be approximate. The maximal tolerated dose, MTD, of IFN-α in man is between 5 and 10 x $10^6 U/m^2$. Using the formula of Freireich et al (7), in which differences in surface area/volume ratios between mouse and man are taken into account, the doses of murine IFN that inhibited transplantable tumor growth are within the range of 5-40 x $10^6 U/m^2$ in man.

It is also important to compare the effects of IFN therapy with established chemotherapy. In one such comparison (4) murine IFN therapy was compared to the published activity of 11 different commonly used chemotherapeutic drugs against the L1210 and P388 lymphomas. Partially pure murine IFN-α/β given at doses slightly higher than the MTD in man had moderate activity and gave equal or better results than five of the eleven agents (4). Similarly in AKR mice with established lymphoma, murine IFN gave better results than seven of ten commonly used chemotherapeutic agents from a variety of chemical classes (4).

One criticism of these murine tumor models is that they use a different end point from clinical studies in man (1). Whereas a positive result in a clinical trial is recorded when a tumor completely disappears or regresses by more than 50%, a positive result in a murine tumor model is inhibition of tumor growth or increase in lifespan. It is important to note that IFN therapy of transplantable murine tumors rarely caused regression or cure.

Effects on experimental and spontaneous metastasis

Metastasis in experimental animals is studied in two ways. In experimental metastases tumor cell suspensions are injected intravenously into mice and colonize the lungs, in spontaneous metastases the spread of tumor cells from a subcutaneous tumor is measured. IFNs or IFN inducers have been shown to inhibit metastases and prolong survival in both types of models (reviewed in 4,8,9, Ramani & Balkwill unpublished results). Doses used were similar to those reported above for the primary tumors.

THE THERAPEUTIC POTENTIAL OF IFNS AS ASSESSED ON HUMAN TUMOR XENOGRAFTS IN NUDE MICE

Administration of human IFNs to the nude mouse can inhibit the growth of some human tumor xenografts derived either from cell lines or primary tumor material. Stasis, after treatment with human IFN-α, of subcutaneous xenografts derived from tumors or breast, bowel, bladder, bone and kidney has been reported. In some cases complete regression and cure of mice occurred (reviewed in 4, 10-15 and Balkwill unpublished results). Xenografts derived from lung cancer cell lines were unresponsive (14). In the studies with osteosarcoma it is of interest that all IFN-α treated tumors were partially replaced by normal appearing bone and bone marrow by the end of the experiment (11).

A wide range of xenografts were responsive to human IFN-α, none responded to human IFN-γ (4,15 and Balkwill unpublished results). The reason for this is not clear but as these two IFNs have different pharmacokinetics properties in nude mice (15) it may be that insufficient IFN-γ was able to reach the tumors. However, recent experiments have shown that intratumoral injection of IFN-γ is equally ineffective (F. Balkwill unpublished results).

Although human IFN-γ did not inhibit the growth of the subcutaneous tumors it was able to have other effects. We have recently found that systemmic therapy with human IFN-γ enhanced or induced Class II MHC expression in the tumor cells at the subcutaneous site (16).

Human IFN-γ is, however, active against xenografts growing in other sites. Intraperitoneal administration of human IFN-γ to intraperitoneal ovarian cancer xenografts resulted in some cures but only when therapy began immediately after tumor cell injection (Balkwill & Ward submitted for publication). Similarly human IFN-γ inhibited the development of pulmonary metastases of a human melanoma cell line injected intravenously when therapy started one day later (Ramani & Balkwill submitted for publication).

The doses of IFN used in some of the studies reviewed above are within the range that could be safely administered to man, but in other studies tumor stasis was only achieved with higher doses. It is important to note that tumor regression or cure, end points that would be scored as responses in clinical trials, were only noted in a small number of tumors.

However the data clearly show that human tumors possess receptors for, and can respond to, IFNs. Moreover, the change to normal bone histology in the osteosarcoma xenografts indicates that IFN therapy can cause differentiated changes in tumor which may result in slower tumor progression and increased lifespan.

THE POTENTIAL OF COMBINING IFNS WITH OTHER CANCER THERAPIES

Interactions between IFNs and chemotherapy have been studied in both animals tumor models and human tumor xenografts. The activity of the following agents have been enhanced by murine IFN-α/β therapy in a variety of animal tumor models: BCNU, cyclophosphamide, cisplatinum, methotrexate and α-difluoromethylornithine, α-DMFO, (reviewed in 4). Enhancement of activity was either additive or synergistic. Similarly in human tumor xenograft models human IFN-α enhanced the activity of cyclophosphamide, doxorubicin, ifosfamide, cisplatinum and α-DMFO (reviewed in 4, 17, 18).

However other studies have shown a lack of, or even negative, interactions between IFNs and chemotherapy. The activity of doxorubicin, cytosine arabinoside, cyclophosphamide and 6-mercaptopurine against L1210 leukaemia was not enhanced by simultaneous administration of murine IFN-α/β (19). Moreover when IFN therapy was started 2 days before the administration of cyclophosphamide it inhibited antitumor activity against a rat liposarcoma (20).

These negative results serve to highlight the potential complexity of interactions between IFN and chemotherapy. Little is known about these interactions. Because IFNs show cell cycle progression they could enhance the activity of cycle specific drugs. A combination of IFN and cyclophosphamide was found to significantly prolong progression of cells through S phase of the cell cycle (21).

IFNs could also influence drug activation and break down because they have been shown to alter the activity of some enzymes in the hepatic cytochrome P450 and glutathione S transferase enzyme systems (21).

Thus the interactions between IFNs and chemotherapy are potentially complex and require a greater understanding - an understanding that can only be achieved with the use of animal models and in vitro systems. Animal models are of paramount importance in studying those drugs that requre in vivo activation, and in studying scheduling of the two agents.

IFNs constitute one part of the cascade of lymphokines and cytokines that are produced during immune responses. Synergy occurs between some cytokines, particularly IFN-γ and tumor necrosis factor, TNF, or the related cytokine lymphotoxin (22). Cell lines that are insensitive or weakly sensitive to the cytostatic or cytotoxic activity of either agent alone are highly sensitive when the two agents are combined (22). We have recently found synergy between human IFN-γ and TNF when treating human ovarian cancer xenografts growing intraperitoneally in nude mice, and additive interactions between human IFN-α and TNF against subcutaneous human tumor xenografts (Balkwill & Ward submitted for publication).

MECHANISMS OF ANTITUMOR ACTION OF IFN

IFNs are complex regulatory molecules that have potential to influence both the tumor and any host response to it. To find out how they act is much more difficult than with many established cancer therapies. IFNs specifically enhance or inhibit the production of over 100 different mRNAs and proteins in the cell (23) and this change in protein synthesis can profoundly alter cellular function. Thus in response to an IFN, whether this is injected or formed naturally in the body, many different cells of an animal bearing an experimental tumor are likely to change their behaviour. The ways in which IFNs could exert antitumor activity are shown in Table 1 and can be divided into two groups: direct effects on tumor cells and indirect effects altering the host relationship with the tumor. There is evidence for both direct and indirect effects from animal tumor models.

Table 1

Indirect antitumor effects of IFNs	Direct antitumor effects of IFNs
Stimulation of host immune system: e.g. enhance effector cell cytotoxicity; enhance production of other cytokines and lymphokines	Cytostasis or cytotoxicity
Depletion or alteration of levels of a crucial host nutrient/ growth factor required by the tumor	Alteration of cell surface antigens
	Change in cytoskeleton
	Change to differentiated phenotype
	Inhibition of oncogene product
	Cure of viral DNA

Evidence for direct antitumor actions of IFNs

Studies in nude mice provide the most obvious evidence for a direct effect of IFNs. As described above, treatment of human tumor xenografts with human IFN-α resulted in tumor stasis and in some cases tumor regression (4). The human IFN-α used in these studies had no measurable effect on the nude mouse host as evidenced by levels of IFN induced enzyme in the murine tissues, NK cell activity, peripheral blood white cell count, or systemmic toxicity, but did raise levels of an IFN induced enzyme in the human tumor tissue (4). Moreover in osteosarcoma xenografts normal bone differentiation was seen after IFN treatment (11). This evidence suggests that IFNs were having a direct effect on the tumor. This direct effect could be manifested in 3 ways:- 1) direct cytostatic activity sometimes accompanied by differentiation, 2) alteration of the tumor ability to respond to the host environment e.g. down regulation of growth factor receptors - a phenomenum reported in vitro (24) or 3) changes in the cell surface antigenicity of the tumor cells thus altering host recognition. Of interest to the latter possibility is the finding that treatment of xenografts with human IFN-γ, while having no effect on tumor growth, strongly enhanced surface expression of Class II HLA (16). While this direct effect would not be expected to alter the response of the nude mouse to a human tumor it may be important in a situation where tumor and host are homologous. Moreover, it demonstrates that IFNs can modulate the surface of cells in solid tumors.

Direct effects of IFNs have also been demonstrated on metastatic cells. In a model of human metastasis DX3 human melanoma cells grow as tumor nodules in the lungs of Balb-c nude mice when injected intravenously. Therapy with human IFNs α and γ strongly inhibited the number of lung tumor nodules while having no effect on the nude mouse host (Balkwill & Ramani submitted for publication).

In a different model system murine IFN-α/β pretreatment of a murine sarcoma line in vitro resulted in a decrease in its metastatic ability in immunocompetent Balb-c mice but not nude mice (25). The authors interpretation of their findings was that IFN pretreatment altered the surface antigen expression of the tumor cells in such a way that they could be recognised and eliminated by the host T cell system.

Not all direct effects of IFNs on tumor cells are of potential benefit. We have recently found that pretreatment of Balb-c COLO 26 adenocarcinoma cells with recombinant murine IFN-γ, but not IFN-α, significantly enhances their metastatic activity (Ramani & Balkwill unpublished results). This enhancement is seen in immunocompetent and nude Balb-c mice, but not in mice depleted of, or deficient in, NK cells. The reasons for this enhancement are not clear. However, they do not seem to be related to changes in surface MHC expression since both IFN types cause similar enhancement of class I MHC and neither IFN affects class II MHC expression (Ramani & Balkwill unpublished results).

Another way to obtain evidence for a direct effect of IFNs on tumors is to block the functions of the immune system or do experiments in immune deprived mice. Treatment of tumor-bearing mice with antilymphocyte serum or x-rays did not prevent the action of murine IFN on Erlich ascites or L1210 leukaemia. Abrogation of mononuclear phagocytes by silica injection did not inhibit antitumor activity of IFNs but did reduce effectiveness (26).

Antimetastatic effects of IFNs against COLO 26 adenocarcinoma cells were found when cells were injected into Balb-c mice, Balb-c nude mice (T cell deficient) beige nude mice (T and NK cell deficient) triple beige nude mice (T, NK and B cell deficient) and in Balb-c mice treated with antiasialo GM1 serum to abrogate NK cells (27), (Ramani & Balkwill unpublished results). However a role for the host mononuclear phagocyte system in the antimetastic effects of IFN-α has not yet been discounted.

Indirect antitumor effects of IFNs

One of the most obvious ways of studying host mediated effects of IFNs on tumor cells is to generate tumorogenic cell lines that are resistant to IFNs *in vitro*. Several different resistant cell lines have been used: L1210 leukaemia, F9 teratocarcinoma, Meth-A sarcoma and Friend leukaemia cells (reviewed in 28). The common finding with all of these experiments has been that IFN therapy works just as well with IFN resistant cells growing as tumors *in vivo* as with the IFN sensitive cells, and this has been interpreted as evidence for a host role in the effects of IFN. In a recent review of data with IFN resistant Friend leukaemia cell, FLC, tumors, Ion Gresser discussed all the evidence for host mediated mechanisms and could find no compelling data suggesting involvement of cells of the host immune system, be they macrophages, NK cells or T cells (28). The effects of IFNs on i.p. FLC tumors could not be transferred by peritoneal cells and no host cell infiltrates were seen in subcutaneous FLC tumors. He suggested that the IFN therapy may cause a deficiency of some necessary substance in the surrounding normal tissues or inhibit the development of tumor vasculature (28).

In other primary transplantable tumors an effect on the host system has been implicated. The activity of IFN-α against a Meth A sarcoma was abrogated when mice were pretreated with antithymocyte globulin, and antitumor effects were also less pronounced in nude mice (29). Circumstantial evidence for an effect of IFN via host macrophages was found in other experiments with Meth A sarcoma (30).

In experimental metastasis the number of lung nodules is inversely correlated with NK cell function. Enhancement of NK function by IFNs or IFN inducers correlated well with inhibition of metastasis in several different models (8,31). However in another model system where IFN-α was given after injection of the metastatic cells strong inhibition of the development of lung nodules was seen in mice totally devoid of NK cells (Ramani & Balkwill unpublished results).

As discussed above the effect of human IFNs on human tumour xenografts provided evidence for direct effects of IFNs on tumors. However other data from this model system suggest a host role in IFNs antitumor effect. Highly purified murine IFN also inhibited the growth of 3 different human tumor xenografts which showed varying sensitivity to human IFN-α (32). No evidence was found for a direct effect of the murine IFN on the human cells and antitumor activity of the murine IFN was also found in beige nude (NK deficient) mice and did not correlate with stimulation of NK cells. Also no host cell infiltrate was found in the tumors. The suggestion from these experiments is that the murine IFN was affecting some non immune host control of the tumor.

LESSONS TO BE LEARNT FROM ANIMAL MODELS

In the animal model systems described here IFNs undoubtedly have antitumor activities at doses that are attainable in man. However IFNs alone have rarely resulted in cure of primary tumors or established metastasis.

The experiments have so far indicated that favourable response is inversely related to low tumor load, that frequent dosing is necessary, but that responses can be seen in a wide range of tumors growing at different sites. The data from the human tumor xenograft studies show that a spectrum of human tumors will respond to direct growth inhibition of IFN-α and that IFN-γ can modulate cell surface antigenicity.

There is much confusing data on mechanismsof antitumor action of IFN and the complexity reflects the complexity of IFNs regulatory effects on cells. Mechanisms are probably dependent on the individual tumor model, the tumor site, the antigenicity of the tumor cell and the type of IFN. However it is still possible that some novel mechanism will be discovered which acts in all the experimental situations.

Some of the most promising therapeutic effects have come from the combination of IFNs with chemotherapy or other biologicals such as tumor necrosis factor and further study in both models and man are warrented.

REFERENCES

1. Martin, D.S., Balis, M.E., Bernard-Fisher, B., Frei, E., Freireich, E.J., Heppner, G.H., Holland, J.F., Houghton, J.A., Houghton, P.J., Johnson, R.K., Mittelman, A. Cancer Res., 46, 2189-2192, 1986.
2. Giovanella, B.C., Fogh, J. Advances in Cancer Research, 44, 69-120, 1985.
3. Giovanella, B.C., Stehlin, J.S., Shephard, R.C., Williams, L.J. Cancer, 52, 1146-1152, 1983.
4. Balkwill, F.R. In: Interferon, vol.4: In vivo and clinical studies, (Eds. N.B. Finter and R.K. Oldham), Elsevier Science Publishers B.V., 1985.
5. Gresser, I., Belardelli, F., Maury, C., Tovey, M.G., Maunoury, M-T. Int.J.Cancer, 38, 251-257, 1986.
6. Gresser, I., Belardelli, F., Maury, C., Manunoury, M-T., Tovey, M.G. J.Exp.Med., 158, 2095-2107, 1983.
7. Freireich, E.J., Gehan, R.A., Rall, D.A., Schmidt, L.A., Skipper, H.E. Cancer Chemother.Rep., 50, 219-244, 1966.
8. Brunda, M., Rosenbaum, D., Stern, L. Int.J.Cancer, 34, 421-426, 1984.
9. Ramani, P., Hart, I.R., Balkwill, F.R. Int.J.Cancer, 37, 563-568, 1986.
10. Yamauoka, T., Takada, H., Yanagi, Y., Kataoka, T., Sakurai, Y. Cancer Chemother.Pharmacol., 14, 184-187, 1985.
11. Brosjo, O., Bauer, H.C.F., Brostrom, L., Nilsonne, U., Nillsson, O.S., Reinholt, F.P., Strander, H., Tribukait, B. Cancer Res., 45, 5598-5602, 1985.
12. Hubbell, H.R., Kvaines-Krick, K., Carter, W.A., Strayer, D.R. Cancer Res., 45, 2481-2486, 1985.
13. Hofmann, V., Groscurth, P., Morant, R., Cserhati, M., Honegger, H.P., Von Hochstetter, A. Eur.J.Cancer Clin.Oncol., 21, 859-863, 1985.

14. Twentyman, P.R., Workman, P., Wright, K.A., Bleehen, N.M. Br.J.Cancer, 52, 21-29, 1985.
15. Balkwill, F.R., Goldstein, L., Stebbing, N. Int.J.Cancer, 613-617, 1985.
16. Balkwill, F.R., Stevens, M.H., Griffin, D.B., Thomas, J.A., Bodmer, J.G. Eur.J.Immunol., 1987 in press.
17. Carmichael, J., Fergusson, R.J., Wolf, C.R., Balkwill, F.R., Smyth, J.F. Cancer Res., 46, 4916-4920, 1986.
18. Heston, W.O.W., Fleischmann, J., Tackett, R.E., Ratlift, T.L. Cancer Res., 44, 3220-3228, 1984.
19. Slater, L.M., Wetzel, M.W., Cescrio, T. Cancer, 48, 5-9, 1981.
20. Marquet, R.L., Schellekens, H., Westbroek, D.L., Jeekel, J. Int. J.Cancer, 31, 223-226, 1983.
21. Balkwill, F.R., Mowshowitz, S., Seilman, S.A., Moodie, E.M., Griffin, D.B., Fantes, K.H., Wolf, C.R. Cancer Res., 44, 5355, 1984.
22. Fransen, L., Van der Heyden, J., Ruysschaert, R., Fiers, W. Eur.J.Cancer Clin.Oncol., 22, 419-426, 1986.
23. Revel, M., Chebath, J. TIBS 11, 166-170, 1986.
24. Zoon, K.C., Karasaki, Y., Zur Nedden, D.L., Hu, R., Arnheiter, H. Proc.Natl.Acad.Sci., 83, 8226-8230, 1986.
25. Wivel, N.A., Pitha, P.M. Int.J.Cancer, 30, 649-654, 1982.
26. Gresser, I., Bourali-Maury, C. P.S.E.B.M., 144, 896-900, 1973.
27. Ramani, P., Hart, I.R., Balkwill, F.R. Int.J.Cancer, 37, 563-568, 1986.
28. Gresser, I. Interferon, 6, 93-126, 1985.
29. Kataoka, T., Matsuura, N., Oh-hashi, F., Suhara, Y. Cancer Res., 45, 3548-3553, 1985.
30. Uno, K., Shimizu, S., Ido, M., Naito, K., Inaba, K., Oku, T., Kishida, R., Muramatsu, S. Cancer Res., 45, 1320-1327, 1985.
31. Hanna, N. Biochimica et Biophysica Acta, 780, 213-226, 1985.
32. Balkwill, F.R., Proietti, E. Int.J.Cancer, 38, 375-380, 1986.

24

EFFECTS OF POLYNUCLEOTIDES ON MONKEYS AND MAN

HILTON B. LEVY

Molecular Virology Section, National Institute of Allergy and Infectious Diseases, National Institutes of Health, Bethesda, MD 20982

ABSTRACT

Interest in the biological effects of nucleic acids as immune modulators developed about simultaneously and almost independently of the realization that some of the nucleic acids could act as interferon inducers. A symposium held in 1970 reported studies, dating from about 1967, on poly adenylic·polyuridylic acid (poly A·poly U) and poly inosinic·poly cytidylic acid as compounds that were able strongly to augment antibody responses in mice to some antigens and to restore immune competence to mice that had become immune compromised for a variety of reasons (1). It was not long after the discovery of interferon that a number of people became aware of the potential use of interferon as a broad spectrum antiviral agent in the treatment of human disease. However, until recently, with the development of DNA recombinant techniques for large scale production of human interferon, obtaining enough interferon to do valid clinical trials was extremely difficult to do. A number of non replicating entities were found that could cause the host (rodents in most studies) to produce interferon. In 1967 it was reported that the most effective of these was a synthetic double stranded RNA, polyinosinic- polycytidylic acid, or polyI·polyC (2). This polynucleotide also was shown to be able to act as an immune adjuvant, acting to augment strongly the production of antibodies to a number of antigens, under conditions where interferon did not do so. Poly A·poly U also was a good immune adjuvant in mice, but induced the formation of only very small amounts of circulating interferon. PolyA·PolyU was a good antitumor agent. It was tested much less extensively than polyI·polyC as an antiviral, but where tested was

M. Revel (ed.), CLINICAL ASPECTS OF INTERFERONS.

less effective (1).

PolyI·polyC proved to be only very slightly active in monkeys and chimpanzees (3). Clinical trials in man as an antitumor agent were very disappointing, because, while the material showed very little toxicity in man it induced the formation of only very low levels of interferon, and showed no antitumor action (4,5). This was later associated with the finding that there is present in primate serum a relatively high concentration of enzymes that hydrolize and inactivate polyI·polyC (6).

A derivative of polyI·polyC, made by complexing the ds RNA with polylysine and carboxymethylcellulose, called polyICLC proved much more resistant to hydrolysis than was polyI·polyC. It was able to induce the formation of large quantitites of interferon in monkeys, chimpanzees and humans (7). The remainder of this chapter will be devoted to an examination of studies that have been done with this stabilized derivative of polyI·polyC, called polyICLC, with polyA·polyU, and with another polynucleotide called Ampligen (8).

The material will be presented in three parts; 1) a brief summary of preclinical studies in rodents, 2) immune modulating effects in primates, including man, and 3) clinical studies.

PRECLINICAL STUDIES IN RODENTS

Extensive studies by Levy (9), Chirigos and Levy (10), Talmadge (11), and Hartmann (12) in mice have demonstrated that PolyICLC exerts immune modulating actions on a number of parameters. Since this review deals with effects in primates, only a brief summary of the data in the above mentioned articles will be given in the next paragraph. The effects of polyA·polyU on immune reactivity has been mostly on production of humoral antibody, particularly using the Jerne plaque test with sheep red blood cells as antigen, where it is a good immune adjuvant. An extensive review is given in (1).

There is a strong augmentation of antibody production when polyICLC is given along with most, but not all antigens (9). There is strong enhancement of NK cell, macrophage, and cytotoxic T lymphocyte activities, without ready development of hyporesponsiveness. In mice there is, in addition, an increase in the number of macrophages in spleen, peritoneal cavity and lung (10,11). These effects are manifest as very low drug levels, well below 1 μg/mouse, and reach a peak at 1 to 10 μg/mouse. Administration of higher doses results in a lower degree of augmentation and can actually be associated with inhibition. Toxicity, of course, increases with increasing dose. Both polyICLC and polyA·polyU increase NK cell activity in peripheral blood, but polyICLC is more active, both with regard to the amount of effect with a given dose of drug, and to the maximum degree of enhancement obtainable with that drug. Figure 1.

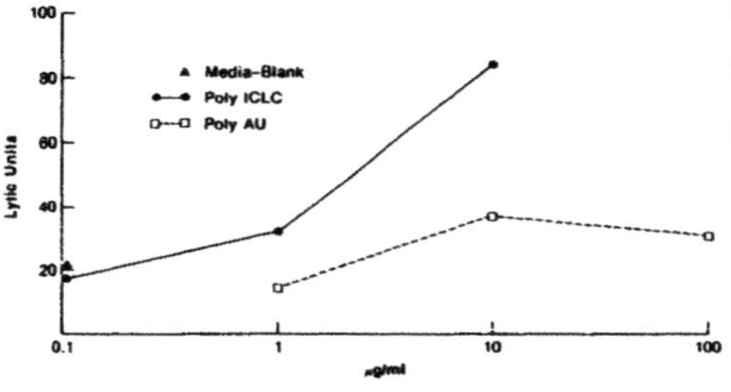

Fig. 1. Effect of poly ICLC, poly A-poly U in vitro on NK cell activity

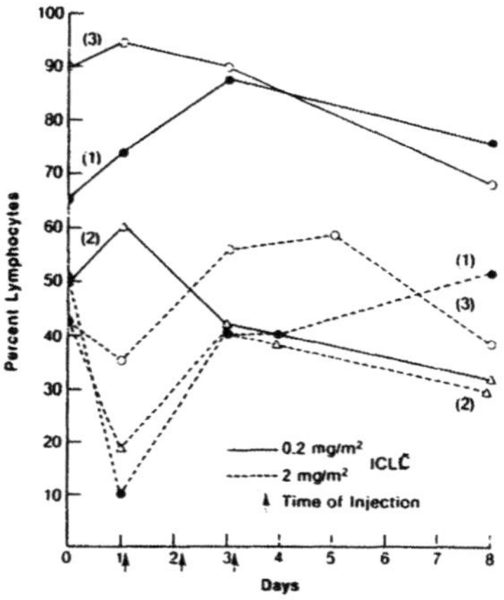

Fig. 2. Effect of two different doses of poly ICLC on % lymphocytes in monkeys.

IMMUNE MODULATING EFFECTS IN PRIMATES, INCLUDING MAN

A number of alterations in cellular activities are seen in primates that have received polyICLC. In general these effects are dose related in primates as they were in mice, in that there is an optimum dose, below and above which there is less effect. At the higher doses there may actually be an inhibition, where at a lower dose there was enhancement.

In monkeys given polyICLC at low doses (0.2 mg/m^2) there is a transient enhancement of the percent lymphocytes in the blood, followed in a few days by a return to normal. At a higher dose (2.0 mg/m^2), there was a transient decrease in the percent lymphocytes, again followed by a return to normal (Levy, H.B., unpublished data). At the higher dose the decrease in lymphocytes is followed by a strong increase in the absolute number of granulocytes. Figure 2.

In man there is a rise in the number of monocytes in the circulation. Table 1 (13). In patients, upon administration of polyICLC, at 100 μg/Kg body weight there is an increase in the number of leu 11+ cells, reflecting a rise in large granular lymphocytes (14). Table 2. At low doses (1 mg/Kg body weight) there was an increase in NK activity as well, but at higher doses there was either no effect or an inhibition (13).

In patients with malignancies or with neurologic diseases who received very low doses of polyICLC (1-3 mg/m^2) there was a small but reproducible increase in the ratio of T helper to T suppressor cells, as determined by FACS analysis. Tables 3 and 4 (13, 14 & Tyndall R. & Levy, H.B., unpublished).

CLINICAL STUDIES

PolyA·PolyU

Based on the observations in animal models discussed above, a trial of polyA·polyU as an adjuvant in operable breast cancer was performed at the Institut Gustave-Roussy in Villejuif, France (15). All patients received standard treatment, which was surgery alone when axillary nodes were negative, or surgery plus cobalt radiation when they were invaded. Patients were divided into two groups. One received the above treatment plus polyA·polyU, one injection per week of 30 mg for 6 weeks; the other group received the standard therapy plus a placebo, and served as control. There were 300 patients about evenly matched and evenly divided as to number. In 1983, when the mean time of follow up was 87 months, it was found that those patients that had received polyA·polyU had an actuarial survival rate of 71%, while the placebo group had a rate of 57%, with a p value for the difference being $< .05$. The difference was even greater in that subgroup of patients where there was evidence of disease in the lymph nodes, 67% vs. 48% with a p value of <0.02. No

Table 1 - Effect of PolyICLC Treatment on % Monocytes in Blood

Patient		0 Time	24 Hrs	% Increase at 24 Hours
F.G.	3/13/83	9.54	15.22	89%
	6/22/83	9.44	8.95	-5%
	7/19/83	7.95	6.39	-20%
	2/6/84	13.1	17.8	36%
V.T.	2/17/83	7.29	32.3	349%
	3/9/83	8.29	31.5	280%
A.P.	2/16/83	6.07	26.0	328%
	3/15/83	19.8	24.7	25%
	4/13/83	7.48	26.5	254%
	12/6/83	3.30	14.7	354%
C.W.	1/9/84	3.70	15.0	305%
R.R.	7/12/83	13.2	22.9	73%
T.B.	7/6/83	15.1	46.2	206%

Table 2 - Effect of PolyICLC Treatment on % Leu 11$^+$ Cells in Blood

		% Leu 11$^+$		
Patient	Date	0 Time	24 Hrs	% Increase at 24 Hours
C.W.	2/15/84	8.26	23.6	186%
F.G.	2/6/84	8.38	15.5	85%
	10/18/83	9.93	30.6	208%
C.W.	1/9/84	6.42	30.5	375%
A.P.	1/10/84	7.65	2.43	204%
	12/6/83	9.96	20.4	105%
	11/1/83	17.0	17.9	5%
T.B.	11/8/83	16.2	37.4	131%
	2/27/84	19.8	42.8	116%
V.T.	3/6/84	15.3	33.2	117%

Table 3 - Effect of PolyICLC on Helper/Suppressor Ratios

	Date of Treatment	0 Time	24 Hrs	% Increase at 24 Hours
F.G.	3/30/83	2.66	7.28	174%
	6/22/83	3.12	4.32	39%
	7/19/83	2.78	4.89	76%
V.T.	3/9/83	2.36	2.50	6%
A.P.	2/16/83	2.05	3.10	51%
	3/15/83	3.26	2.54	-22%
	4/13/83	2.01	2.19	9%
H.T.	5/2/83	1.46	2.26	55%
R.R.	7/12/83	2.81	3.35	19%
T.B.	7/6/83	1.65	3.61	119%

Table 4 - Effect of Repeated Doses of PolyICLC on One Patient on Helper/Suppressor Ratio

Patient J.C.L. Time (hr)	Helper/Suppressor
0	2.0
8	1.4
24	2.6
8	3.3
24	5.14
0	2.4
8	5.14
0	1.7
8	5.2
24	5.1
0	1.2
8	5.2
24	5.1

interferon was found in the serum of patients receiving the polyA·polyU, although there was an increase in two enzymes that are associated with the presence of interferon, namely protein kinase and 2'-5' A synthetase. NK cell activity was also frequently increased in those patients receiving the duplex. The mean percentage of E-rosette-forming cells also was increased in those patients.

There was virtually no toxicity in any of the patients that could be attributed to the drug, except for an occasional mild fever. Currently there are ongoing comparable adjuvant trials with polyA·polyU in colonic and colorectal cancer, melanoma, and in resectable gastric carcinoma.

Poly ICLC

Several small open ended trials in a variety of diseases have been performed with polyICLC. Initial phase I studies were done by Levine et al in refractory cancer patients (16). The standard oncologist approach was used, which was to determine the maximum tolerated dose. As has been suggested, and will be mentioned again, this is probably a bad approach to use with biologic response modifiers. The regimen used in that study was as follows: a given dose of polyICLC was administered i.v. to 3 patients and the patients were observed for a week. If no untoward reactions were noted that dose was given daily for 2 weeks. If that dose was tolerated, three different patients received the next higher dose according to the same schedule. In that study, it was found that the MTD was 12 mg/m^2. At this level of drug, a mean peak serum level of interferon of about 2000 I.U. per ml was found. Subsequent studies indicated that most patients could not tolerate 12 mg/m^2. In investigations with the Childrens Cancer Testing Group in terminally ill leukemic children, it was seen that the MTD was between 6 and 9 mg/m^2 (17). There were no cures in these refractory children, but, there were some anti-leukemic effects in that in some marrows all tumor cells disappeared. The children died before it could be seen whether a more normal marrow would have grown back.

On the other hand, in relatively healthy children, with severe juvenile laryngopapilloma at Johns Hopkins Hospital and U. of North Carolina, 12 mg/m^2 was tolerated and even 15 mg/m^2 was acceptable,

Table 5 - PolyICLC in Patients with Multiple Myeloma

M-Component Type	Comments
kappa light chains	67% decrease in B-J excretion, plus correction of hypercalcemia and disease stabilization with first period of treatment. 44% decrease in B-J excretion with polyICLC restarted 2 months later. Normocalcemic for 5 months.
IgG kappa	M-component decrease from 5.2 g to 3.9 g. Improved bone pain and performance status (became ambulatory) for 2-3 months.
IgG lambda	Trial stopped because of toxicity (malignant hypertension).
IgM kappa	Plasmaphoresis requirement decreased from q 14 days to q 28 days.
IgG kappa	Stable disease parameters for 2 months.
lambda light chains	Died one week after iniation of polyICLC.
IgG kappa	50% decrease in serum IgG M-component, plus symptomatic benefit.

although side effects were severe. There were clear-cut benefits in most children (18). Figure 3 shows the change in frequency of the need for surgical intervention in one of the children before and after the institution of treatment with polyICLC.

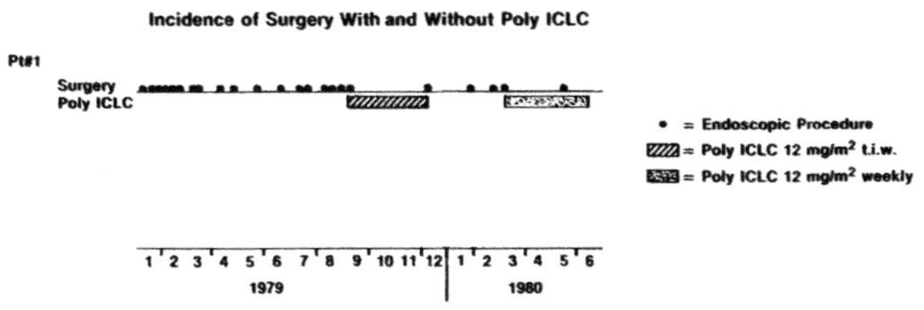

Fig. 3. Effect of poly ICLC treatment of a patient with juvenile laryngo papilloma

There were no beneficial effects in a limited study in women with metatastic breast cancer (19), nor in malignant melanoma (20). It may be that the dose of drug used in these studies was too high to develop maximum immune enhancing capability. There were definite beneficial effects in multiple myeloma as seen in studies at the U. of Arizona (21). Table 5.

Of particular interest are the strong suggestions of benefit in some patients with neurologic disease of putative immunological origin, namely chronic polyneuropathy (Guillain-Barre), and multiple sclerosis. In 12 peripheral neuropathy patients, perhaps half have shown moderate to dramatic improvement, while others have shown none (22). There patients have an extreme sensitivity to fever. Their muscular weakness is strongly worsened by fever, and it was so with polyICLC. However, this was transient and they were back to pretreatment strength within 24-36 hours. Frequently, by the third day, an objectively demonstrated improvement could be seen. Similar side effects were seen in a study with multiple sclerosis patients. With these patients vigorous treatment with antipyretics, and anti-emetics was desirable, and where needed a cooling blanket was used. A group of ten M.S. patients have been treated with monthly infusion of polyICLC for from 10 to 18 months. All initially stabilized or improved. Three had improvement of DSS (disability status score), and four of AI (ambulatory index). Four deteriorated after 5, 10, 14, and 17 months of treatment. An example of one patient who improved is shown in Fig. 4. It was concluded that polyICLC can be safely given to patients with M.S., with aggressive management of fever and that a controlled trial is warranted (14).

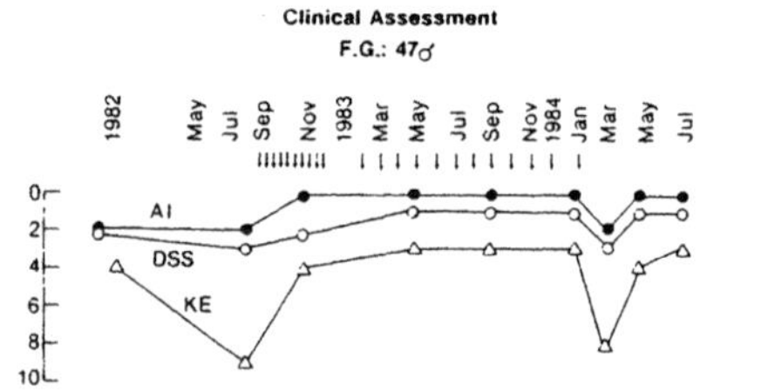

Fig. 4. The clinical courses of a patient (F.G.) who improved during treatment is shown.

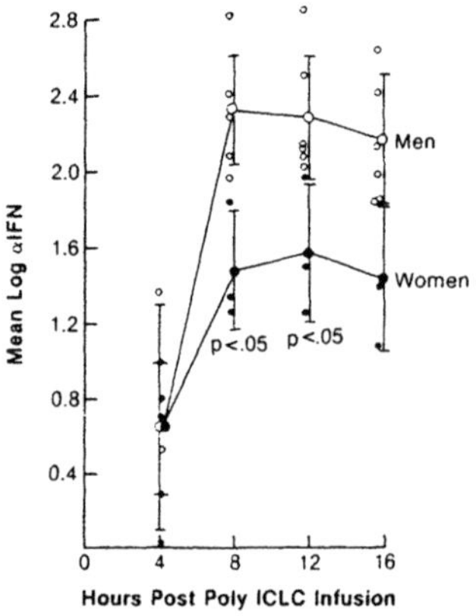

Fig. 5. Gender difference in interferon production in response to poly ICLC

It was noted as part of the M.S. study that women made significantly less interferon than did men in response to polyICLC (Fig. 5). The same type of difference in response was later noted in male and female Rhesus monkeys (23). There did not appear to be any demonstrable difference in the characteristics of the patients that might account for this difference in interferon production, nor was there any difference in the degree of clinical improvement. There may have been somewhat less discomfort in the female patients.

As has been mentioned several times, it appears that maximum immune enhancement is not found with maximum tolerated dose of polyICLC. Rather, the MTD is frequently found to give much less enhancement than the optimum biological dose, which may be 10 to 50 times less (11). At the higher doses toxicity is greater, of course. Current clinical studies are being developed around the idea of using much lower doses, and using the intramuscular route.

AMPLIGEN

A totally different approach to the use of double stranded RNAs has been proposed by T'so and Carter (8). These investigators working on the concept that polyI·polyC has proven too toxic for use in man, have developed complexes of polyribonucleotides that are more readily hydrolizable than is polyI·polyC. The reasoning was that such complexes would be stable just long enough to produce the biological change desired, and then would be inactivated by hydrolysis. They did this by inserting a molecule of uridylic acid for every 12th cytidylic acid in the poly C strand. This produced a more easily hydrolizable duplex which they call Ampligen. There have been a few fragmentary reports on clinical effects of Ampligen (25). There clearly is very little toxicity. There is also little or no interferon made. A few of the 10 patients studied may have shown a slight rise in NK cell activity. The limited amount of data do not allow for an evaluation of clinical effectiveness. Further studies are being comtemplated at this time.

While the idea of producing drugs with lower toxicity is, of course, desirable, toxicity was never a problem with polyI·polyC. Rather, ineffectiveness in primates was the problem. PolyICLC is very much more effective in primates than is polyI·polyC, but it is indeed more toxic. Derivatives of polyICLC can be prepared that are much more easily hydrolized than is the standard preparation, by using lower molecular weight components. Such derivatives are much less toxic than is the standard preparation made with higher molecular weight components, but unfortunately the more hydrolizable complexes are ineffective in primates. That this may prove to be the case with Ampligen is suggested by the observation that Ampligen does not induce interferon in man. On the other hand, neither does polyA·polyU, and it appears to have beneficial action in adjuvant therapy. It will be interesting to follow the course of clinical studies with Ampligen.

CONCLUSION

As has been noted, polyICLC has several activities in primates, including man. In addition to inducing the synthesis of significant quantities of interferon, it enhances NK cell and macrophage activity, increases the ratio of T helper to T suppressor cells, and augments antibody formation to a number of antigens. The cell-associated immune enhancing actions occur maximally at low doses of the drug, where side effects are very mild. The induction of large quantities of interferon (= or > 1000 u/ml of serum) require larger doses, ca.5-9 mg/m^2. Under these conditions immune enhancement is less than maximal, NK cell activity is usually decreased, and side effects such as fever and myalgia can be strong. Levels of interferon of 200-500 u/ml can be achieved with drug doses of ca 3-4 mg/m^2, where side effects are less and some immune enhancement is still found.

There are three different disease situations in which the use of polyICLC has potential:

I) - For the treatment of virus diseases, where interferon probably plays the major role, the use of the drug at 3-4 mg/m^2 is indicated. The significant side effects of intravenous administration of this level of drug would probably preclude its use for the treatment by this route of minor illnesses. Intramuscular use appears to give much milder adverse effects, and should be considered for treatment of viral diseases. Studies with rabies in monkeys indicated that i.m. treatment was very effective in post exposure prophylaxis (25).

II) - For the treatment of cancer, where immune modulation may play a major role, low levels of the agent, given i.m. should be investigated. Protocols for such studies are currently being written.

III) - For the treatment of AIDS, low doses should be assessed. Several factors suggest the use of polyICLC in this disease; - double stranded RNAs have long been reported as restoring immune competence to mice that had lost competence for a variety of causes (1). In addition, the improved ratio of T helper/T suppressor cells would be corrective of a problem in AIDS. The general augmentation of cellular immunity and antibody production would point in the same direction, while the interferon produced might be expected to exert some antiviral action.

REFERENCES

1. Biological Effects of Polynucleotides (Eds. R.F. Beers and W. Braun), Springer Verlag, N.Y., 1971.
2. Field, A.K., Tytell, A.A., Lampson, G.P., and Hilleman, M.R. Proc. Natl. Acad. Sci. U.S. 58:1004-, 1967.
3. Levy, H.B. Texas Rep. Biol. and Med. 35:91-98, 1977.
4. Young, C.W. Med. Clinics North America 55:721-728, 1971.
5. Robinson, R.A., DeVita, V.T., Levy, H.B., Baron, S., Hubbard, S. P. and Levine, A.S. J. Natl. Canc. Inst. 57:599-602, 1976.
6. Nordlund, J.J., Wolff, S.M. and Levy, H.B. Proc. Soc. Exp. Biol. Med. 133:439-444, 1970.
7. Levy, H.B., Baer, G., Baron, S., Buckler, C.E., Givvs, C.J., Iadarola, M.J., London, W.T. and Rice, J. J. Invect. Dis. 132: 434-439, 1975.
8. Tso, P.O.P., Alderfer, J.L., Levy, J., Marshall, L.W., O'Malley, 299-312, 1976.
9. Levy, H.B. and Riley, F. In: Lymphokines (Ed. E. Pick), Academic Press, Vol. 8, 1983, pp. 303-322.
10. Chirigos, M.A., Papademetriou, V., Bartocci, A., Read, E. and Levy, H.B. Immunopharmac. 3:329-337, 1981.
11. Talmadge, J.E., Adams, J., Philips, H., Collins, M., Lenz, B., Schenider, M., Schlick, E., Ruffman, R., Wiltrout, R.H., and Chirigos, M.A. Cancer Research 45:1058-1065, 1985.
12. Hartmann, D., Adams, J.S., Meeker, A.K., Schneider, M.A., Lenz, B.F., and Talmadge, J.A. Cancer Research, 46:1331-1338, 1986.
13. Maluish, A.E., Reid, J.W., Crisp, E.A., Overton, W.R., Levy, H. B., Foon, K.A., and Herberman, R.B. J. Biol. Res. Modifiers, 4:656-663, 1985.
14. Bever, C.T. Jr, Salazar, A.M., Neely, E., Ferraraccio, B.E., McFarlin, D.E., and Levy, H.B. Neurology, 36:494-498, 1986.
15. Lacour, J. J. Biol. Res. Modifiers, 4:538-543, 1985.
16. Levine, A.S., Siviluch, M., Wiernik, P.H., and Levy, H.B. Cancer Research 39:1645-1650, 1979.
17. Lampkin, B.C., Levine, A.S., Levy, H.B., Krivit, W. and Hammond, D. Cancer Research 45:5904-5909, 1986.
18. Leventhal, B.G., Kashiman, H., Levine, A.S., and Levy, H.B. J. Pediatrics, 99:614-616, 1981.
19. Theriault, J. and Hortobagyi, M. Personal communication.
20. Hawkins, M.J., Levin, M., and Borden, E.C. J. Biol. Res. Mod. 4:664-668, 1985.
21. Durie, B.G.M., Levy, H.B., Voakes, J., Jett, J.R. and Levine, A.S. J. Biol. Res. Mod. 4:651-655, 1985.
22. Engel, W.K., Cuneo, R. and Levy, H.B., Lancet, 1978.
23. Bever, C.T., Jr., McFarlin, D.E., MacFarland, H., and Levy, H.B. J. Int. Res. 5:423-428, 1985.
24. Brodsky, I., Strayer, D.R., Krueger, L.K. and Carter, W.A. J. Biol. Resp. Mod. 4:669-675, 1985.
25. Baer, G.M., Shaddock, J.H., Moore, S.A., Yaeger, P.A., Baron, S. and Levy, H.B. J. Inf. Dis. 136:58-62, 1977.

25

DELIVERY OF INTERFERONS TO TARGET CELLS BY MONOCLONAL ANTIBODIES: POTENTIAL FOR THE SPECIFIC TARGETING OF OTHER LYMPHOKINES

H.K. HOCHKEPPEL AND S.S. ALKAN, PHARMACEUTICALS RESEARCH DIVISION, CIBA-GEIGY LIMITED, CH-4002, BASEL, SWITZERLAND

INTRODUCTION

The successful cloning, production and purification of interferons (IFN) and other lymphokines (LKs) have revived interest in their therapeutic potential as both antitumor and immunomodulatory molecules. However, clinical studies with conventionally administered IFNs have produced ambiguous results. It was soon realized that both natural and recombinant IFNs have a very short half-life and at higher doses produced severe side effects. This is not surprising, because LKs are induced and probably also produced locally, released into microenviroments and they mostly act on neighboring cells (1). All the IFNs which are of about 20 Kd molecular weight act through their specific cell surface receptors. Since various kinds of cells in the body bear IFN receptors and almost all IFNs exert multiple biological activities, natural target cell specificity of IFNs is considered to be very limited. Consequently, systemically administered IFNs can be expected to produce unwanted side effects with only little therapeutic benefit because of inefficient and improper body distribution. All this necessitates novel approaches for the clinical application ofIFN or other LKs (2, 3, 4). One possible approach to overcome the above problems would be to deliver IFN to desired sites by antibodies (5, 6). As one example we shall summarize here our experimental results using monoclonal antibodies (mAb) specific for either viral antigens, tumor antigens or idiotypes as carriers of human interferons.

M. Revel (ed.), CLINICAL ASPECTS OF INTERFERONS.

Targeting IFNs via mAbs:

Details of experiments were described previously (5). MAb purified human recombinant IFNα/D subtype was coupled to mAbs specific either for Epstein Barr virus membrane antigen (EBV-MA), tumor antigens or an idiotype using a bifunctional cross-lingking reagent, N-succinimidyl 3-(2-pyridiyl dithio) propionate (SPDP) (5, 6). Uncoupled IFN was separated from IFN-mAb conjugates by gel chromatography. The conjugates retained specific antibody activity as tested by binding to their respective target cell lines. The IFN-mAb conjugates also retained the full biological activity of IFN as tested by standard antiviral and antiproliferative activity on several target cell lines as described previously (5).

Figure 1

r IFNα-D targeted by anti-EBV MAbs
(15 minutes pulse at 4 °C)

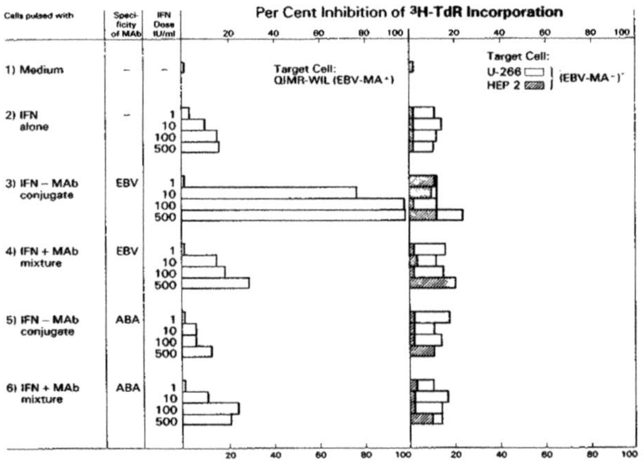

To demonstrate the specific targeting of IFN via mAb a series of in vitro pulsing experiments were performed at 4°C for only 15 minutes, a condition that did not allow free IFN to bind to its receptor, but without effecting the mAb binding to target cells. As can be seen in Fig. 1, a minimum amount of IFN-mAb conjugate (5-10 units/ml) exerted a pronounced antiproliferative action on a cell line (QIMR-WIL) which is EBV-MA$^+$, but not on EBV-MA$^-$cells (U266,

Hep2). Free IFN, IFN/mAb mixture or an irrelevant IFN-mAb conjugate had no effect. In a similar way the antiviral action of IFN-mAb conjugates was demonstrated (5). Only EBV-MA$^+$ target cells pulsed with IFN-mAb, but not with free IFN, were protected from lysis with mengo virus. Recently, we performed pulsing experiments with radiolabelled conjugate as described above to quantitatively compare the efficiency of the IFN-mAb conjugate with that of free IFN. In these experiments the cells were incubated continuously for 2 days with various doses (measured by cpm and not units) of free or mAb coupled IFNs. Results in Table 1 demonstrate that mAb linked IFN exerts an antiproliferative activity already at much lower doses of IFN on EBV-MA$^+$ target cells than free IFN. Also the antiviral activity of IFN-mAb conjugate was clearly superior to free IFN. Currently, we are utilizing also other mAb-target cell systems such as anti-idiotypic mAbs as carrier of IFN to idiotypic positive hybridoma cells or anti-glycosphigolipid mAbs as carrier of IFN to colon carcinoma cells and obtained similar results as obove. In vivo experiments are in progress (Alkan, Hochkeppel and Towbin, to be published).

We have recently extended our studies to other modes of targeting interferons by mAb. It is now possible to prepare bispecific mAbs either by means of biological (hybrid hybridomas) (7) or by chemical (8) methods. Such $(Fab')_2$ molecules with double specificities (one for IFN another for a tumor antigen) would not only deliver exogenous but also endogeneous to desired sites. Our preliminary experimental results with bispecific mAb-LK targeting are encouraging.

Fig. 2 shows, among others, our currently available bi-specific mAb conjugates which have already been tested for their ability to transport IFN to mouse hybridoma target cells possessing idiotypic mAb on their cell surface as specific tumor markers. In all cases mAb delivered IFN exerted a clearly stronger antiproliferative activity than untreated IFN on target cells.

Table 1

Comparison of anti-proliferative effects of IFN and IFN-MAb conjugate

	QIMR-WIL CELLS (EBV+)			U-266 CELLS (EBV−)	
Cells incubated with	IFN dose (cpm)	Viable cells/ml ($\times 10^{-4}$)	% inhibition of growth	Viable cells/ml ($\times 10^{-4}$)	% inhibition of growth
Medium	0	110	0	104	0
IFN alone	10^1	108	2	108	0
	10^2	109	1	110	0
	10^3	110	0	93	11
	10^4	114	0	92	11
Medium	0	128	0	92	0
IFN-MAb conjugate	10^1	92	28	88	4
	10^2	80	38	89	3
	10^3	50	61	92	0
	10^4	39	70	88	4

Figure 2

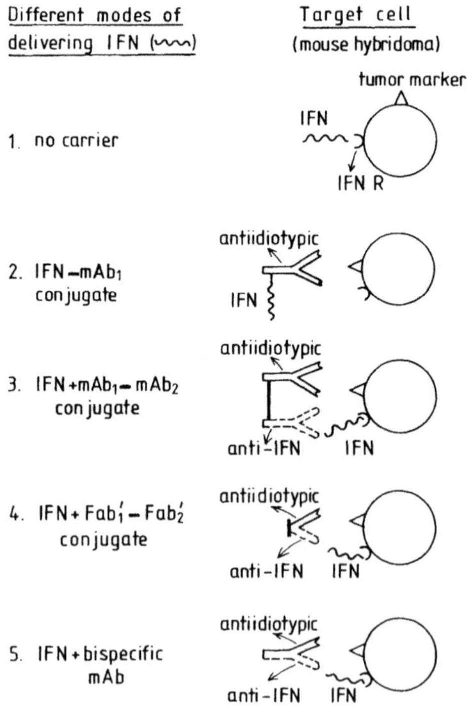

Table 2

BINDING AFFINITY OF ANTIBODY CONJUGATED RADIOLABELLED IFNα BDDD OR IFNα BDDD TRANSPORTED BY DOUBLE ANTIBODY CONJUGATE TO ID+ 10K44 TARGET CELLS IS INCREASED IN COMPARISON TO FREE IFNα BDDD

CELLS INCUBATED WITH	IFN DOSE (CPM)	EXCESS OF INHIBITOR ADDED (COLD IFNα)	TIME OF INCUBATION AT R.T. (MIN.)	CPM BOUND	% REPLACEMENT
1. IFNα ALONE	5×10^4	0	60	3265	0
	"	20	"	940	73.3
	"	200	"	720	77.9
2. IFNα-MAB$_1$ CONJUGATE	"	0	"	6364	0
	"	20	"	3789	40.5
	"	200	"	3390	46.8
3. IFNα + MAB$_1$-MAB$_2$ CONJUGATE	"	0	"	5976	0
	"	20	"	3362	43.7
	"	200	"	3043	49.9

In addition, antibody targeted IFN had a higher affinity to the IFN receptor on mouse hybridome cells (Table 2). The bi-specific antibody conjugates were specifically developed for IFN or other LKs which are too labile for direct crosslinking.

Proposed advantages of LK-targeting

The mechanism of IFN-mAb action outlined above is unknown. However, some of the characteristics and the differences between LK-targeting and conventional drug-targeting are obvious. First, IFN and other LKs are biomolecules all of which act through specific surface receptors on target cells. One important consequence of this is that once targeted, a LK will exert its effect via its receptor, without the need of antibody-mediated endocytosis. In contrast, a drug-Ab conjugate has to enter into every target cell which is in practice almost impossible to achieve. If a target cell modulates the surface antigen and/or sheds it, a drug-Ab conjugate loses its potential effect totally. In contrast, a LK-Ab conjugate will also be active on cells in the vicinity of the target site. Secondly, even mAbs are not strictly specific for tumor/target cells, they can best be classified as semi-specific. This creates a great problem when toxic substances are being targeted via Abs. However, since LKs, especially at lower doses, are not toxic, semi-specific mAbs should not cause the same serious problems as in drug-targeting. Thirdly, most LKs exert multiple biological effects which may allow combination therapy with a single LK. For instance, IFNs stimulate both specific (T and B lymphocytes) and non-specific (MØs, NK cells) host defence mechanisms. In experimental models IFN-mAb conjugates have been shown to stop the growth of target tumor cells (5) and also activate NK cells (6). Fourthly, conjugation of LKs might stabilize the LKs, and in vivo prolong their half-life which is normally very short. Lastly, since mAb will increase the efficacy of LKs, lower doses will be required for therapy which in turn reduces their side effects. In summary, although the LK-targeting method might also suffer from some of the classical problems of drug-delivery (unspecific uptake by liver, RES, antigenicity of Abs) we believe that it brings distinct advantages over the conventional concept of drug/toxin-targeting.

References

1. Bocci, V. Immunomodulators as local hormones: New insight regarding their clinical utilization J. Biol. Resp. Modif. 4:340-352, 1985.

2. Gregoriadis, G. Targeting of drugs with molecules, cells and liposomes. TIPS 4:304-307, 1983.

3. Koff W.C.,Fidler, I.J., Showalter, M.K., Chakrabarty, M.K., Hampar, B. Ceccorulli, L.M. and Kleinermann, E.S. Human monocytes activated by immunomodulators in liposomes lyse herpesvirus-infected but not normal cells. Science 224:1007-1009, 1984.

4. Eppstein, D.A., Mersh, V., Van Der Pas, M., Felgner, P.L. and Schreiber, A.B. Biological activity of liposome encapsulated murine interferon γ is mediated by a cell membrane receptor. Proc. Natl. Acad. Sci USA 82:3688-3692, 1985.

5. Alkan, S.S., Miescher-Granger, S., Braun, D.G. and Hochkeppel, H.K.,Antiviral and antiproliferative effects of interferons delivered via monoclonal antibodies, J. Interferon Res. 4: 355-363, 1984.

6. Pelham, J.M., Gray, J.D., Flannery, G.R., Pimm, M.V. and Baldwin R.W., Interferon conjugation to human osteogenic sarcoma monoclonal antibody 791T/36, Cancer Immunol. Immunother. 15:210-216, 1983.

7. Milstein, C. and Cuello, A.C. Hybrid hybridomas and their use in immunochemistry. Nature 305:537-540, 1983.

8. Brennan, M., Davison, P.F. and Paulus, H, Preparation of bispecific antibodies by chemical recombination of monoclonal IgG1 fragments. Science 229:81-83, 1985.

26

IMMUNOMODULATORY ACTIONS OF INTERFERONS

H.K. Hochkeppel and S.S. Alkan

Ciba-Geigy Ltd, CH-4002 Basel, Switzerland

INTRODUCTION

The growth of scientific knowledge on interferons (IFN) has been accelerated by hybridoma and recombinant DNA technologies. In the past, among many biological effects of IFNs, anti-viral and anti-tumor activities received much attention and only recently have publications started to accumulate emphasizing the immunoregulating abilities of IFNs. While we refer the reader to earlier extensive reviews (1-4) in this short review we wish to summarize very recent data which deal with the effect of IFNs on various immunocytes with a special emphasis on IFNγ. The immune system which has evolved to protect the individual from infectious agents has to solve four major problems: The first is to discriminate self from non-self. The second is to mount a proper class of immune response so that different kinds of antigens can be efficiently eliminated. The third is regulation and orchestration of its own responses. The fourth is the generation of diversity in order to recognize an almost infinite number of antigens (each specifically). Available evidence indicate that IFNs influence the immune system in solving the first three problems. Here the T lymphocytes seem to be crucial cells, because they can help or suppress the functions of other lymphocytes. It is known that T cells can only recognize foreign antigens in association with self-antigens whose expression is regulated by IFNs. Among the cells affected by IFNs are antigen presenting cells (macrophages, dendritic cells, B cells etc.) and T lymphocytes which play a central role in regulating immunity because they produce, beside IFNs, other lymphokines (LK), such as interleukin 1 and 2. Consequently, the actions of IFNs on various cells of the immune system are imbedded in a network of LK actions with positive/negative feedback systems and interaction cascades.

M. Revel (ed.), CLINICAL ASPECTS OF INTERFERONS.

I. REGULATION OF MACROPHAGE ACTIVITIES BY IFNs

Endogeneous interferons protect macrophages against viral infections (5) and blocking of endogeneous IFNs by antibodies reduces the resistance against murine tumors (6). IFNα, ß and γ can render macrophages tumoricidal for various tumor cells or cell lines in vitro (7-12), however IFNγ seems to act at a dose one hundred times lower than IFNα or ß (7) and also has been shown to act in vivo (11) (Table 1). In this context, the arguments as to whether or not IFNγ and macrophage activation factor (MAF) are identical (8,10), must await more specific assay systems. In addition, there may exist other LKs with IFN-like activities which may effect either different subpopulations or differentiation stages of monocytes.

Table 1

Immunomodulatory actions of interferon

action (in vitro)	IFNα	IFNß	IFNγ
monocyte/macrophage:			
- induction of tumoricidal activity	+	+	++
- induction/increase of unspecific phagocytosis	+	+	+
- induction/increase of FcR mediated phagocytosis	(+)	(+)	++
- induction of ADCC	?	?	++
- induction and/or increased production of IL1	+	+	++
- increased production of PA	-	-	+
- increased production of TNF	-	-	+
- suppression of CSF activity	+	+	+
- induction of MHC class I (H-2/HLA) expression	+	+	++
- induction of MHC class II (I-A) expression (also: endothelial, mast, neuronal, fibroblast and various tumor cells)	-	-	++

Apparently all interferons increase unspecific phagocytosis (13-16) of macrophages, whereas the Fc receptor mediated phagocytosis is mainly induced by IFNγ by augmenting the number and density of Fc receptors on the surface of macrophages (3,17-20). Via this receptor, IFNγ, and perhaps also IFNα or ß, induce antibody dependent cellular cytotoxicity (ADCC) (21-23). Recently it has been shown that there are two different Fc receptors, a low and a high affinity form, and only the high affinity receptor is responsible for the ADCC. IFN appears to act by converting the low affinity into the high affinity form (24). In a very recent study it has been demonstrated that mouse lung macrophages, which are normally unresponsive to activation for tumor cytotoxicity by lipopolysaccharide (LPS), become responsive to LPS following treatment with IFNγ (25).

IFNγ - and probably also IFNα or ß - augment the production of interleukin 1 (IL1) or even directly induce this lymphokine at higher doses (12, 26). IL1 induces interleukin 2 receptor (IL2 R) expression on T- helper (Th) cells (27). This receptor is responsible for T and perhaps B cell proliferation (28-32). An antibody against this receptor can block the induction of IFNγ by mitogens (33). In addition, IFNγ increases the production of tissue plasminogen activator (PA) and tumor necrosis factor (TNF) in macrophages (34, 35) whereas IFNα, ß and γ suppress the activity of colony stimulating factor (CSF) (36). This latter effect, however, is questionable since experiments concerning the influence on CSF activity were performed with crude natural IFNs.

All three IFN species induce MHC class I antigen expression, that is H-2 in mice and HLA - A, B, C in man (37-40). As with the tumoricidal activity IFNγ acts at much lower doses than IFNα or ß (39) and also only IFNγ induces significant expression of MHC class II antigens (41-48), that is Ia in mice and HLA-DR in humans. As well as its effect on monocytes, IFNγ is able to induce MHC class II antigen on endothelial, mast, neuronal, fibroblast and various tumor cells (49-53). In a recent study it was demonstrated that IFNγ modulates HLA-DR and DC antigens of cultured thymic epithelial cells (54). The above findings collectively indicate that IFNγ is not only able to regulate the Ia of antigen presenting cells (peripheral effects), but also has the potential to influence the thymus with important consequences in terms of the education of immunocytes for self-recognition (central effect). IFNα was recently shown to antagonize IFNγ in the induction of MHC class II antigens (78).

II. EFFECT OF IFNs ON NATURAL KILLER (NK) CELLS AND LYMPHOCYTES

IFNα, ß and γ activate NK cells to be tumoricidal <u>in vitro</u> (55-57), however at the present time little is known about the <u>in vivo</u> effect of IFNs on NK cells (<u>Table 2</u>). There is however, growing evidence suggesting that one of the functions of NK cells is to control hematopoiesis. It has been shown that IFNγ synergistically acts with NK cell derived colony inhibiting activity (58), and recent data indicate that interferons might be mediators of hematopoietic suppression in aplastic anemia <u>in vivo</u> (59).

Table 2

Immunomodulatory actions of interferon (continued)

action	IFNα	IFNβ	IFNγ
in vitro			
NK cells:			
- induction of cytotoxicity	+	+	+
lymphocytes:			
- induction of B-cell differentiation	-	-	+
- augmentation or inhibition of antibody production	+	+	+
in vivo			
- augmentation of antibody production	-	-	+
- increase of Th cell activity	-	-	+

Figure 1

Interaction of IFNγ with immunocytes

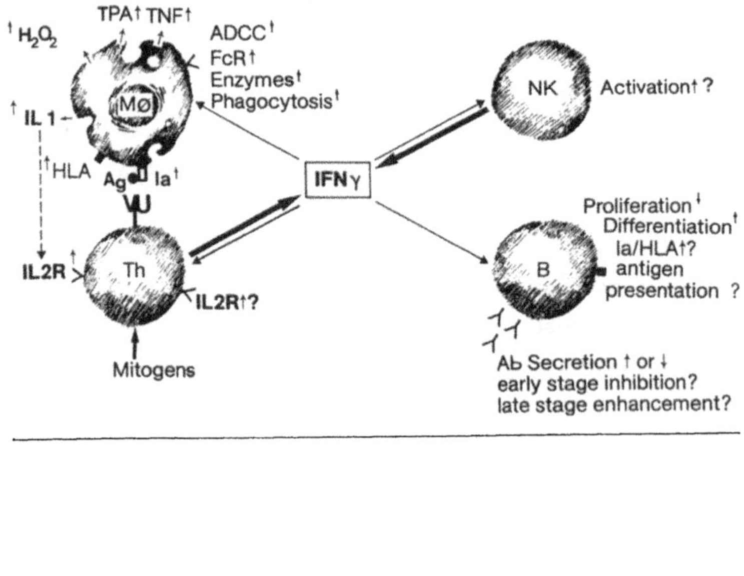

Also Interferon accelerates autoimmune diseases in mice (60). IFNγ seems to be one of the B cell differentiations factors (61-65) and there are also several reports showing that IFNα, β or γ can either inhibit or enhance antibody production (66-70). These effects seems to be very dependent on the differentiation stage of the B cells affected and on the dose of IFN applied (71). In vivo, at least IFNγ seems to enhance antibody formation if added simultaneously with the antigen (69, 72) and there are indications that this augmentation occurs via direct stimulation of T-helper cells.

III. IFNγ AND THE LK/IMMUNOCYTE NETWORK

In Fig. 1 we have schematically illustrated the interactions of IFNγ with the various cells of the immune system. In the context of these interactions several questions remain to be resolved, for example the direct effect of IFNγ on Th cell activity, IL2 R expression and Ia expression and antigen presentation by B cells. Is IFNγ also the mediator responsible for these effects? As already mentioned above IFNγ induces IL1 in macrophages which directly effects the Th cells to express IL2 R and to produce IL2 itself which would then induce IFNγ generation (32). IFNγ induces expression and augmentation of Ia on macrophages, and thereby might regulate antigen presentation which is a crucial event for effective cellular and humoral immunity. An open question is at what stage IFNγ augments or inhibits the antibody production of B cells as it is still very difficult to isolate pure B cells which are at specific differentiation stages. There are recent indications that IFNγ is probably one of the late stage B cell differentiation factors (62-65).

According to G.Wong et al., (unpublished results, 1985 TNO-ISIR Meeting) IFNγ might directly induce the production of TNF and lymphotoxin (LT), and TNF as well as LT induce the production of IFNγ in Th cells. This positive feedback might ensure that IFNγ and TNF or LT, which seem to act synergistically (70), are induced simultaneously. If these results hold, IFNγ could indirectly induce its own production a) via the IL1-IL2 pathway and b) via TNF or LT induction. Taken together the above results clearly demonstrate the complexity of LK interactions and therefore, in order to use IFNγ or other LKs for the correction of immunological disorders, one might think of delivering specific LKs to desired organs, tissues or cells via carrier molecules such as monoclonal antibodies (73) or liposomes.

IV. ROLE OF IFNγ IN APC-T-B CELL INTERACTIONS

If one looks at the cellular interactions occuring during an immune response, one can differentiate two kinds of APC-T-cell interrelation. The interaction of antigen presenting cells (APC) with T-killer cells (Tk) is MCH class I restricted and is regulated by all three species of IFNs (α, β, γ) (37-40). In contrast, the interaction of APC with Th cells is MHC class II restricted and mainly regulated by IFNγ, and not by IFNα or β (49-53). The question as to whether the interaction of T-helper with B cells, a process which is also MHC class II restricted, is regulated by IFNγ is currently being investigated and a recent study showed that murine IFNγ enhances helper T-cell activity both in vitro and in vivo (72). An adjuvant activity of human IFNα during hepatitis B vaccination has also been reported (74). In addition, alterations in the expression of selected class I and II histocompatibility antigens in a variety of tissues of mice after treatment with recombinant IFNγ have been reported in detail (75). IFNγ caused a five to fifteen fold increase in MHC I (H-2K) and MHC II (I-A) antigens throughout the body. It was concluded that the dynamic and selective induction of class I and II antigen expression by IFNγ is likely to exert an effect on the immune response at many tissue sites. Whether or not IFNγ can influence the type and the class of immune response remains to be determined.

V. EFFECT IFNγ ON B CELL DIFFERENTIATION

The development of B cells from stem cells can be divided into two major differention stages, an antigen independent pre-B cell and an antigen dependent B cell stage. During the antigen independent phase precursor B cells are generated which finally develop into resting B cells. In a second stage, resting B cells are activated leading to FcR expression and Ia-antigen augmentation and subsequent differentiation into plasma cells. Early studies showed that IFNα/β can inhibit the antibody response in vitro (71) and a recent study demonstrated that murine IFNα/β inhibited the cell growth and antibody secretion of plasmocytomas and hybridomas (76). In contrast IFNγ seems to act at a rather late stage of B cell development in combination with several other factors (i.e. TRF, BSF1) (77) driving the B cell towards final differentiation and antibody production (62-65).

CONCLUSIONS

IFNs, and specifically IFNγ, seem to play a central role within the network of lymphokine interactions regulating immunocyte actions. Whereas all three IFNs exerts multiple effects on macrophages, mainly IFNγ seems to have a direct effect on T and B cells. Since IFNγ seems to be interlinked with the induction and action of other lymphokines it is likely that it plays a critical role in the body's defense mechanism against foreign antigens, and a failure of IFNγ production (high or low) may therefore significantly imbalance the whole immune system. Because there exist positive/negative feedback regulations between interferons and interleukins we feel that management of certain autoimmune diseases (such as systemic lupus erythematosis or rheumatoid arthritis) should be possible by treatment with LKs or LK antagonist molecules.

Abbreviations:

ADCC:	Antibody Dependent Cellular Cytotoxicity
APC:	Antigen Presenting Cells
CSF:	Colony stimulating Factor
IFN:	Inferferon
IL1:	Interleukin 1
IL2:	Interleukin 2
LK:	Lymphokine
LT:	Lymphotoxin
MAF:	Macrophage Activating Factor
MHC:	Major Histocompatibility Complex
NK:	Natural Killer Cells
Th:	T-helper cell
Tk:	T-killer Cell
TNF:	Tumor Necrosis Factor
PA:	Tissue Plasminogen Activator

REFERENCES

1. Schultz, R.M. Lymphokine Rep. 1:63-97, 1980.
2. Friedman, R.M. and Vogel. S.N. in: Advances in Immunology (eds. Dixon F.J., and Kunkel, H.G), pp. 97-140, Academic Press, N.York, London 1983.
3. Trinchieri, G. and Perussia, B. Immunology Today 6, 131-136, (1985)
4. DeMaeyer-Guignard, J. and DeMaeyer, E. Interferon (London) 6: 69-91, 1985.
5. Belardelli, F., Vignaux, F., Proietti, E. and Gresser, I. Proc. Natl. Acad. Sci. USA 81:602-606, 1984.
6. Gresser, I., Belardelli, F., Maury, Ch., Maunoury, M.-T. and Tovey, M.G. J. Exp. Med. 158:2095-2107, 1983.
7. Varesio, L., Blasi, E., Thurman, G.B., Talmadge, J.E., Wiltrout, R.H and Herberman, R.B. Cancer Res. 44:4465-4469, 1984.
8. Nathan, C.F., Prendergast, Th.J., Wiebe, M.E., Stanley, E.R., Platzer, E., Remold, H.G., Welte, K., Rubin, B.Y. and Murray, H.Y. J. Exp. Med. 160:600-605, 1984.
9. Celada, A., Gray, P.W., Rinderknecht, E. and Schreiber R.D. J. Exp. Med. 160:55-74, 1984.
10. Peters, P.M., Svedersky, L.P., Shalaby, M.R. and Palladino, jr. M.A. Lymphokine Res., 4:265-273, 1985.
11. Murray, H.W., Spitalny, G.L. and Nathan, C.F. J. Immunol. 134: 1619-1622, 1985.
12. Arenzana-Seisdedos, F.A., Virelizier, J.L. and Fiers, W.J. Immunol. 134:2444-2448, 1985.
13. Rollag, H. and Sonnenfeld, G. Scand. J. Immunol. 20:149-155. (1984)
14. Rollag, H. and Degrê, M. Acta Pathol. Microbiol. Scand. Sect. B, 89:153, 1981.
15. Degrê, M., Rollag, H., Melby, K., Midtredt, T., Morland, B., Sonnenfeld, G. and DeLey, M. In: The Biology of the Interferon System (DeMaeyer, E., Galasso, G. and Schellekens, H. eds.); Elsevier North-Holland, New York, page 161, 1981.
16. Wang, E., Michl, J., Pfeffer, L., Silverstein, S., and Tamm, I. J. Cell Biol. 91:238 a (Abstract 13026), 1981.
17. Perussia, B., Dayton, E.T., Lazarus, R., Fanning, V. and Trinchieri, G. J. Exp. Med. 158:1092-1113, 1983.

18. Vogel, S.N., English, K.E., Fertsch, D. and Fultz, M.J. J. Interferon Res. 3:153, 1983 a.

19. Vogel, S.N., Finbloom, D.S., English, K.E., Rosenstreich, D.L. and Langreth, S.G. J. Immunol. 130:1210, 1983 b.

20. Fertsch, D. and Vogel, S.N. J. Immunol. 132:2436-2439, 1984.

21. Hokeland, P. and Berg, K. J. Immunol. 127:1585-1588, 1981.

22. Lang, N.P., Ortaldo, J.R., Bonnard, G.D. and Herbermann, R.B. J. Natl. Cancer Res. Inst. 69:339-343, 1982.

23. Akiyama, Y., Lubeck, M.D., Steplewski, Z. and Koprowski, H. Cancer Res. 44:5127-5131, 1984.

24. Gergely, Y. 7th European Immunology meeting of E.F.I.S., Jerusalem, Abstract S03, p. 35, 1985.

25. Akagacsa, K. and Tokunaga, T. J. Exp. Med. 162:1444-1459, 1985.

26. Baraschi, D., Censini, St. and Tagliabue, A. J. Immunol. 133:-764-767, 1984.

27. Mizel, S.B. Immunol. Rev. 63:51-72, 1982.

28. Zubler, R.H., Lõwenthal, J.W., Erard, F., Hashimoto, N., Devos, R. and MacDonald, H.R. J. Exp. Med. 160: 1170-1183, 1984.

29. Bettens, F., Kristensen, F., Walker, C., Bonnard, G.D. and DeWeck, A.L. Cell. Immunol. 86:337-346, 1984.

30. Bismuth, G., Leclercq, L., Duphot, M., Moreau, J.L. and Theze, J. Cell. Immunol. 86:90-100, 1984.

31. Reem, G. and Yeh, N. J. Immunol. 134:953-958, 1985.

32. Vilcek, J., Henriksen-Destefano, D., Siegel, D., Klion, A., Robb, R.J. and Le J. J. Immunol. 135:1851-1856, 1985.

33. Croll, A.D., Wilkinson, M.F. and Morris, A.G. Cell. Immunol. 92: 184-189, 1985.

34. Jones, C.M., Varesio, L., Herbermann, R.B. and Pestka, S. J. Interferon Res. 2:377-386, 1982.

35. Svedersky, L.P., Nedwin, G.E., Goeddel, D.V. and Palladino, M.A. J. Immunol., in press, 1985.

36. Klimpel, G.R., Fleischmann, W.R. and Klimpel, K.D. J. Immunol. 129: 76, 1982.

37. Fellous, M., Nir, U., Wallach, D., Merlin, G., Rubinstein, M. and Revel, M. Proc. Natl. Acad. Sci. USA, 79:3082-3086, 1982.

38. Houghton, A.N., Tomson, T.M., Gross, D., Oettgen, H.F. and Old, L.J. J. Exp. Med. 160:255-269, 1984.

39. Ball, E.D. Guyre, P.M., Glynn, J.M. Rigby, W.F.C., and Fanger, M.W. J. Immunol. 132:2424-2428, 1984.

40. Ozato, K., Wan, Y.J. and Orrison, B.M. Proc. Natl. Acad. Sci. USA, 82:2427-2431, 1985.

41. Basham, T. and Merigan, Th. J. Immunol. 130:1492-1494, 1983.

42. King, D.P. and Jones, P.P. J. Immunol. 131:315-318, 1983.

43. Kelley, V.E., Fiers, W. and Strom, T.B. J. Immunol. 132:240-245, 1984.

44. Sztein, M.B., Steeg, P.S., Johnson, H.M. and Oppenheim, J.J. J. Clin. Invest. 73:556-565, 1984.

45. Koch, N., Wong, G.H. and Schrader, J.W. J. Immunol. 132:1361-1368, 1984.

46. Becker, S. Cell. Immunol. 91:301-307, 1985.

47. Capobianchi, M.R., Ameglio, F., Tosi, R. and Dolei, A. Human Immunol. 13:1-11, 1985.

48. Ling, P.D., Warren, M.K. and Vogel, S. J. Immunol. 135:1857-1863, 1985.

49. Ameglio, F., Capobianchi, M.R., Dolei, A. and Josi, R. Infect. Immun. 42:122-125, 1983.

50. Houghton, A.N., Johnson, T.M., Gross, D., Oettgen, H.F. and Old, L.J. J. Exp. Med. 160:255-269, 1984.

51. Schwartz, R., Momburg, F., Moldehauer, G., Dörken, B. and Schirrmacher, V. Int. J. Cancer 35:245-250, 1985.

52. Carrel, S., Schmidt-Kessen, A. and Giuffrê, L. Eur. J. Immunol. 15:118-123, 1985.

53. Anböck, J., Niederwieser, D., Romani, N., Fritsch, P. and Huber, C. Arch. Dermatol. Res. 277:270-275, 1985.

54. Berrih, S., Arenzana, Seisdedos, F., Cohen, S., Devos, R., Charron, D. and Virelizier, J.-L. J. Immunol. 135:1165-1171, 1985.

55. Weigent, D.A., Langford, M.P., Fleischmann, W.R.jr. and Stanton, G.J. In: Human Lymphokines (Khan, A. and Hill, N.O. eds.), pp. 539-550, Academic Press, New York and London, 1982.

56. Herberman, R.B., Ortaldo, J.R., Mantovani, A., Hobbs, D.S., Kung, H.F. and Pestka, S. Cell. Immunol. 67:160-167, 1982.

57. Kirchner, H. Progress Clin. Biochem. and Med. 1, 192-193, 1984.

58. Degliantoni, G., Murphy, M., Kobayashi, M., Francis, M.K., Perussia, B. and Trinchieri, G. J.Exp. Med. 162:1512-1530, 1985.

59. Zoumbos, N.C., Gascon, P., Dijeu, J.Y. and Young, N.S. Proc. Natl. Acad. Sci. USA 82:188-192, 1985.

60. Adam, C., Thoua, Y., Ronco, P., Verroust, P. and Tovey, M. Clin. exp. Immunol. 40:373-382, 1980.

61. Nakamura, M., Manser, T., Pearson, G.D.N., Daley, M.J. and Gefter, M.L. Nature (London) 307:381-382, 1984.

62. Leibson, H.J., Gefter, M., Zlotnik, A., Marrak, Ph. and Kappler, J.W. Nature (London) 309:799-801, 1984.

63. Sidman, C.L., Marshall, J.D., Shultz, L.D., Gray, P.W. and Johnson, H.M. Nature (London) 309:801-803, 1984.

64. Falcoff, R.J.M., Butler, J.L., Dinarello, C.A. and Fauci, A.S. J. Immunol. 133:692-696, 1984.

65. Nakagawa, T., Hirano, T., Nakagawa, N., Yoshizaki, K. and Kishimoto, T. J. Immunol. 134:959-966, 1985.

66. Murray, P.D., Swain, S.L. and Kagnoff, M.F. Fed. Proc. 44:1293 (Abstract), 1985.

67. Shalaby, M.R., Weck, P.K., Rinderknecht, E., Harkins, R.N., Frane, J.W. and Ross, M.J. Cell. Immunol. 84:380-392, 1984.

68. Pelton, B.K. and Denman, A.M. Clin. exp. Immunol. 59:398-404, 1985.

69. Aune, T.M. Int. J. Immunopharmac. 7:65-71, 1985.

70. Neubauer, R.H., Goldstein, L., Rabin, H. and Stebbing, N. J. Immunol. 134:299-304, 1985.

71. Gisler, R.H., Lindahl, P. and Gresser, I. J. Immunol. 113:438-444, 1974.

72. Frasca, D., Adorini, L., Landolfo, S. and Doria, G. J. Immunol. 134:3907-3911, 1985.

73. Alkan, S.S., Miescher-Granger, S., Braun, D.G. and Hochkeppel, H.K. J. Interferon R. 4:355-363, 1984.

74. Grob, P.J., Joller-Jemelka, H.I., Binswanger, K., Zaruba, K., Descoeudres, C. and Fernex, M. Eur. J. Clin. Microbiol. 3:195-198, 1984.

75. Skoskiewicz, M.J., Coloin, R.B., Schneeberger, E.E. and Russell, P.S. J. Exp. Med. 162:1645-1664, 1985.

76. Bhayani, H. and Hudson, L. J. Interferon Res. 4:41-50, 1984.

77. Vitetta, E.S., Ohara, J., Myers, C., Layton, J., Krammer, P.H., and Paul, W.E. J. Exp. Med. 162:1726-1731, 1985.

78. Ling, P.D., Warren, M.K. and Vogel, S.N. J. Immunol. 135:1857-63, 1985.

27

SYNERGISTIC ACTIONS OF INTERFERONS AND TUMOR NECROSIS FACTOR

R. FEINMAN and J. VILČEK

Department of Microbiology, New York University Medical Center, 550 First Avenue, New York, NY 10016

INTRODUCTION

Although the clinical application of interferons has resulted in some notable therapeutic successes, most common forms of cancer are proving to be resistant to interferon therapy. Failure of single-agent therapy with interferons in many neoplastic diseases has led to attempts to combine interferon with chemotherapeutic agents or with other cytokines. A cytokine widely considered as a potential anti-tumor agent is tumor necrosis factor (TNF). Although clinical trials with TNF have started only recently (1), various combinations of TNF and interferons have already been evaluated quite extensively in preclinical studies, both in tissue culture and in animals. Results of these studies will be reviewed in this article.

Tumor necrosis factor (TNF), is a monocyte/macrophage-derived protein, first detected in the sera of animals injected sequentially with Bacillus Calmette-Guerin (BCG) and endotoxin (2). Sera from such animals caused a hemorrhagic necrosis of Meth A sarcoma and some other transplanted tumors _in vivo_ and inhibited the growth of some transformed cells in culture (3). Recently, human TNF was purified to homogeneity (4) and its cDNA was cloned and expressed in _E. coli_ (5-7). TNF is functionally and structurally related to lymphotoxin (LT), a lymphocyte-derived protein (8-10). (Some authors have proposed that LT be designated TNF-beta and the "original" TNF be termed TNF-alpha.) Interestingly, the structural genes for both TNF and

M. Revel (ed.), CLINICAL ASPECTS OF INTERFERONS. Copyright 1988. Kluwer Academic Publishers, Boston. All rights reserved.

LT were shown to be linked to the major histocompatibility complex genes on the short arm of human chromosome 6 (11) or murine chromosome 17 (12).

Until a few years ago TNF and LT were considered selective and specific inhibitors of tumor cell growth. However, when homogeneous preparations of recombinant *E. coli*-derived TNF became available, it has become apparent that TNF is a multifunctional cytokine mediating a wide range of biological actions. (The realization that TNF is more than a tumor necrosis factor is reminiscent of the earlier recognition that interferons are not the strictly specific antiviral agents they had once been considered.) The first surprising activity found to be mediated by TNF was its ability to increase serum triglyceride levels in animals and to inhibit lipoprotein lipase activity in cultured cells. These activities were earlier ascribed to an LPS-induced, macrophage-derived protein, termed "cachectin"; it is now clear that cachectin and TNF are identical (13). Cachectin/TNF appears to be an important mediator of endotoxin-induced shock in Gram-negative infections (14). The long list of other biological actions mediated by TNF includes a pyrogenic action *in vivo*, activation of granulocyte, monocyte and T cell functions, regulation of hemostatic properties in endothelial cells, stimulation of collagenase and prostaglandin E_2 production in fibroblasts, and stimulation of the expression of several cellular genes, including HLA-A,B,C and HLA-DR in some tumor cell lines. These and other actions of TNF have been reviewed elsewhere (3,14,15). Although TNF is cytolytic or cytostatic for some tumor cells, in several normal cell lines and in some other tumor cells TNF was found to exert a growth factor-like mitogenic effect (16,17). This multitude of TNF's biological actions indicates the importance of TNF in physiological and pathophysiological processes, but it also points to potential problems in attempts to use it therapeutically as an antitumor agent.

CYTOTOXIC/CYTOSTATIC AND ANTI-TUMOR ACTIONS OF TNF

Like the interferons, TNF affects tumor cell growth both directly and indirectly. The cytotoxic/cytostatic activities seen in cultured tumor cells are the result of a direct action of TNF on the target cell. In contrast, the hemorrhagic necrosis seen in Meth A sarcoma-carrying mice injected with endotoxin or with TNF (16) and the action of TNF on some other tumors in experimental animals is apparently the result of an indirect action of TNF.

That TNF and LT can be directly cytostatic or cytotoxic for tumor cells has been known for many years, and yet, the mechanism of this action is poorly understood. It is apparent that not all tumor cells are susceptible to an inhibitory action of TNF; in fact, approximately 60% of the tumor cell lines examined are resistant (6,16). In most cell lines resistance is not due to an absence of TNF receptors or a lower binding affinity for the ligand, because most TNF-sensitive and resistant lines were found to have a similar number of cell surface binding sites (ranging from a few hundred to several thousand), with an apparent Kd in the range of 200 pM (18,19). Interaction of TNF with the cellular receptor appears to be necessary for the cytotoxic action, because microinjection of TNF directly into the cytoplasm or nucleus of TNF-sensitive L929 cells failed to produce cytotoxicity or growth inhibition (20). It has been suggested that the cytostatic/cytotoxic action might be mediated by a TNF-induced release of prostaglandins, proteases or free radicals (21). Others have suggested DNA fragmentation (22) labilization of lysosomal membranes with an ensuing release of lysosomal proteases, and phospholipase or protease activation as the basis of TNF cytotoxicity (21).

Metabolic inhibitors such as actinomycin D and mitomycin C greatly increase the susceptibility of cells to TNF cytotoxicity (23-25). A similar increase in TNF cytotoxicity was seen in tumor cells incubated at 39°-40° C, instead of 37° C (26-28). It appears that the greatest enhancement of cellular susceptibility to the cytotoxic action of TNF is seen in the presence of inhibitors of protein synthesis, e.g., cycloheximide (23). Nonmalignant cells are generally resistant to the cytotoxic and cytostatic actions of TNF, however, in the presence of cycloheximide even normal human fibroblasts undergo rapid lysis by TNF (29). Since cytotoxicity can be demonstrated in the presence of inhibitors of cellular RNA or protein synthesis, it is obviously not mediated by TNF-induced activation of cellular gene products. In addition, the fact that cytotoxicity can be increased in the presence of metabolic inhibitors suggests that damage produced as a result of TNF action in most cells can be repaired if cellular RNA and protein synthesis are allowed to proceed normally. The idea that inhibition of a repair process was responsible for increased cytotoxicity in the presence of metabolic inhibitors was proposed many years ago by Rosenau et al. (30) on the basis of their work with LT.

Several investigators analyzed the relationship between the cell cycle and the cytotoxic/cytostatic action of TNF. Lee et al. (31) found that treatment of B16 melanoma cells with LT increased the proportion of cells in the G0/G1 stage, decreased the number of cells in S, and increased the number of cells in the G2+M phases of the cell cycle. In TNF-sensitive exponentially growing murine L929 cells TNF transiently arrested cell cycle progression in the G2 phase before inducing cell lysis (32). Trinchieri et al. (33) found that treatment of the myeloid cell line HL-60 with TNF initially increased ^{3}H-thymidine incorporation and the proportion of cycling cells; this initial stimulatory effect was followed by a reduction in DNA synthesis and a decrease in the proportion of cells in the S/G2/M phases of the cell cycle. Details of the inhibitory action of TNF on cultured cells can vary depending on the target cell and conditions. Sugarman et al. (34) found that the presence of some growth factors can antagonize the cytostatic and cytotoxic action of TNF. In some other cell lines TNF or LT not only fail to inhibit, but actually stimulate cell growth in a growth factor-like manner (16,17,31).

In contrast to the inhibitory actions on tumor cells in culture, it is likely that -- unless TNF is administered intratumorally -- much of the antitumor activity seen with TNF in experimental animals is indirect and not related to the susceptibility of the tumor cells themselves to the action of TNF. This conclusion is supported by numerous recent observations. Haranaka et al. (35) suggested that host T cells are important in the antitumor activity of TNF because TNF was more effective in Balb/c nu/+ mice than in the thymus deficient Balb/c nu/nu mice. There is increasing evidence that the primary lesion responsible for the characteristic TNF-induced hemorrhagic necrosis of the Meth A sarcoma is vascular, probably due to an action on the endothelial cell (36). It was also shown that the sensitivity of tumor cells to TNF action in culture does not correlate with the sensitivity of the tumors to TNF action in syngeneic mice (36,37). Finally, Asher et al. (38) showed that recombinant human TNF was more effective against murine tumors that were immunogenic than against tumors that were non-immunogenic.

SYNERGISTIC ENHANCEMENT OF TUMOR CELL CYTOTOXICITY WITH TNF AND INTERFERONS

Many studies have shown that interferons can enhance the susceptibility of tumor cells to the cytotoxic actions of TNF or LT. Williamson et al. (39) reported that LT and IFN-alpha or IFN-gamma were synergistic in the cytolysis of breast and lung carcinomas and melanomas. TNF and IFN-gamma also showed a marked synergistic cytotoxicity in a wide array of other human tumor cell lines, particularly breast, cervix and colon carcinomas (40,41). Tumor cells naturally resistant to the cytostatic/cytotoxic action of TNF could be rendered susceptible to TNF action by an exposure to interferons, especially IFN-gamma. In our own experiments, the human colon carcinoma cell line, HT-29, was virually resistant to the cytotoxic effects of either TNF or IFN-gamma alone; incubation with IFN-gamma, but not with alpha nor beta IFN, sensitized HT-29 cells toward the cytotoxic action of TNF (28). Other investigators (16,42-44) have reported similar findings.

In an attempt to explain the synergism between IFN-gamma and TNF, we and others have examined the effect of various interferons on TNF binding to cell surface receptors (18,19, 45-47). Incubation of several different cell lines with IFN-gamma caused a significant increase (30-100%) in TNF binding. Generally, this increase was less marked and less regularly seen with IFN-alpha or IFN-beta. Increased binding of TNF in cells incubated with IFN-gamma was shown to be due to an increased expression of TNF receptors, without a demonstrable change in the binding affinity. IFN-gamma also caused a stimulation in the synthesis of the TNF receptor protein, probably resulting from a direct stimulation of transcription (45,47). In the case of HT-29 cells, the results seem to suggest a causal relationship between enhanced TNF binding and increased cytotoxicity (18, 47). However, in several tumor cell lines that are resistant to TNF action, IFN-gamma caused an increase in TNF binding without a concomitant increase in the sensitivity to TNF cytotoxicity (18). Other experiments suggested that TNF receptor modulation does not appear to be a major mechanism of synergism between TNF and interferons (47).

The synergism between TNF and interferons may be related to the ability of both agents to slow progression through the cell cycle. In suppressing cell growth TNF and interferons probably activate separate inhibitory pathways whose combined actions are greater than the sum of their separate actions. The potentiating action of interferons on TNF cytotoxicity might also be related to a somewhat similar action of various inhibitors of RNA and protein synthesis, mentioned earlier. Like metabolic inhibitors, interferons might interfere with a putative repair process protecting cells from TNF cytotoxicity.

Using a molecular approach to dissect the mechanisms by which TNF alone and in combination with IFN-gamma inhibits tumor cell growth, Yarden et al. (48) showed that both IFN-gamma and TNF reduce c-myc protooncogene expression in Hela cells, and that c-myc expression was more strongly inhibited in Hela cells treated with TNF and IFN-gamma together than with either one cytokine alone. Their findings also suggested that IFN-gamma and TNF inhibit c-myc expression through different molecular mechanisms. Downregulaton of protooncogenes implicated in cell cycle control, such as c-myc, might play a role in the inhibitory action of TNF and its synergistic enhancement by interferons. However, a more complete understanding of the mechanisms involved in the synergism between TNF/LT and interferons in their cytostatic/cytotoxic actions on tumor cells will require much additional work.

SYNERGISTIC ANTI-TUMOR ACTIONS OF TNF AND INTERFERONS

Balkwill et al. (49) examined the combined action of recombinant human TNF and recombinant human interferons (IFN-alpha and IFN-gamma) on human tumor xenografts in thymus-deficient nude mice. In the nude mouse model the antitumor actions of TNF and of the interferons are probably due largely to direct effects on the tumor cells, rather than indirect immunomodulatory mechanisms. This conclusion is supported by the demonstration that systemic (intraperitoneal, i.p.) administration of human TNF showed little or no therapeutic effect against a variety of subcutaneous (s.c.) human tumor xenografts. In contrast, intratumoral administration of human TNF was highly effective against the same tumors. Human TNF given i.p. was also effective against human ovarian cancer grown i.p. in the nude mice. Both human IFN-alpha and IFN-gamma acted synergistically with TNF against s.c. and i.p. tumors. Combined i.p. treatment with TNF and IFN-gamma was particularly effective against i.p. grown ovarian cancer.

Synergism between TNF and IFN-gamma was also seen in syngeneic tumors in mice. Brouckaert et al. (50) examined the actions of recombinant human or murine TNF, with or without recombinant murine IFN-gamma, in C57BL/6J mice inoculated s.c. with B16BL6 melanoma cells. In this model human TNF or murine IFN-gamma alone given i.p. showed very little activity, but combined i.p. treatment was highly effective. The authors also showed that murine TNF was much more effective against the B16 melanoma tumors (and also more toxic) than human TNF. In fact, murine TNF alone was so effective that the addition of IFN-gamma was not necessary to achieve a complete inhibition of tumor growth. The greater efficacy of murine TNF in the C57BL/6J mice can be ascribed to some degree of species specificity.

Results of animal experiments are sufficiently encouraging to consider the use of combined therapy with TNF and IFN-gamma in human cancer. A combination of TNF and IFN-gamma is also being considered for the treatment of patients infected with the human immunodeficiency virus (HIV). The rationale for such combination therapy is said to be the ability of the two agents to reduce HIV infection in cell culture systems (51). Toxicities resulting from the administration of TNF and IFN-gamma might pose a major obstacle in devising suitable therapeutic protocols.

Lysis of tumor cells by mixtures containing interferon and TNF or LT is not merely a laboratory phenomenon or the result of an artificial therapeutic intervention. Some years ago we observed that crude preparations of natural human IFN-gamma lysed some tumor cells very efficiently, whereas pure IFN-gamma did not (44). The difference in the actions of these preparations was due to the presence of TNF and LT in crude preparations of IFN-gamma; only when both IFN-gamma and TNF or LT were present did tumor cell lysis occur. TNF or LT and IFN-gamma or IFN-gamma are often made by the same cells in response to the same stimuli and natural mixtures of these cytokines are likely to be produced in the intact organism. In addition, we have shown that IFN-gamma can sensitize tumor cells to lysis by peripheral blood monocytes (28). Since TNF is an important mediator of monocyte cytotoxicity for tumor cells, this action represents another form of a synergistic interaction between IFN-gamma and TNF, likely to be relevant for anti-tumor defenses in the body.

REFERENCES

1. Blick, M., Sherwin, S.A., Rosenblum, M. and Gutterman, J. Cancer Res. 47:2986, 1987.
2. Carswell, E.A., Old, L.J., Kassel, R.L., Green, S., Fiore, N. and Williamson, B. Proc. Natl. Acad. Sci. USA 72:3666, 1975.
3. Old, L.J. Science 230:630, 1985.
4. Aggarwal, B.B., Kohr, W.J., Hass, P.E., Moffat, B., Spencer, S.A., Henzel, W.J., Bringman, T.S., Nedwin, G.E., Goeddel, D.V. and Harkins, R.N. J. Biol. Chem. 260:2345, 1985.
5. Pennica, D., Nedwin, G.E., Hayflick, J.S., Seeburg, P.H., Derynck, R., Palladino, M.A., Kohr, W.J., Aggarwal, B.B. and Goeddel, D.V. Nature 312:724, 1984.
6. Wang, A.M., Creasey, A.A., Ladner, M.B., Lin, L.S., Strickler, J., Van Arsdell, J.N., Yamamoto, R. and Mark, D.F. Science 228:149, 1985.
7. Shirai, T., Yamaguchi, H., Ito, H., Todd, C.W. and Wallace, R.B. Nature 313:803, 1985.
8. Williams, T.W. and Granger, G.A. Nature 219:1076, 1968.
9. Ruddle, N.H. and Waksman, B.H. J. Exp. Med. 128:1267, 1968.
10. Aggarwal, B.B., Henzel, W.J., Moffat, B., Kohr, W.J. and Harkins, R.N. J. Biol. Chem. 260:2334, 1985.
11. Spies, T., Morton, C.C., Nedospasov, S.A., Fiers, W., Pious, D. and Strominger, J.L. Proc. Natl. Acad. Sci. USA 83:8699, 1986.
12. Muller, U., Jongeneel, C.V., Nedospasov, S.A., Lindahl, K.F. and Steinmetz, M. Nature 325:265, 1987.
13. Beutler, B., Greenwald, D., Hulmes, J.D., Chang, M., Pan, Y.-C. E., Mathison, J., Ulevitch, R. and Cerami, A. Nature 316: 552, 1985.
14. Beutler, B.A. and Cerami, A. Nature 320:584, 1986.
15. Le, J. and Vilček, J. Lab. Invest. 56:234, 1987.
16. Sugarman, B.J., Aggarwal, B.B., Hass, P.E., Figari, I.S., Palladino, M.A. and Shepard, H.M. Science 230:943, 1985.
17. Vilček, J., Palombella, V.J., Henriksen-DeStefano, D., Swenson, C., Feinman, R., Hirai, M. and Tsujimoto, M. J. Exp. Med. 163:632, 1985.
18. Tsujimoto, M., Yip, Y.K. and Vilček, J. J. Immunol. 136:2441, 1986.
19. Aggarwal, B.B., Eessalu, T.E. and Hass, P.E. Nature 318:665, 1985.
20. Nitsu, Y., Watanabe, N., Sone, H., Neda, H., Yamauchi, N. and Urushizaki, I. Jpn. J. Cancer Res. 76:1193, 1985.
21. Old, L.J. Nature 326:330, 1987.
22. Ruddle, N.A. Immunol. Today 6:156, 1985.
23. Hahn, T., Toker, L., Budilovsky, S., Aderka, D., Eshhar, Z. and Wallach, D. Proc. Natl. Acad. Sci. USA 82:3814, 1985.

24. Matthews, N. and Watkins, J.F. Br. J. Cancer 38:302, 1978.
25. Ruff, M.R. and Gifford, G.E. In: Lymphokines, Vol. 2 (ed. E. Pick), Academic Press, New York, 1981, pp235-272.
26. Matthews, N. Br. J. Cancer 44:405, 1983.
27. Tsujimoto, M., Yip, Y.K. and Vilček, J. Proc. Natl. Acad. Sci. USA 82:7626, 1985.
28. Feinman, R., Henriksen-DeStefano, D., Tsujimoto, M., and Vilček, J. J. Immunol. 138:635, 1987.
29. Kirstein, M. and Baglioni, C. J. Biol. Chem. 261:9565, 1986.
30. Rosenau, W., Goldberg, M.L. and Burke, G.C. J. Immunol. 111:1128, 1973.
31. Lee, S.H., Aggarwal, B.B., Rinderknecht, E., Assisi, F. and Chiu, H. J. Immunol. 133:1083, 1984.
32. Darzynkiewicz, Z., Williamson, B., Carswell, E.A. and Old, L.J. Cancer Res. 44:83, 1984.
33. Trinchieri, G., Kobayashi, M., Rosen, M., Loudon, R., Murphy, M. and Perussia, B. J.Exp. Med. 164:1206, 1986.
34. Sugarman, B.J., Lewis G.D., Eessalu, T.E., Aggarwal, B.B., Shepard, H.M. Cancer Res. 47:780, 1987.
35. Haranaka, K., Satomi, N. and Sakurai, A. Int. J. Cancer 34:263, 1984.
36. Palladino JR., M.A., Shalaby, M.R., Kramer, S.M., Ferraiolo, B.L., Baughman, R.A., Deleo, A.B., Crase, D., Marafino, B., Aggarwal, B.B., Figari, I.S., Liggit, D. and Patton, J.S. J. Immunol. 138:4023, 1987.
37. Gresser, I., Belardelli, F., Tavernier, J., Fiers, W., Podo, F., Federico, M., Carpinelli, G., Duvillard, P., Prade, M., Maury, C., Bandu, M.-T. and Maunoury, M.-T. Int. J. Cancer 38:771, 1986.
38. Asher, A., Mule, J.J., Reichert, C.M., Shiloni, E. and Rosenberg, S.A. J. Immunol. 138:963, 1987.
39. Williamson, B.D., Carswell., E.A., Rubin, B.Y., Prendergast, J.S. and Old, L.J. Proc. Natl. Acad. Sci. USA 80:5397, 1983.
40. Fransen, L., Van Der Heyden, J., Ruysschaert, R. and Fiers, W. Eur. J. Cancer Clin. Oncol. 22:419, 1986.
41. Schiller, J.H., Bittner, G., Storer, B. and Willson, J.K.V. Cancer Res. 47:2809, 1987.
42. Scheurich, P., Ucer, U., Kronke, M. and Pfizenmaier, K. Int. J. Cancer 38:127, 1986.
43. Williams, T.W., Bellanti, J.A. J. Immunol. 130:518, 1983.
44. Stone-Wolff, D.S., Yip, Y.K., Kelker, H.C., Le, J., Henriksen-DeStefano, D., Rubin, B., Rinderknecht, E., Aggarwal, B.B., Vilček, J. J. Exp. Med. 159:828, 1984.
45. Ruggiero, V., Tavernier, J., Fiers, W. and Baglioni, C. J. Immunol. 136:2445, 1986.
46. Tsujimoto, M., Vilček, J. J. Biol. Chem. 261:5384, 1986.
47. Tsujimoto, M., Feinman, R., Vilček, J. J. Immunol. 137:2272, 1986.
48. Yarden, A. and Kimchi, A. Science 234:1419, 1986.
49. Balkwill, F.R., Lee, A., Aldam, G., Moodie, E., Thomas, J.E., Tavernier, J. and Fiers, W. 1986. Cancer Res. 46:3990, 1986.

50. Brouckaert, P.G.G., Leroux-Roels, G.G., Guisez, Y., Tavernier, J. and Fiers, W. Int. J. Cancer 38:763, 1986.
51. In: AmFAR Directory of Experimental Treatments for AIDS and ARC, Vol. 1 (Eds. M. Gottlieb, M. Grieco, L.N. Painter, and S. Bernstein) Mary Ann Liebert, Inc., New York, 1987, p. 86.

INDEX

GPSR Compliance

The European Union's (EU) General Product Safety Regulation (GPSR) is a set of rules that requires consumer products to be safe and our obligations to ensure this.

If you have any concerns about our products, you can contact us on ProductSafety@springernature.com

In case Publisher is established outside the EU, the EU authorized representative is:

Springer Nature Customer Service Center GmbH
Europaplatz 3
69115 Heidelberg, Germany

Batch number: 10755284

Printed by Printforce, the Netherlands